OXIDES AND OXIDE FILMS

Volume 1

THE ANODIC BEHAVIOR
OF
METALS AND SEMICONDUCTORS SERIES

Edited by

John W. Diggle

RESEARCH SCHOOL OF CHEMISTRY
THE AUSTRALIAN NATIONAL UNIVERSITY
CANBERRA, AUSTRALIA

Oxides and Oxide Films, Volume 1
Edited by John W. Diggle

ADDITIONAL VOLUMES IN PREPARATION

OXIDES AND OXIDE FILMS

Volume 1

Edited by

John W. Diggle

RESEARCH SCHOOL OF CHEMISTRY
THE AUSTRALIAN NATIONAL UNIVERSITY
CANBERRA, AUSTRALIA

1972
MARCEL DEKKER, INC., New York

MARCEL DEKKER, INC.
95 Madison Avenue, New York, New York 10016

LIBRARY OF CONGRESS CATALOG CARD NUMBER 72-83120
ISBN 0-8247-1143-2

PRINTED IN THE UNITED STATES OF AMERICA

INTRODUCTION TO THE SERIES

The aim of this series is to present invited authorative articles by scientists in the general field of anodic behavior of materials. The topics of interest will thus be such as all aspects of anodic oxide films, corrosion reaction films in general, anodic dissolution of metals together with the related areas such as electropolishing and electrochemical machining, anodic oxidation of organic compounds, both from synthetic and electrode kinetic points of view, anodic gas evolution reactions and a variety of related areas as well as their industrial ramifications. The anodic behavior of metals and non-metals in solid state electrolytes and in ionic liquids as well would thus merit inclusion in this series also. The emphasis will be upon an interdisciplinary approach, attempting to interweave electrochemical approaches with the consideration of materials science or other overlapping fields (depending on the subject under review) and to present the expositions at a sufficiently advanced level in order to bring out the salient points in great depth.

Successive volumes will be devoted, whenever possible, to a given subject and theme in order to create various books each of which—it is hoped—will assist in the advancement of knowledge in a particular area, e.g., some aspects of anodic oxide films or anodic dissolution etc. In several cases, e.g., anodic oxide films, a number of contributions will be included from peripheral areas even though they may not address themselves to anodic, or even electrochemical matters. Thus, the several volumes on anodic oxide films will perhaps be best labeled as treatments of oxides and oxide films in general with strong emphasis, however, on anodic or related aspects, at least in the majority of the chapters. This kind of approach is considered essential in order to place the anodic behavior of materials in an inter-disciplinary perspective, especially with regard to materials science - this aspect must be stressed as the most pivotal point of the present series. For example, any volumes on anodic oxide films and their electrochemical (or electrocatalytic) behavior would be less than complete, if appropriate chapters on subjects such as defects in oxides, heterogeneous catalytic behavior of oxides, space charges in oxides, radiation damage especially in relation to defect structure, glassy characteristics of oxides etc. are not included. Similar comments also pertain, of course, to subjects other than oxide films to be covered in this series.

It should be pointed out that a deliberate attempt has been made by the editor to encourage contributors to express opinions as well as facts. In this way it is hoped not only to increase the readability of these contributions and thus heighten the readers interest, but also to provoke controversy for, quoting the words of William Hazlitt, what ceases to be subject of controversy ceases to be a subject of interest.

John W. Diggle

Canberra, Australia
September 1972

PREFACE

Over a decade has passed since the publication of Young's
"Anodic Oxide Films", and in this time contributions to the field
of oxides and oxide films have been so significant that it is
doubtful that no one single publication, and perhaps no one
single author, can do justice to the subject. It was from such
an opinion that the initial volumes in this series were conceived.
The present volume initiates this series on oxides and oxide films
by considering four fundamental aspects — the phenomena of
passivity, ionic conductivity, electronic conductivity and oxide/
electrical double-layer processes.

The first chapter (Brusić's) concerns the general phenomena
of passivation and passivity. Although it deals more or less
specifically with the passivation of iron, since it has been the
one metal most extensively studied, this chapter does elucidate
the complexities with which one is faced when attempting to
fully comprehend the processes involved in both the active and
passive states.

The second chapter (Dignam's) deals with the much studied,
but still not fully understood, ion-transport phenomena through
oxide films. Over the last few years the theories proposed by
Dignam for ion conduction in growing oxides have injected fresh
life into an area of oxide films which was inherently anchored
to a crystalline lattice concept, despite the largely amorphous
nature of anodic oxides. Although there are still many argu-
mentative aspects concerning the conduction of ions across oxides,
significant progress has been made and further progress can be
expected.

The third chapter (Mead's) concerns itself with the equally important electron transport phenomena through dielectric layers. In his contribution, Mead has attempted to show that, in the ideal case, after comprehending all the parameters involved, the phenomena of electron transport can be accounted for in terms of rather simple mechanisms.

The final chapter in this volume (Ahmed's) reviews and critically examines present knowledge concerning the electrical double layer at an oxide-electrolyte interface. This area of investigation has been marked by an understandable preference to experiment with bulk oxides and hence the applicability of such work to oxide films on metals has been precisely defined here.

John W. Diggle

Canberra, Australia
September 1972

CONTRIBUTORS TO VOLUME 1

VLASTA BRUSIĆ, IBM, Thomas J. Watson Research Center, Yorktown
    Heights, New York

M. J. DIGNAM, Department of Chemistry, University of Toronto,
    Toronto, Ontario, Canada

C. A. MEAD, California Institute of Technology, Pasadena,
    California

SYED M. AHMED, Department of Energy, Mines and Resources,
    Mineral Sciences Division, Mines Branch, Ottawa, Canada

CONTENTS OF VOLUME 1

3 ELECTRONIC CURRENT FLOW THROUGH IDEAL DIELECTRIC FILMS

C. A. Mead

4 ELECTRICAL DOUBLE LAYER AT METAL OXIDE-SOLUTION INTERFACES

Syed M. Ahmed

CONTENTS OF OTHER VOLUMES

OXIDES AND OXIDE FILMS

Volume 1

Chapter 1

PASSIVATION AND PASSIVITY

Vlasta Brusić

IBM

Thomas J. Watson Research Center

Yorktown Heights, New York

I. INTRODUCTION

Every national economy would greatly benefit if all iron and other metal constructions were as resistant to the influence of time and setting as the famous 1600-year-old iron pillar of Delhi. Its corrosion rate above ground, estimated from comparison with the corrosion rate of similar steel in the same dry climate, corresponds to a loss of 5μ/yr [1], or 0.1 Å/min or 1.6×10^{12} atoms of metal per square centimeter per second. Such an estimate is probably too high because it uses linear, rather than the more applicable parabolic, extrapolation [1]. Yet this rate is many times less than the theoretical rate of reaction based on the kinetic-theory equation for the number of collisions of reactive gas molecules (air, 25°C) with a flat surface. The theoretical rate gives 8.5×10^{22} atoms of iron per square centimeter per second [2]. The estimated rate is also several orders of magnitude less than the corrosion rate of most metallic constructions in most of the natural, and especially industrial, environments.

Corrosion still imposes an unfortunate and appreciable burden on industry in the amount of more than 5 billion dollars per year in the United States and similarly 5 to 6 billion rubles per year in the Soviet Union. Yet technology predominantly depends on those corrosive, nonnoble metals that owe much of their usefulness to the existence of the passive state. This state is also one of the results of the reactivity of such metals or of their tendency to react with the oxygen and oxygen-containing compounds of the environment to form a protective film on the surface and to thereafter become less reactive to the influence of ambience than would be expected from their standard electrode potential.

From the time the passivation phenomenon was first observed and described [3-7] up to today the complexity of the problem on the

one side and the practical importance of its application on the other
have drawn the attention of many scientists. (see, for example, the
excellent collection of work in Refs. [8-11]).

The present discussion draws particularly on studies that have
been published in the last 10 years and are concerned with metal
electrodes under conditions producing passivity, which are the ones
that best clarify the passivation process.

II. PASSIVATION PHENOMENA

A. Definition of Passivity

Passivation can be defined as a constant slowing down of any
action, process, or reaction. In this respect the decrease in the
oxidation rate with increasing potential on an "inert" metal surface
(Pt) as occurs with organic compounds [12] or hydrogen [13] is
described by Gilroy and Conway [12] and Schuldiner [13] as "passivation in electrochemical reaction. " Similarly the term "passivity"
may be applied to any improved corrosion resistance of a metal in
an electrolyte (or air), whatever the cause. However, the general
opinion expressed in scientific literature clearly narrows down the
use of the term. The term is applied only to metals thermodynamically unstable under given conditions, the improved corrosion resistance of which is accompanied by a simultaneous shift of the
electrode potential in the positive direction; that is, the term is
applied to metal oxidation [14-16]. Thus Wagner [17] has given
the following phenomenological definition of passivity:

"A metal may be called passive when the amount of metal consumed by a chemical or electrochemical reaction at a given time is
significantly less under conditions corresponding to a higher affinity

of the reaction (i. e., a greater decrease in free energy) than under conditions corresponding to a lower affinity."

Such a definition enables us, without hypothesizing about the nature of the oxidized state, to distinguish between "the passivity" on one hand and "the inhibition" (and diminution of the corrosion rate by paints or galvanoplastic deposits) on the other. In the latter the adsorption of the active species from the solution influences the reaction kinetics at the metal-solution interface without appreciably influencing the potential difference.

An electrochemical example given by Wagner [17] is that the steady rate of anodic dissolution of a metal in a given environment is lower at a more noble single electrode potential than it is at a less noble potential; that is, a typical curve of current i against potential V shows the appearance of a negative di/dV slope (see Fig. 1). A chemical example of passivity (such as the classical case of iron in nitric acid) can be exhibited by a metal in solution if the partial anodic and cathodic currents (i. e., the current of metal dissolution and the current of reduction of the oxidized form of an oxidation-reduction system that is present in the solution) become equal at a potential sufficiently higher than that of reversible metal dissolution [15-19]. In this case the metal displays an open-circuit potential, which is in the passive region of the i-V curve (Fig. 1).

The comparison of such a steady-state potential of the metal with its corrosion potential in a specific medium is defined by Tomashov and Chernova [15] as the "degree of its passivity." The same authors have given another approximate calculation of the degree of passivity, using as the determining factor the ratio of anodic and cathodic polarization at a certain potential. This has

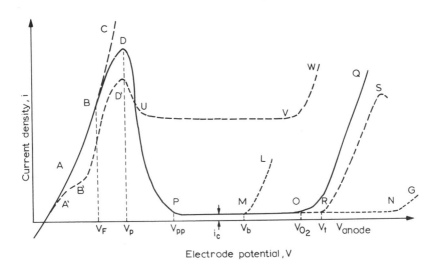

Electrode potential, V

Fig. 1. Schematic representation of the anodic processes on metals. Curve ABDPOQ: typical active, passive, and transpassive i-V curve for metal in water (Fe, Ni); curve ABC: free dissolution (Fe in alkali); curve ABDML: i-V curve in the presence of active anions, with possible salt formation in the active region (Fe in chloride solution); curve ABDPRS: active, passive, and transpassive behavior of metals that may dissolve in transpassive region as high-valency cations, with possible appearance of secondary passivity (Cr); curve ABDUVW or A'B'D'UVW: i-V curve for metals forming less protective film (Cu, Zn) or iron in electropolishing solutions; curve ABDPNG: i-V curve for metals with low electron conductivity, with oxide growth possible at high potentials (Al, Ti, Ta). Reprinted from Ref. [15] by courtesy of Plenum Press.

produced the following series in decreasing order of passivity: Zr, Ti, Ta, Nb, Al, Cr, Be, Mo, Mg, Ni, Co, Fe, Mn, Zn, Cd, Sn, Pb, Cu.

B. Passivation as an Electrochemical Problem

Although passivity is a general phenomenon and applies to all metals, the discussion of the experimental findings as well as theoretical considerations is often focused on iron, for which the

difference between the active and passive states is especially pronounced. If the electrode is oxidized potentiostatically, a current-
potential curve of the type shown schematically in Fig. 1 is
obtained.

In the active state metal ions go into the solution

$$M \rightarrow M^{2+} + 2e^- , \qquad (1)$$

undergoing hydration. At still higher potentials the polarization
curve deviates from simple logarithmic dependence (Tafel law, metal
dissolution) because the dissolution is hindered by another anodic
process, that is, the formation of a protective film [15] through a
reaction of the type

$$M + H_2O \rightarrow M-O + 2H^+ + 2e^- . \qquad (2)$$

At the passivation potential V_p a limiting passivation current i_p is
reached, and the acceleration of metal dissolution is equal to the
retardation of the process. With further increase in potential the
rate of metal dissolution is increasingly hindered by the process,
which evidently terminates at V_{pp}. Thereafter the dissolution rate
(i. e., corrosion rate) is independent of potential. The increase in
current at some higher potential is determined by the overpotential
of oxygen discharge (or the presence of some oxidation-reduction
system). The region is termed "transpassive."

The name "Flade potential," V_F, has a variety of connotations.
Many investigators, following Flade's original definition, assign
this name to the potential at which an already passivated metal begins to lose passivity [20-22]. Some authors, however, prefer to
assign this term to the passivation potential V_p [23] at which the
current is maximal or even to V_{pp}, the potential of "complete
passivity" [24]. According to Tomashov and Charnova [15], V_F is
the thermodynamic potential at which the formation of a metal oxide

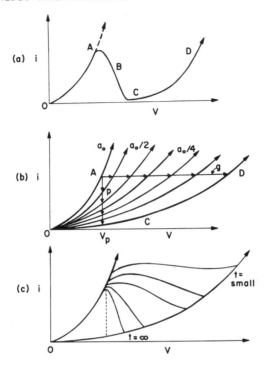

Fig. 2. Schematic representation of galvanostatic (g) and potentiostatic (p) i-V curves.

becomes possible, as shown in Fig. 1. As much as possible in this chapter all discussions refer to the values given in this figure.

What kind of processes may be responsible for such an experimental current-potential curve?

Let us reexamine the curve, this time as shown in Fig. 2(a). The curve has a maximum and a minimum, with a region AC where dV/di is negative and regions OA and CD, which can be separately described by the ordinary nonlinear resistance of a simple electrode process. Points on the line AC are not galvanostatically accessible and would not be potentiostatically accessible if this were a simple electrode process. Any such process can be described in terms of a reaction resistance and a capacitance lag (time delay), that is, in terms of $i = f(V)$ or $V = f(i)$ and $dV/di = R(e)$, where resistance, capacitance, and differential resistance are real and positive. The present system must be at least double; that is, it consists of at least two individual systems coupled together. How can the coupling be expressed? In coupling one can imagine two possible states with a number of elements in between, such that the process OA produces the state CD and the process CD produces the state OA. For example, the process OA can be described by

$$i = a(i, t) f(V) , \qquad (3)$$

where a is a reacting material, a bare electrode surface, which is changed as the current passes; that is,

$$a = a_0(1 - ki_t t) , \qquad (4)$$

with k being the proportionality constant.

If one works galvanostatically, the slope di/dV will change as shown in Fig. 2(b): at the beginning $a = a_0$. With increasing time a becomes smaller and the potential will move up (see Eq. (3)) either ad infinitum or until the potential of another reaction is reached (CD).

If one works potentiostatically, a is again proportional to the current, which varies with time; it decreases rapidly at first and then decays slowly, even more slowly approaching the curve CD (vertical line, Fig. 2(b)). The initial decrease in the current will be proportional to the change in a, which in turn decreases more rapidly as the current (i.e., the potential) increases. Thus at potential V_1 after $t = t_1$, a may decrease to $a_0/2$, but at V_2 $(V_2 > V_1)$ a may decrease to $a_0/4$, possibly resulting in the current at time t_2, i_{t_2}, which is smaller than the current at some earlier time t_1, i_{t_1}; that is, a current-potential curve with a negative dV/di slope can be found.

The reaction by which a free electrode surface diminishes
(Eq. (4)), that is, film formation, may not occur below a certain
potential V_F, the standard electrochemical potential for the reaction.
Thus points OA below V_F belong to another, separate reaction, such
as free dissolution. Also the current at the potential $V > V_F$ may
be only partly due to the film-forming reaction and partly may still
be due to metal dissolution, hydrogen evolution, and similar reactions; thus points above V_F will be a result of the competition between reactions that diminish a and reactions that either do not
change a or enlarge it. In the ideal case, when metal dissolution
at the potentials above V_F is negligible, the resulting current-
potential curve would look as shown schematically in Fig. 2(c),
giving the active-passive transition depending on the potential at
which "film" formation is thermodynamically possible, on the relative rates of metal dissolution and "film" formation, and on the time
one waited before reading.

C. Problems in Characterizing the Passivation Process

The first theory of passivity was suggested over a century ago
when Faraday [7] stated, "my strong impression is that the surface
is oxidized. . . ." However, the investigation of the nature of the
oxidized state introduced the traditional polemics between those
supporting the adsorption theory and those supporting the oxide-film
theory of passivity. The controversy is at least partly due to seman-
tic ambiguities in the description of passivating films. The terms
most often used are the following: adsorbed (chemisorbed) oxygen
and oxygen-containing species [20, 25-33], monomolecular oxide
[34], monolayer (or less) of oxide [14, 35-44], ordered, monomolecular, two-dimensional film of a definite chemical phase [45], three-
dimensional film [19], oxide not related to any known oxide [46],

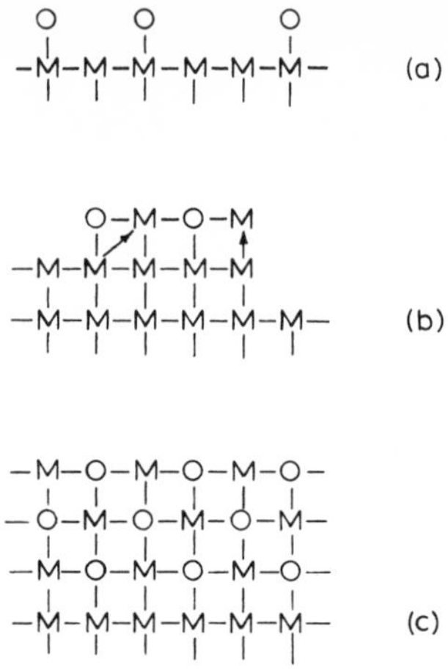

```
     O      O           O
     |      |           |
  -M-M-M-M-M-M-      (a)
   |  |  |  |  |  |
```

```
        O-M-O-M
        |/ |  |  ↑
   -M-M-M-M-M
    |  |  |  |  |
  -M-M-M-M-M-M-      (b)
   |  |  |  |  |  |
```

```
  -M-O-M-O-M-O-
   |  |  |  |  |  |
  -O-M-O-M-O-M-
   |  |  |  |  |  |
  -M-O-M-O-M-O-
   |  |  |  |  |  |
  -M-M-M-M-M-M-      (c)
   |  |  |  |  |  |
```

Fig. 3. Characterization of passivating films: (a) adsorbed layer, (b) monomolecular (or less) oxide, and (c) three-dimensional oxide.

nonstoichiometric oxide [47], and duplex oxide [21, 23, 48-54]. It is not always clear what is meant by "surface oxides" or "surface adsorption, " especially when coverage with oxygen is less than a monolayer. In the opinion of this author the term "adsorbed layer" should be reserved for the situation shown in Fig. 3a, characterized by the formation of preferentially unidirectional bonding between adsorbate (oxygen) and substrate (metal) without metal ions leaving their lattice site.

In contrast, it is suggested that in the monomolecular (or less) oxide layer, metal atoms have left their regular position in the lattice (at kink sites and any other sites) to enter together with oxygen atoms into a new, alternating arrangement of oxygen and metal, which, even if two-dimensional, resembles the arrangement in the formation of the nuclei of the respective oxides (i. e. , lateral bonds are formed, Fig. 3b) [55].

In the three-dimensional oxide there is a repeated distance and symmetry relationship in a vertical dimension also, with either a more or less continuous relationship with the substrate (as in epitaxial film) or with a complete misorientation, as in nonepitaxial, deposited film.

When possible, in the following text the characterization of the passivating film will be given accordingly.

There are also scientific difficulties in approaching the problem of the entities formed in passivation. This is particularly so in the region of low electrode coverage, where there is no possibility of clearly distinguishing between the adsorbed or (monomolecular or less) oxide layer without an experimental method that is sufficiently sensitive. Experimental approach to the problem was for a long time connected with the measurement of electrochemical parameters, which may have been insufficient and ill-defined [42]. These experimental approaches usually measured an average current density and thus gave little information about the local current distribution, the nature and thickness of the passive film, and the film distribution over the surface. Some of the other difficulties involved in this type of experiment include the following:

1. Characterization of the surface and surface topography prior to, during, and after the passivation process. In the earlier experiments the electrode was probably oxidized prior to oxidation, as shown for aluminum [56] and titanium [57]. Also, a metal surface can be populated by energetically different sites [58, 59], as schematically shown in Fig. 4, where the surface energies for the different positions increase in the order

$$E(3) < E(1) < E(0.5) < E(2) < E(6) < E(7) < E(4) < E(5).$$

Since the most active sites dissolve first, one can expect two kinds of effect: (a) the electrode may be roughened, increasing the electrode area and thus diminishing the current density (thus at constant potential the total current may even increase) and (b) with time the number of the most active sites may decrease, and the current may

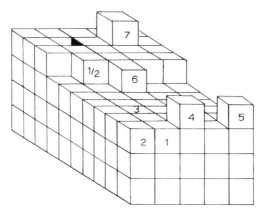

Fig. 4. Possible characteristic lattice sites on the surface of a cubic lattice. Reprinted from Refs. [58] and [59] by courtesy of Pergamon Press.

decrease with time. In any case the experiments on the "old" electrode could be different from those on a freshly prepared surface.

Similarly the reaction behavior of the metal atom depends on crystallographic orientation, as shown by Piontelli [60] for the anodic dissolution of nickel in sulfuric acid as well as in chloride solution, by Foley et al. [61] for the passivation behavior of iron, by Despic et al. [62] for the dissolution of iron, and by Pickering and Frankenthal [63] for film formation and reduction on Fe 24% Cr alloys.

2. The electrochemical behavior of the main reaction partner, the metal atom, depends also on the electrode pretreatment, which often varies from author to author. For example, Despic and Bockris [64] and Bockris and Damjanovic [65] have shown that surface adion concentrations at the reversible potential on silver vary for more than an order of magnitude, depending on surface preparation. These concentrations are smallest for electrodes polarized after preparation by quenching in helium and are largest for electrodes prepared in situ by anodic pulsing. Moreover, different mechanisms pro-

posed for iron dissolution may be related to the state of the surface:
The Tafel slope obtained on well-annealed, zone-refined iron is
40 mV and can be explained by the mechanism in which the second
electron transfer is rate determining [66]. In contrast, the slope
obtained on cold-worked, strongly deformed iron is 30 mV and can
be explained by the two-electron transfer as the rate-determining
step [67]. The evidence that the pretreatment influences kinetic
parameters is not always strong, as the results of Hoar [14] show
that the free energy for dissolution is 10.6 kcal/mole for both
annealed and cold-worked nickel.

3. The quantitative evaluation of the kinetic parameters is
influenced by change in electrode resistance caused by the formation of a passive layer, as well as by change in the activity of the
ionic reactants at the surface, both of which are often unknown.

4. The Tafel behavior of the reaction is often limited to a
relatively narrow region of potential, with the general assumption
that the transfer coefficient α does not depend on potential. There
is some indication that this assumption is doubtful [68].

Still both galvanostatic and potentiostatic measurements have
been very useful, especially since they have been supplemented by
many other techniques, such as potentiodynamic (single- and
multiple-sweep) and ac-impedance investigations, coulometry, and
in particular nonelectrochemical methods like ellipsometry, electron
microscopy and diffraction, and (possibly) Mössbauer techniques.
Some of the difficulties, such as the preparation and characterization of the initial state of the electrode surface, apply equally to
nonelectrochemical methods. Furthermore, the results of the nonelectrochemical approach may be somewhat questionable because
(a) the specimen to be examined by electron diffraction and microscopy is usually removed from the initial surroundings and can

undergo additional changes that are not characteristic of the original system; (b) study of the Mössbauer effect requires application of radioactive isotopes; and (c) although ellipsometry has the advantage of being sensitive, nondestructive, and applicable in situ during the electrochemical experimentation, the evaluation of measured parameters is not always simple and straightforward.

However, the assumptions applied in the evaluation of the results of the electrochemical and nonelectrochemical measurements usually rely on the different parameters of the system, and similar conclusions support the validity of the findings.

D. Information on the Passivation Process

The mechanism involved in the phenomenon of passivity is basically investigated in two ways: (a) study of the kinetics of the electrochemical processes taking place on passive metals and alloys, and (b) study of the nature and structure of the passive film, with little direct experimental evidence about the processes in the active potential region. The Appendix gives the essential literature data on the subject of theories of passivity, with the author, method used in the investigation of the problem, type of passivating film suggested, thickness of the film (according to the model and, if given, by the experimental evidence) at V_p or V_{pp}, and the most important thoughts about the passivation process and the cause of passivity.

E. Passivation Theories

1. The Adsorption Theory

The adsorption theory has been developed by Uhlig et al. [20, 25–29, 69], Frumkin et al. [70], Kabanov et al. [36–38], and Kolotyrkin [30] (see Appendix). According to Kolotyrkin [30],

passivity is described as a specific case of the widespread phenomenon of the change in the kinetics of an electrodic reaction due to
the activated adsorption of the oxygen of water; the dissolution rate
is generally represented by

$$i = k_1 c' \exp\left(\frac{\alpha_1 F}{RT}\right) V,$$
(5)

where α is the transfer coefficient equal to βz; when the valency
of the dissolving ion, z, is unity, α is equal to the symmetry factor
β, or to the fraction of the potential available to lower the energy
barrier for dissolution.   The quantity c' is the number of atoms per
square centimeter, and if with the superimposition of adsorption,
c' changes with potential in such a way that

$$c' = c_0' \exp\left(-\frac{\alpha_2 F}{RT}\right) V$$
(6)

the total current becomes

$$i = k_1 c_0' \exp\left[\frac{(\alpha_1 - \alpha_2)F}{RT}\right] V.$$
(7)

Providing that in the first stage of the passivation process (at
$V_p < V < V_{pp}$) $\alpha_2 >> \alpha_1$ and in the second stage (at $V > V_{pp}$)
$\alpha_2 = \alpha_1$, this equation can explain the experimentally observed drop
of the current with potential and its independence of potential in the
passive region.   The basic anodic processes, even in the passive
potential region, are direct dissolution of the metal as one and the
competitive adsorption of anions as another.

In Uhlig's earlier view, one of the factors determining passivation (and chemisorption) is the ratio of the work function to the
enthalpy of sublimation ΔH_s.   If this ratio is less than unity, conditions are favorable to passivation because the electron would
escape more readily than the atom, favoring the chemisorption of
substances like oxygen.   A passive film is composed, then, from
chemisorbed atomic and molecular oxygen (supplemented perhaps by
OH and H_2O).   The formation of chemical bonds satisfies the surface

affinities of the metal without metal atoms leaving their lattice site.

Recently this model was discussed by Uhlig [69] in the light of the results reported by Germer and MacRae [71-73]. These studies (LEED) show that when oxygen atoms (ions) are chemisorbed, a specific number of metal atoms enter the approximate plane of adsorbed oxygen atoms to form a very stable structure of negative oxygen ions and positive metal ions, in which the ratio of one to another depends on oxygen pressure. This structure can be detected on the metal even when heating to higher temperatures eliminates the diffraction pattern of oxide (NiO).

It is argued that on typical transition metals (Ni, W, Cr, Ti, Ta) the formation of such a layer (i. e., $M \cdot O \cdot O_2$) proceeds with more favorable free energy of formation than the oxide formation, as they have unfilled d electron energy levels leading to the formation of strong chemical bonds between oxygen and the metal. These are the metals that typically exhibit passivity. For nontransition metals with filled d levels, such as copper or zinc, the heats of oxygen adsorption are expected to be lower, and the formation of oxides is more favorable. Such metals do not exhibit thin-film passivity.

Correlation of the observed onset of Wagner's passivity on alloys like Ni-Cu, Ni-Zn-Cu, and Cu-Ni-Al to the occupancy of the d levels of the alloys is given in support of the theory [29]. According to the theory, the same type of passive film (i. e., $M \cdot O \cdot O_2$) is formed in solutions, interposing a stable barrier between metal and electrolyte, displacing adsorbed H_2O and increasing the activation energy for the hydration and dissolution of the metal lattice. Such films are assumed to decrease the exchange-current density i_0 and thus to increase the polarization of the metal in the noble direction, where more oxygen can be adsorbed, which in turn forms nucleation of metal oxide. Thus in the passive state a thicker oxide film may be detected.

Briefly, the theory explains the onset of passivity by the formation of a thin adsorbed layer that either shifts the electrode potential
in the double layer or influences the kinetics of the anodic process;
that is, the important happenings occur in the small interface region
between the metal and the solution.

If, however, such a theory (e.g., Kolotyrkin's [30]) should
apply, one would expect a pure logarithmic increase in the current
at potentials below V_p, a very sharp peak, a pure logarithmic
decrease in the current until V_{pp} is reached, and the presence of
the adsorbed film in dimensions smaller than monomolecular ones
even at high passive potentials. This is in disagreement with
experimental evidence (see below). On the other hand, a chemisorbed film of $M \cdot O \cdot O_2$ composition in the electrolyte is not very
likely for the following reason: Most metals (Cr [74, 75], Ti [74,
75], Fe [51, 76], Ni [77]) can be passivated in a solution saturated
with deoxygenated argon — that is, with water oxygen — as long as
some water is present in the solution [77]; in such solutions, the
formation of molecular oxygen at low cathodic potentials is impossible, and without its presence the total number of millicoulombs
determined in the film (0.01 C/cm^2 according to Uhlig, or 1 to 2
mC/cm^2 according to Nagayama and Cohen [51] and Brusic [76]
at V_{pp}) and recalculated into film thickness would exceed reasonable limits for the adsorbed monolayer of oxygen ions. In the usual
correlation between the number of millicoulombs and film thickness
it is assumed that 0.5 mC/cm^2 corresponds to the monolayer of O^{2-}
ions if each metal ion adsorbs one oxygen ion. If the adsorbed
layer has a structure that contains both metal and oxygen ions (as
suggested by Germer and MacRae [71] for oxygen adsorption and
equally for the formation of the initial film formed with the adsorption of H_2O [78]), the resulting passivating film is several layers
thick — that is, a thin oxide is more likely than an adsorbed multi-

layer of O^{2-}. Finally, factors that favor adsorption may also favor the formation of oxides.

2. The Oxide-Film Theory

The oxide-film theory describes the state of improved corrosion resistance through the formation of a protective film on the metal substrate; this consists of the reaction products of the metal with its environment. Such a film is a definite new phase, even if it is as thin as a single monolayer [45] or thick enough to be stripped away from the surface and examined separately [19]. Electron diffraction [45, 51, 52, 56] and ellipsometric [76, 79-90] studies give the experimental evidence for the theory. In this case the physicochemical properties of the metal relative to a corrosive medium depend to a large degree on the properties of the protective film. The properties of the film, however, are not uniquely determined.

Adherents to this theory have different opinions on the potential at which the film forms, its thickness, the mechanism of formation, and, most important, the "cause" of passivity. In the earlier theories it was postulated that the passivation follows the formation of a "primary layer" of small conductivity, with porous character, which is sometimes due to precipitation of metal salt on and near the electrode [91]. In the pores the current increases, and by polarization at an "Umschlagspotential" ($V_F = V_p$, Fig. 1) an actual passive layer is formed. Thus the essential concept of the passivation process is connected with the change of the properties (chemical or physical) of the primary film at a certain potential. The passive film is free from pores and presents a barrier between the metal and the environment. It is electronically conductive and slowly corrodes in solution [21, 23, 46, 91].

These general ideas were further developed in detail by Sato and Okamoto [92, 93], Bockris, Reddy, and Rao [79], and Pavlov and Popova [94, 95].   Pavlov and Popova have studied the behavior of lead in sulfuric acid solution.   They have observed the formation of well-recognized crystals of $PbSO_4$ at lower potentials and have concluded that the passivation and depassivation phenomena take place in the intercrystalline spaces.   If these spaces are on the order of magnitude of the ionic diameters, the film behaves as a permeable membrane with the amount of lead ions formed exceeding the flux of SO_4^{2-} moving toward the lead surface.   Thus the pH in the area increases and basic sulfate of lead and lead oxide are formed, and the electrode is stably passivated.   If the intercrystalline spaces are of larger dimension, passivation can take place only on anodic polarization.

Sato and Okamoto have studied the i-V curve for nickel in acid solution and have calculated the passivation potential as a potential for NiO/Ni_3O_4 transition.   This was in fair agreement with the experimentally observed value.   The higher valence oxide was thought to have a self-repairing ability; that is, a higher valence oxide theory of passivity was suggested.   In contrast to this higher valence oxide film, a surface oxide film NiO has the same valency as the ions of active dissolution; thus it has no self-repairing ability and cannot passivate the metal.   Because anodic dissolutions at active potentials proceed at a high rate, Sato and Okamoto have suggested that the primary film forms by a dissolution-precipitation mechanism, although no direct experimental evidence was offered.   Bockris, Reddy, and Rao [79] have studied the passivation of nickel in acid solution by a combination of electrochemical methods and in situ ellipsometry.   In their view the primary oxide is relatively thick

($\sim 60\overset{\circ}{A}$), is formed by the precipitation, is poorly conducting, and is transformed into well-conducting, nonstoichiometric NiO. This transformation is thought to be the essential step in the passivation process [79].

This theory, however, is not generally accepted. As pointed out by Hoar [14], some of the best passivating films, as those on aluminum or tantalum, are poor electron conductors, whereas many poorly passivating films, such as CuO or PbO_2, are excellent electron conductors. Also, it may very well be that the experimental results, in particular the ellipsometric data, of Bockris et al. are in error due to the possible influence of the continuous electrode roughening in the active potential region (see, for example, Refs. [76 and 83]). Such an error may be the reason that the reported refraction index for the first passivating film is unusually high (about 4.0), that the changes of ellipsometric and coulometric data in the passive potential region are not consistent, and that the main indicator of the prepassive-film transition into electronically conducting film is not trustworthy. The latter was derived from the variation of the ellipsometrically determined change of the extinction coefficient with potential.

De Gromoboy and Shreir [96] argued in their experimental study of nickel in sulfuric acid that higher oxides may form directly from the metal and at the metal surface; the observed "passivation potentials" (determined from anodic charging curves) were found to correspond closely with the potentials calculated for $Ni \rightarrow NiO$, $Ni \rightarrow Ni_3O_4$, $Ni \rightarrow Ni_2O_3$, and $Ni \rightarrow NiO_2$, the phenomenon being sharply dependent on the presence or absence of small concentrations of thiourea. Similarly Arnold and Vetter [35] suggest the direct formation of NiO at the electrode. From the anodic charging curves these authors have concluded that the film reaches a monomolecular dimension at the potential of complete passivity.

This and many other similar hypotheses can be said to form the monolayer-oxide theories. Here the onset of passivity is due to the formation of a single phase, an oxide, which is very thin in this potential region (at V_p, $\theta < 1$) and grows to a monolayer at V_{pp} [14, 34, 35, 38-44]. Its formation is a potential-dependent reaction involving the direct oxidation of the metal by the discharge of water, which, once thermodynamically possible, is suggested to be kinetically easier in its crucial primary stage than is metal dissolution. However, metal or oxide dissolution can still be simultaneous electrode processes [39-42, 96]. The decrease in the anodic current is described as a result of a decrease in film-free area caused by film formation and its influence on the kinetics of the anodic reaction.

Krasil'shchikov [97] has suggested that the decrease in the anodic current may depend on the localization of the potential difference at the metal-solution interface. Experimental results have shown that the activation energy and the rate of dissolution on stainless steel depend very little on the temperature. Thus he has concluded that the process responsible for the high degree of passivity is connected with the temperature-independent process, such as the presence of the potential barrier in the metallic surface or, better, in the metal oxide. This barrier may be formed as the result of the O^- adsorption at the oxide surface and of the penetration of these ions into the semiconducting layer. The reaction of O^- with the cations of an oxide, such as occasional Ti^{3+} in TiO_2 or Ni^{3+} in the passivating film on nickel, may lead to a decrease in the current-carrier concentration and a decrease in the reaction rate.

In this respect and regarding the discussion on the initial passivations, as pointed out by an increasing number of writers [14, 15, 32, 33, 84], the oxide-film and adsorption theories do not contradict but rather supplement one another. In looking at the primary act of passivation as the formation of a tightly held monolayer containing oxide or hydroxide anions and metal cations, Hoar [14] concurs with "adsorptionist" Uhlig as well as "filmists" Schwabe and Arnold and Vetter. In the formation of such a film, adsorption may, however, play an important role [98]. Adsorption is the first stage of the process; in an oxygen atmosphere molecular oxygen is adsorbed, dissociated, and chemisorbed, whereas in an aqueous solution, water molecules or OH^- will be adsorbed and chemisorbed (see Fig. 3a). The second stage is the intrusion of metal ions from the lattice into the adsorbed layer (see Fig. 3b), which in the third stage may grow in the third dimension. With time and potential (at $V > V_{pp}$) oxide growth is a function of the film's ionic and electronic conductivity. The thickness of the film will then be further responsible for the retardation of the anodic processes, both growth of the oxide and anodic dissolution, since the ions have to pass through the protective film. Thus one may speak of a combined oxide-film-adsorption theory of passivity [15].

III. CHARACTERISTICS OF ACTIVE AND PASSIVE STATES

As far as the electrode material is concerned, both anodic dissolution in its simplest formulation

$$M \rightarrow M^{z+} + ze^- \tag{1}$$

and anodic film formation by a reaction of the type

$$M + \frac{z}{2} H_2O \rightarrow MO_{z/2} + zH^+ + ze^- \tag{2}$$

are dissolution processes.

A more realistic formulation for ordinary aqueous corrosion conditions is in fact

$$M + xH_2O \rightarrow M^{z+} \cdot xH_2O + ze^-, \tag{8}$$

or

$$M + (x+y)H_2O \rightarrow [M(OH)_x O_y]^{z-x-2y} + (x+2y)H^+ + ze^- . \qquad (9)$$

When a complexant C is present,

$$M + yC^{n-} \rightarrow (MC_y)^{z-yn} + ze^- , \qquad (10)$$

or for a nonaqueous solvent

$$M + xSol \rightarrow M^{z+} \cdot Sol_x + ze^- . \qquad (11)$$

In all cases the dissolution process (Eqs. (8) through (11)) requires less energy than that required for the removal of the metal from the lattice as a "bare" cation M^{z+}; this is due to the large negative free energy of hydration (solvation) or of the complexing of the cation.

Similarly water molecules in the Helmholtz double layer at a metal surface under "anodic" conditions — that is, at potentials more positive than the potential of zero charge (pzc) — tend to orientate with their oxygen toward the metal, presenting the possibility of the following reactions:

$$M \cdots O \overset{H}{\underset{H}{\diagup}} \rightarrow M-OH + H^+ + e^- , \qquad (12)$$

$$M \cdots O \overset{H}{\underset{H}{\diagup}} \rightarrow M-O + 2H^+ + 2e . \qquad (13)$$

Again the formation of a surface monolayer of "solid" hydroxide or oxide requires a much smaller energy because of the large negative free energy for cation association with surrounding oxide ions. Thus both processes, Eqs. (1) and (2), are similar; anions (molecules) from the environment replace the metal electrons and are the "binders" for the metal cations.

Which of the two processes prevails — the formation of the solvated ions (Eqs. (8) through (11)), that is, dissolution with an appreciable rate, or the formation of the solid film (Eqs. (12) and (13)),

which may or may not slow down the dissolution — depends on the
potential, the composition of the solvent, the specimen geometry
and microstructure and, to a great extent, on the properties of the
solid film. The last, if it forms, separates the reacting substances,
and further reaction is possible only if some of the reacting species
can diffuse through the layer. Thus the reaction depends on its
thermodynamic stability, crystal structure and habit (which determine the adhesion between layer and metal), thickness, conductivity, phase-boundary reactions, and diffusion and transport processes
to and from the boundary. All strongly influence the kinetics of
thermodynamically possible reactions.

A. Metal Dissolution

1. Active Dissolution

Active dissolution occurs with the majority of metals when they
are continuously made the anode in a solution in which they can form
readily soluble products. This applies even for noble metals, such
as palladium or silver in orthophosphoric acid [99].

The following experimental observations have been made:

1. The lattice cations immediately concerned are those at sin-
gularities, such as kink sites, as indicated by the nonrandom dissolution of the metal surface to produce crystallographic facets. From
these kink sites mobile adions may be formed as intermediate states
[100].

2. Dissolution rates on low-index planes are normally slower
than those on high-index planes [62], although in the potential region where the film-formation process becomes possible the opposite
may be true [63].

3. The process may involve the intermediate formation of lower valence cations, as in the dissolution of copper [101], anodic dissolution of beryllium in anhydrous media [102], or the dissolution of indium [103, 104] and bismuth [103].

4. The dissolution may involve the hererogeneous disproportionation reaction of the low-valency intermediate, as shown for the anodization of indium and concentrated indium amalgams and bismuth amalgam in very concentrated solution [103].

5. The intervention of anion catalysts (hydroxyl or others) occurs in the process, as shown for the dissolution of iron [41, 42, 66, 105, 106], nickel [74, 77], and other metals.

These are complicated factors determining, to a greater or lesser extent, the kinetics and mechanism of anodic dissolution.

The net rate at which ions are removed from the crystal is given by an expression of the type

$$i = i_0 [\exp(\frac{\alpha_a F\eta}{RT}) - \exp(-\frac{\alpha_c F\eta}{RT})] , \qquad (14)$$

which for high anodic η gives

$$\eta = \frac{RT}{\alpha_a F} \ln i - \frac{RT}{\alpha_a F} \ln i_0 , \qquad (15)$$

where i is the current density at anode overpotential η, α is the anodic or cathodic transfer coefficient, i_0 is the exchange-current density for the dissolution $\rightleftarrows$ deposition reaction, and the other symbols have their conventional meanings.

Using i_0 as the indicator, Piontelli [107] has made a useful empirical classification of metals, later modified by Vermilyea [9], which is given in Table 1. According to this, iron, cobalt, nickel, and the transition metals in general exhibit a very sluggish dissolution reaction. Hoar [98] has recently argued that this may be a

Table 1

Piontelli's Classification for the Electrodeposition of Metals[a]

Normal metals $(i_0 \approx 10\text{-}10^{-3} \text{ A/cm}^2)$	Intermediate $(i_0 \approx 10^{-3}\text{-}10^{-8} \text{ A/cm}^2)$	Sluggish $(i_0 \approx 10^{-8}\text{-}10^{-15} \text{ A/cm}^2)$
Lead	Copper	Iron
Tin	Zinc	Cobalt
Mercury	Bismuth	Nickel
Thallium	Antimony	Transition metals
Cadmium		Noble metals
Silver		

[a]Reproduced from Ref. [107] by courtesy of R. Piontelli.

function of the incomplete d band in nickel and other transition metals, which may hinder the separation of metal cations from electrons in the lattice, that is, may hinder metal dissolution. Therefore an easy passivation of those metals may occur not as a result of altered adsorption properties as argued by the adsorption passivation theory (see Section II.E.1), but as a result of a slow dissolution, thus making it easier for the metal → metal-oxide potential to be reached [98].

From the variation in i_0 with the pH and the concentration of Fe^{2+}, as well as from the Tafel slope for the steady-state anodic log i-V curve (determined as $dV/d \log i = 0.040$ V) and other kinetic parameters, Bockris, Despic, and Drazic [66] have suggested the following mechanism for iron dissolution:

$$Fe + OH^- \rightleftharpoons FeOH + e^- , \tag{16}$$

$$FeOH \xrightarrow{rds} FeOH^+ + e^- , \tag{17}$$

$$FeOH^+ \rightleftharpoons Fe^{2+} + OH^- , \tag{18}$$

where "rds" is the rate-determining step.

The presence of the $FeOH_{ads}$ was indeed confirmed by Weissmantel, Schwabe, and Hecht [108] by mass spectroscopy, although
more than one mechanism is suggested to follow this initial step.
For example, Heusler [105, 109] and Schwabe [42] emphasize the
possibility of the following mechanism:

$$Fe + H_2O \rightleftarrows FeOH + H^+ + e^- , \qquad (19)$$

$$Fe + H_2O \rightleftarrows FeOH^- + H^+ , \qquad (20)$$

$$FeOH_{cat} + FeOH^- \overset{rds}{\rightleftarrows} FeOH^+ + 2e^- + FeOH_{cat} , \qquad (21)$$

$$FeOH^+ + H^+ \rightleftarrows Fe^{2+} + H_2O , \qquad (22)$$

where FeOH acts catalytically in the reaction.

Another mechanism, similar to that given by Eqs. (16) through
(18), is suggested by Sato and Okamoto [93] for nickel in acid solution. They have emphasized the possibility of the rate-determining
step being changed from the first to the third step as the $NiOH^+$
ion concentration increases in the vicinity of the electrode. The
rate is also influenced by the amount of OH^- ions involved in a
reaction cycle. Also, as shown for iron [66], the speed of the reaction $Fe^{2+} + 2e^- \rightleftarrows Fe$ depends on the anion present in the solution; i_0 decreases in the order $ClO_4^- > SO_4^{2-} > Cl^- > CH_3COO^- >$
NO_3^- , and this (except for NO_3^-) is the order of the specific adsorbabilities of these ions on mercury [66].

Some evidence for the importance of the adsorption in the dissolution reaction is given in the work of Marshakov and Altukhov [110].
Anodic polarization curves on iron, cadmium, zinc, copper, and
silver were measured with and without an ultrasonic field. The field
may increase the atomic energy in the metal lattice as well as the
energy of the ions in the solution. In iron dissolution the prevailing
effect is the inhibition of the rate-controlling anion adsorption and
thereby a decrease in dissolution rate. In the case of cadmium,
copper, and silver the dissolution rate increases in the presence of

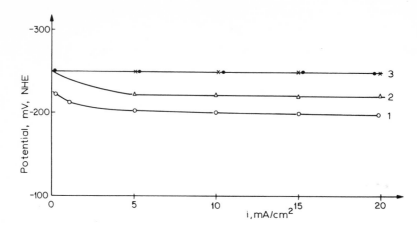

Fig. 5. Anodic polarization curves for iron in 1 N H_2SO_4.
Curve 1: no scouring; curve 2: with scouring at 500 rpm; curve
3: with scouring at 1000 and 2000 rpm. Reprinted from Ref. [74]
by courtesy of Pergamon Press.

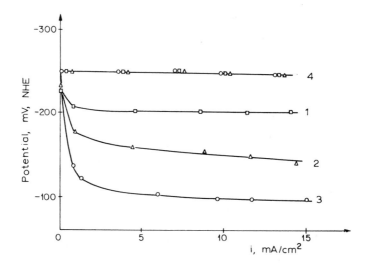

Fig. 6. Effect of varying anion composition of solution on the
anodic polarization curve for iron. Curve 1: without scouring, 1 N
H_2SO_4; curve 2: without scouring, 1 N HCl; curve 3: without scouring, 1 N H_2SO_4 + 0.01 N KI; curve 4: for all solutions with scouring
at 2000 rpm. Reprinted from Ref. [74] by courtesy of Pergamon Press.

the ultrasonic field, either due to an increase in the energy of the metal atoms in a lattice (Cd) or due to the acceleration of the rate-controlling ion transport from the electrode (Ag, Cu).

Direct evidence for adsorption inhibition in the process of iron and nickel dissolution has been reported by Schwabe and Kunze[111-113] and recently by Tomashov and Vershinina [74] through use of the method of continuous surface renewal of a solid metal surface in contact with the solution and through the simultaneous polarization of the electrode by an external source of potential or current[74]. The metal surface was renewed by continuous scouring with a corundum disk rotating at a rate varied from 250 to 4500 rpm. Anodic polarization curves for iron in 1 N H_2SO_4 as well as in HCl and 1 N H_2SO_4 + 0.01 N KI, obtained in the presence of air over the solution with and without scouring, are shown in Figs. 5 and 6.

The overvoltage of the reaction decreases with increasing scouring rate, which in turn determines the completeness of the mechanical removal of the adsorbed anions and molecules (Fig. 5). When accompanied by sufficiently vigorous surface removal, the anion in the solution (SO_4^{2-}, Cl^-, I^-) does not significantly affect the shape of the polarization curve (Fig. 6), and the iron electrode becomes almost completely unpolarizable within the range of current densities used. It was concluded that the active anodic dissolution is slowed by the transfer of Fe^{2+} ions from the surface to the solution over an adsorption barrier, rather than by the slow charge-transfer step. Also the renewal of the electrode surface enlarges the exchange-current density, so that the iron electrode approaches a reversible electrode potential with respect to its ions. Confirmation of these hypotheses is given by the following experiments:

1. In a solution where air is replaced by argon and where Fe^{2+} ions are added in a concentration of 1 M, the theoretical value of -0.44 V for a standard iron electrode was measured.

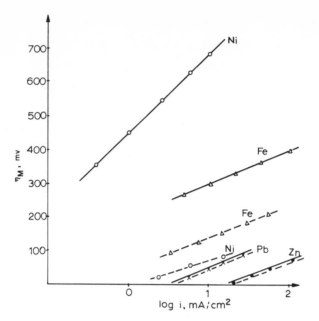

Fig. 7. Overpotential for metal dissolution (η_M) as a function of log current density for nickel and iron in 1 N H_2SO_4, and lead and zinc in 1 N HCl without scouring (solid curves) and with scouring (broken curves) at 2000 rpm. Reprinted from Ref. [74] by courtesy of Pergamon Press.

2. The effect of scouring is shown to be more pronounced if a given metal has a small exchange current (Fig. 7). Iron and nickel show a considerable decrease in the overpotential of metal ionization with scouring; however, lead and zinc ("normal" and "intermediate" metals — see Table 1) show hardly any change in their anodic behavior; anodic dissolution on such metals occurs without marked inhibition even in the absence of surface renewal.

In contrast, it was found [74] that on readily passivating metals, such as titanium and chromium, even scouring at a rate of up to 2000 rpm does not lead to a significantly large Tafel region, but rather to anodic passivation (Fig. 8). However, the open-circuit

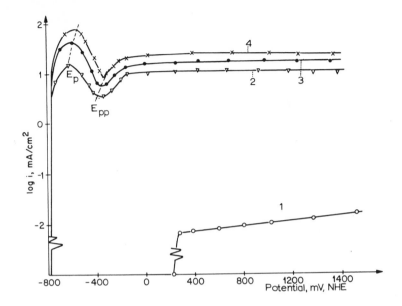

Fig. 8. Anodic potentiostatic curves for titanium in 1 N H_2SO_4, in air. Curve 1: without scouring; curves 2, 3, and 4: with scouring at 500, 1000, and 2000 rpm, respectively. Reprinted from Ref. [74] by courtesy of Pergamon Press.

potential of titanium was displaced in the negative direction by almost 1 V (i.e., to −0.7 V) by continuous surface scouring. Initially a 30-mV shift of the potential from its open-circuit value causes a considerable current to appear, identical for all rates of surface renewal; that is, titanium dissolves anodically because in this potential range the oxygen-adsorption rate is not high enough to passivate the continuously renewed surface.

The mechanism of the anodic dissolution of titanium may involve the direct formation of Ti^{3+} without any solution-soluble intermediates, as suggested by Armstrong et al. [114]; this may explain the absence of a scouring-rate effect. The oxygen-adsorption rate, however, is assumed to increase on further increase in potential, and at some potential V_p, depending on the scouring rate, it becomes

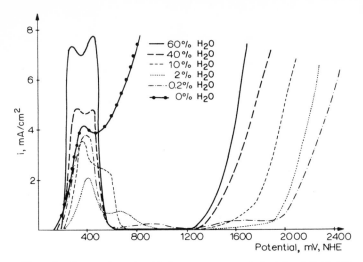

Fig. 9. Current-potential curves on nickel in $2N$ H_2SO_4-DMF-water mixtures. Reprinted from Ref. [77] by courtesy of Pergamon Press.

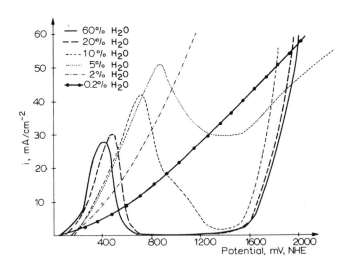

Fig. 10. Current-potential curves on nickel in mixtures of $2N$ H_2SO_4-acetonitrile-water. Reprinted from Ref. [77] by courtesy of Pergamon Press.

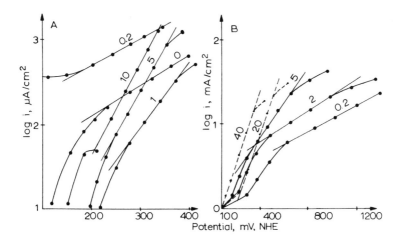

Fig. 11. Tafel slopes in DMF-water (A) and acetonitrile-water (B) mixtures. Numbers on the figure denote the water content in percent. Curve marked 40 in B is displaced by -0.1 V. Reprinted from Ref. [77] by courtesy of Pergamon Press.

equal to that of surface removal, that is, passivation begins. It is postulated [74] that in the presence of scouring the passivating film is initially adsorbed, thickening at higher potentials and covering an increasing portion of the electrode surface. However, it is thinner and less perfect than it is when formed without continuous scouring. Thus the current in the passive potential region is several orders of magnitude higher than it is on the electrode with no surface renewal. The experiments on titanium, as well as on nickel and chromium, which under similar conditions can also undergo passivation, were conducted in deoxygenated argon, and that may be taken as evidence that these metals passivate with water oxygen [74].

The influence of water on the anodic behavior of nickel was examined by Schwabe and Schmidt [77]. Experiments were conducted in mixtures of 2 N H_2SO_4, dimethylformamide (DMF) and water (Fig. 9), and of 2 N H_2SO_4, acetonitrile, and water (Fig. 10). Figures 9, 10, and 11 show the following:

1. The Tafel slope, $d\eta/d \log i$ (Fig. 11) determined from the i-V curve in the active potential range, is highest in the water-free electrolyte, decreasing to a limiting value with an increase in water content of up to 5% in the DMF-water mixtures and 20% in the acetonitrile-water mixtures. Thus an increase in water content increases the anodic dissolution at prepassive potentials.

2. The presence of water is a necessary factor for the appearance of passivation.

3. Again, the passivation reaction depends on the nature of the strongly adsorbable anions and molecules.

The presence of the active, adsorbable anions also influences the dissolution reaction. The effect of halide ions on metal dissolution, as well as on passivation and the breakdown of passivity, has been widely studied. Schwabe [42] has shown that, for otherwise similar experimental conditions, Cl^- ions hinder the anodic dissolution in the lower potential region but strongly accelerate it at higher potentials.

The electrodissolution kinetics of iron in neutral and alkaline solutions in the presence of chloride ions were recently studied by Asakura and Nobe [115, 116] through transient-polarization techniques. It has been observed that the anodic Tafel slopes were 40 mV/decade for fast polarization and 80 mV/decade for slow or steady-state polarization. These data have been discussed in the light of Bockris' mechanism (see Eqs. (16) through (18)). In the fast anodic polarization the data can be explained by the second electron transfer being the rate-determining step, with the assumption that the concentration of H^+ at the electrode is not appreciable and can remain nearly constant. In addition, the reaction of Cl^- with formed Fe^{2+} is assumed to form a complex in the solution and to be at equilibrium

$$Fe^{2+} + mH_2O + nCl^- \rightleftharpoons [Fe(OH)_m(Cl)_n]^{2-n-m} + mH^+ . \qquad (23)$$

Thus the overall reaction can be expressed by

$$Fe + mH_2O + nCl^- \rightleftharpoons FeX + mH^+ + 2e , \qquad (24)$$

where FeX represents the formed complex ion with charge $2-n-m$. It is assumed, however, that in the steady state the concentration of H^+ at the electrode is different from that in short-term experiments, being determined by the diffusion rate of the species from the electrode to the solution; this difference is directly responsible for the appearance of a higher Tafel slope [115, 116].

2. Breakdown of the Film; Activation

Partial or complete breakdown of passivity, with the onset of a high dissolution rate (i.e., corrosion) is brought about by any process that leads to (a) partial or complete removal of the passivating film or (b) its structural change, as observed during the electrolytic breakdown of passive films on aluminum, titanium, and tantalum [117].

In aqueous solution film removal may be effected through electrochemical reduction or oxidation, chemical dissolution, mechanical disruption, and especially by the presence of active anions (chlorides). These may either encourage the activation of the "weak" sites in the film (mechanical flaws, inclusions, etc.) or introduce new ones. Depassivation is usually manifested by a fairly sudden fall in potential to a value slightly more noble than that required for passivation, that is, to the activation or Flade potential.

Metal oxides on nickel [96], chromium [74], and especially aluminum [14] are not readily reducible, hydrogen ions being reduced instead. The vigorous evolution of hydrogen gas usually leads to depassivation. Oxide on iron is, on the other hand, easily reducible

in many solutions, and it is often possible to estimate the film
thickness from the charge required to produce the full activation.
The process may be, however, more complicated, involving the reduction of the passive film to an Fe(II) compound. This compound
may be easily soluble, and its dissolution may produce a negative
error in the film-thickness estimate. Furthermore, if the dissolution
proceeds nonuniformly and produces pitting, the current due to iron
dissolution may prevail and mask the current due to oxide reduction.

Kruger [118] has observed, by optical and transmission electron
microscopy, discrete, noncontinuous cathodic sites on iron that
had been passivated and then placed in a dilute copper sulfate solution until passivity had decayed, as indicated by potential measurements. It was determined indirectly (through copper deposition at
the cathodic sites during the attack on the passive film) that the
breakdown depends on the properties of the passivating film. In
turn these are affected by the metal surface since surface orientation influences the number of breakdown sites [118].

Using transmission electron microscopy, Pickering and Frankenthal [63] have concluded that physical or chemical inhomogeneities
in the passive film, not associated with dislocation termini or grain
boundaries in the alloy, appear to be the most probable sites for the
breakdown of the film on Fe 24% Cr alloy, when the potential is
slowly swept from the passive to the active region. In this case
local breakdown and crystallographic pitting were observed on many
crystallographic planes, indicating, in contrast to the results of
Kruger [118], that the nucleation of film breakdown is independent
of substrate orientation. (Effects of the substrate arising from multi-
atomic steps cannot be ruled out since an indication of possible
mutual alignment of some of the pits does exist.) However, if the
activation were imposed by a fast change in potential, the effects
of film inhomogeneities would be lost, and the film would be

reduced more or less uniformly, at least over each grain, giving the appearance of general etching.   These as well as similar results of Kruger's for the reduction of the passive film on iron [82] have shown that an "aggressive anion" is not necessary for the initiation and initial propagation of pitting.   This refutes the suggestion of Kolotyr-kin [119].

Active anions may, however, produce pitting and high currents in the passive potential region (see Fig. 1).   Their influence on film breakdown, and therefore on increased metal dissolution, is still a subject of much discussion [14].   This phenomenon has been studied on stainless steel [120-129], aluminum [131], zirconium [132], magnesium [133], and nickel, cobalt, titanium, tantalum, as well as iron [14, 130, 134-136].

The observed pitting is not crystallographic, but rather hemi-spherical, with the formation of highly reflecting surfaces.   Break-down sites are usually observed at the grain boundaries — that is, at points where the film is less perfect.   At high anodic potentials and with concentrated solution, the sites may become so numerous that overall electrobrightening of the surface occurs.   From the standpoint of the film theory, the rupture of the film takes place through the onset of gradual displacement of oxygen in the protec-tive film by Cl^- on reaching the breakdown potential V_b (see Fig. 1). Increased anion adsorption should in turn produce increased anion penetration into the film, increasing the ionic conductivity of the oxide.   Thus the oxide becomes able (at certain points) to sustain a high current density and to produce electrobrightening by random removal of cations in the form of a soluble complex [14, 15].

As an alternative possibility, Hoar [14] has postulated a "mechanical" mechanism for anion penetration, in which he sug-gests that the breakdown is caused by the lowering of interfacial tension or interfacial free energy with the progression of anion

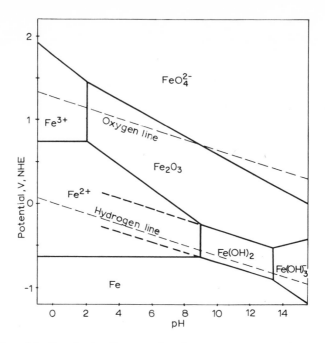

Fig. 12. Pourbaix diagram for the system Fe–H_2O at 25°C, taking into consideration passivation by $Fe(OH)_2$ and Fe_2O_3 only and assuming a total ionic concentration of 10^{-6} M/ℓ. Reprinted from Ref. [137] by courtesy of Pergamon Press.

adsorption. Finally a type of peptization occurs; the adsorbed anions repel one another, with the result that the oxide, to which they are attached, ruptures. Each of these alternative mechanisms might be controlled by the rate of anion adsorption. Thus, as Hoar has suggested, the presence of the anions in solution, or the ratio of the anion/water concentration, should not be neglected in discussing the thermodynamically possible anodic processes on metals.

B. Thermodynamics of the Active–Passive Transition

The factors that determine whether an anode in aqueous solution is active, passive, or brightening are (a) the metal–solution potential difference [14, 24, 137] and (b) the anion/water ratio [14].

1. The Metal-Solution Potential Difference

The metal-solution potential difference, combined with the influence of pH, determines whether the formation of aquocations (Eqs. (8)
through (11)) or oxide formation (Eqs. (12) and (13)) is the thermodynamically prevailing possibility.   Such thermodynamic data, even
if unsupplemented by information of a kinetic nature, are usefully
expressed in the graphical form devised by Pourbaix [137] giving
the domains of thermodynamic stability for a variety of compounds
in potential-pH diagrams.   The Pourbaix diagram for iron is shown
in Fig. 12.   In order that oxide formation may proceed, the anode
potential must be at least as high as that for oxide or hydroxide
formation from the metal and water in the particular solution.

If the oxide forms as a "monolayer, " there is a question as to
whether or not one can apply with some assurance these thermodynamic data, obtained on the bulk oxides.   Some experimental evidence does indicate that for many metal-oxide systems the free
energy of formation of the first monolayer is indeed close to that
observed for the bulk phase [138], and hence no appreciable difference in the reversible potential should be expected.   On the other
hand, Vermilyea [9] has shown that, if the first monolayer forms by
two-dimensional nucleation, the potential of the two-dimensional-
film formation may be lower than that expected from the thermodynamics, the actual potential being dependent on the relative
values for the interfacial energies between metal-film, film-solution,
and metal-solution.

At higher pH and potentials a higher oxide phase, often more
passive, may be formed.   At lower pH aquocations are stable.   However, metals with low exchange-current densities for the dissolution
reaction (see Table 1) can be more readily polarized beyond the
extrapolated line (Fig. 12) to form a metastable oxide (hydroxide).

Consequently, as already discussed, relatively easy passivation may occur for metals having a sluggish dissolution reaction. In neutral and slightly alkaline solutions passivation occurs readily

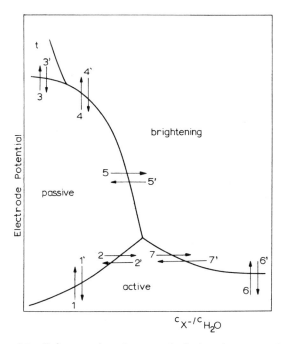

Fig. 13. Scheme showing anode behavior at various potentials in solutions of various anion/water concentration ratios c_{X^-}/c_{H_2O}. The processes are as follows: 1, anodic passivation; 1', activation — the Flade relation; 2, activation by adding "corrosive" anions; 2', inhibition by removing (or sequestering) "corrosive" anions; 3, transpassivation by raising the anode potential; 3', repassivation by lowering the anode potential; 4, breakdown, via pitting, leading to anodic brightening, by raising the anode potential; 4', repassivation by lowering the anode potential; 5, breakdown, via pitting, leading to anodic brightening by adding "corrosive" anions; 5', repassivation by removing "corrosive" anions; 6, anodic brightening by raising the anode potential in concentrated solution; 6', anodic etching of brightening metal by lowering potential in concentrated solution; 7, anodic brightening, from etching, by increasing the anion concentration; 7', anodic etching, from brightening, by decreasing the anion concentration. Reprinted from Ref. [14] by courtesy of Pergamon Press.

for nearly all metals through the formation of stable oxides. Some metals, such as tin and especially tantalum and molybdenum, form stable oxides (i. e., passivate even at low pH values). At very high potentials metals may form oxides that can partially redissolve as hydrolyzed ions, such as CrO_4^{2-} or FeO_4^{2-}, and in the transpassive potential region corrosion may be expected, with possible occurrence of secondary passivity.

2. The Anion/Water Ratio

The second factor is Hoar's [14] extension of the Pourbaix diagram, shown schematically in Fig. 13. It illustrates the joint influence of the measurable anode potential and the anion/water ratio. At a low c_{X^-}/c_{H_2O} ratio the electrode may behave according to the appropriate Pourbaix diagram, showing regions of active, passive (sometimes having more than one oxide with passivating properties), and transpassive behavior. At higher c_{X^-}/c_{H_2O} ratios the electrode brightens (process 7), this becoming more likely at higher potentials (process 6). Since thermodynamic data on ion adsorption on metals and oxides are very scarce, the diagram should be taken only as a qualitative display of experimental results.

Thermodynamic data can, of course, give only information on what is possible, without telling us which oxide or hydroxide is in fact most readily formed from a number of possible candidates that may be thermodynamically stable under a particular set of conditions, and without telling us whether any particular oxide is protective.

C. Properties of Passivating Films

1. Thickness

Film thickness increases with both time and electrode potential, and depends inversely on the corrosion current (see Fig. 1) and the

electron conductivity of the film.   There is some experimental evidence (obtained from coulometry and ellipsometry — see Appendix)
that the thickness varies from monomolecular dimensions in the
active potential region up to dimensions of several hundred angstroms,
but mostly up to dimensions of less than 100 Å , in the passive potential region.   For example, on iron in a slightly alkaline solution
thickness is determined to be about 6 to 10 Å [76] at V_p and between
30 and 50 Å [51, 52, 76, 83] at the potential just below the oxygen-
evolution reaction.   Even when of monomolecular dimensions, the
film has essential passivating properties.

2. Composition and Structure

The results of several experimental approaches, such as anodic
or cathodic charging curves, microchemical analysis, ellipsometry,
electron (and x-ray) diffraction, and Mössbauer spectroscopy, still
do not identify the passivating film uniquely, as can be seen from
the rather complicated but widely studied case of the passive film
on iron.   Nagayama and Cohen [51, 52], using the electron-
diffraction method, have shown that the passive film on iron is a
two-phase oxide of Fe_2O_3 and Fe_3O_4 having a cubic structure with a
mean lattice parameter of 8. 37 ± 0. 04 Å if attached to the metal
surface, or of 8. 3 ± 0. 04 Å if separated from the metal surface, that is,
γ-Fe_2O_3 with or without Fe_3O_4.   Similar results were obtained by
Foley et al. [61].   Since the lattice parameters of these two oxides
are similar and since γ-Fe_2O_3 contains all the reflections of Fe_3O_4,
the results cannot positively identify the presence of Fe_3O_4.   However, it has been suggested [51, 61] that Fe_3O_4 may be near the iron
surface and Fe_2O_3 on the outside.   This is because the structure
$Fe|Fe_3O_4|Fe_2O_3$ is thermodynamically the most stable system [21,
23, 139].

Electron-diffraction investigations on (a) the passivated and then reduced electrode (reduced in the active potential region) and (b) the electrode that is initially oxide-free and then potentiostated in the active potential region show the presence of Fe_3O_4 [67] in the former case and of either Fe_3O_4 or FeOOH, depending on the crystallographic orientation in the latter. This may be taken as indirect evidence for the presence of Fe_3O_4 in the passive film.

On the other hand, chemical analysis of the passive film separated from the metal [140] has shown the presence of only Fe^{3+}, with the amount of Fe^{2+} estimated at less than 2.5%, which suggests that γ-Fe_2O_3 is the essential and most probable constituent of the passive film. More recently, combined electrochemical and ellipsometric methods have led to the same conclusion [76] supported by the results of a Mössbauer study [141], in which no Fe_3O_4 could be detected in the passive film.

Brusic [76] has suggested (on the basis of coulometric and ellipsometric results obtained in situ) that the initial film on iron in slightly alkaline solution at $V < V_p$ is $Fe(OH)_2$, which transforms completely into Fe_2O_3 at higher potentials ($V > V_p$). The identification of $Fe(OH)_2$ by electron diffraction may be particularly difficult because (a) being the component of lower valency iron, it oxidizes easily to a higher oxide or hydroxide and (b) hydroxides in general, when subjected to the influence of an electron beam, may be transformed into anhydrous oxides. The latter has been shown by Armstrong et al. [104] with a passivating film on indium, initially determined $In(OH)_3$ being transformed, within the microscope, into In_2O_3.

Noncrystalline oxides are found on aluminum, tantalum, and titanium when formed at low potentials (i. e. , below the breakdown voltage) and crystalline γ-Al_2O_3, β-Ta_2O_5, and TiO_2 are found at higher potentials [117]. Whether potential or thermal effects induce crystallization is not known.

Tomashov et al. [142] have shown that TiO_2 indeed forms in strongly oxidizing media or at high potentials (> 8.0 V) and that spontaneous passivation or anodic passivation at lower potentials leads to the formation of lower oxides, such as the mixed oxide $Ti_2O_3 \cdot 3$–$4\,TiO_2$ or even Ti_3O_5.

Gellings [143] has summarized the oxidation states most often found in the metallic oxides for many metals, in an attempt to correlate the bond strength in the oxide (i.e., the charge/radius ratio of the metallic cation) to its passivating properties. The most favorable charge/radius ratio seems to be between 4 and 7 in the intermediate range of pH. Accordingly, metals that can form several oxides, once passive, may easily show a breakdown of passivity under reducing conditions if they form lower valency oxides (e.g., Fe or Cr). Alternatively, if a higher valency state is possible (Cr^{6+}, Mo^{6+}, Fe^{6+}), the metals may dissolve in the transpassive region. For example, chromium passivates to form Cr_2O_3 [144] but may dissolve at higher potentials as CrO_4^{2-} [14, 15]. Similarly, electron-diffraction results on iron in the transpassive region [59] show that γ –Fe_2O_3 was not present, the oxide being identified as Fe_3O_4. These results are somewhat surprising and may be in error. They certainly show, however, that the compositions of the passive and transpassive films may not be the same, which, until these results, had not been directly indicated.

Fleischmann and Thirsk [45] have succeeded in determining the composition of very thin films, practically monomolecular, such as $Cd(OH)_2$ on cadmium amalgam and ZnO on zinc. However, the determination of the composition of the initial film on metals, formed in the active potential region, as well as many other parameters of its characterization, requires much more data. In this respect two experimental techniques seem promising: high-energy photoelectron spectroscopy and photoelectric polarization, the latter method being in situ, and the former being direct inasmuch as electron diffraction is direct.

Briefly, photoelectron spectra are generated when photoelectron emission from a surface film, bombarded with x-rays, is analyzed with respect to electrostatic energy. This technique, commonly known as ESCA (electron spectroscopy for chemical analysis), yields information concerning not only the elements being studied but also the chemical environment of the atoms concerned [145]. Such information will be particularly valuable in the study of thin two-phase films.

The in situ photoelectric polarization technique developed by Oshe et al. [146-148] consists of illuminating the surface film with light of an energy appropriate to the optical absorption region of the film. The polarity and magnitude of the photocurrent produced give both the type and degree of nonstoichiometry, positive polarity corresponding to metal excess and negative polarity to metal deficiency.

Applying this technique to the passivation of iron in sodium sulfate, Oshe, Rosenfel'd, and Doroshenko [148] noted that, as the passivation potential was exceeded, the surface film, which was metal excess in character, became more stoichiometric. This, it was suggested, was due to the Fe_3O_4, present in the active region, being oxidized to Fe_2O_3.

3. Conductivity

The complicated phenomenon of conductivity of the anodic oxide film is discussed elsewhere in this volume. Here the subject is mentioned briefly insofar as it is relevant to the growth of the film and the shape of the i-V curve.

a. Electronic Conductivity. Passive layers on iron, nickel, chromium, and other metals are reported to have a good electron (or hole)
conductivity [14, 21, 23, 149]. For electrochemists this is consistent with the establishment of oxidation-reduction reactions, such
as Ce^{4+}/Ce^{3+}, HNO_3/HNO_2 [23] (the latter leading to the chemical
passivation of iron, for example), or oxygen evolution. Measurement of the resistance of the passive film on iron formed in nitric
acid indicates R_{film} < 0.1 ohm/cm^2; that is, if the thickness is
about 100 Å, the conductivity σ is about 10^{-5} ohm^{-1} cm^{-1} [150].
This value is not inconsistent with the value measured on epitaxially
grown single crystals of γ-Fe_2O_3 [151], where, at room temperature,
the conductivity was determined as being between 3×10^{-8} and
5×10^{-3} ohm^{-1} cm^{-1}, depending on the addition of foreign atoms or
departure from stoichiometry. The latter plays a particularly important role in thin anodic films. Oxide films on aluminum and tantalum
do not show the ability to evolve oxygen (i.e., they are poor electron conductors), but they are able, on the other hand, to sustain
fairly large electric fields and to grow to considerable thickness
through ionic movement. An intermediate behavior is shown by the
oxides of tin and titanium, on which oxygen evolution and film
thickening can both occur. Titanium oxide is the more complicated,
showing a strong influence on the semiconducting properties. For
example, Paleolog et al. [152] have studied the oxidation-reduction
reaction

$$Fe(CN)_6^{4-} \rightleftharpoons Fe(CN)_6^{3-} + e^-$$

and observed that the cathodic reaction is somewhat polarized, but
the anodic one is almost absent. Furthermore, the evolution of oxygen occurs only at high potentials (close to that of electrolytic
breakdown). In contrast, reduction reactions proceed easily [153].
Further discussion of this point is outside the scope of the present

work. The role of oxides in electrode reactions is to be dealt with in a subsequent contribution in this series.

b. Ionic Conductivity and Film Growth. Ionic conductivity occurs predominantly through the movement of defects in the lattice [149] and depends on the concentration gradient and on the electric field strength. The potential gradient (i. e., the field E) is assumed to reduce the height of the energy barrier for the ions moving with the field

$$i = const\ n\ exp\left(- \frac{U - \beta\, zeaE}{kT}\right),$$ (25)

and to increase the barrier for the ions moving against the field

$$i = const\left(n + a\, \frac{\partial n}{\partial x}\right) exp\left[- \frac{U + (1 - \beta)zeaE}{kT}\right],$$ (26)

where n is the number of mobile ions per unit volume, a is the jump distance ($\approx$ lattice constant), x is the distance through the oxide, and U is the chemical activation energy. With very strong electric fields, on the order of 10^6 to 10^7 V/cm (typical of oxide films on most metals), the influence of the reverse current can usually be neglected [10]. And as suggested by Verwey [154], Mott [155, 156] and Cabrera [157, 158], the overall current shows an exponential dependence on the field:

$$i = k_0\ exp\left(\zeta\, \frac{V}{L}\right),$$ (27)

with $\zeta = \beta\, zaF/RT$, L being the film thickness.

Expression (27) is the fundamental equation for high-field ionic conductivity and is a key equation in the theory of anodic film growth. It is, however, modified in detail according to what is considered to be the rate-determining step in ionic movement.

According to the model of Verwey [154], the limiting factor is the rate at which ions move from one interstitial position to another. The concentration of mobile species is large and has a value

corresponding to electroneutrality, that is, the value that would
exist in the bulk oxide. The system is under internal (not interfacial)
control. (Aluminum oxidation was examined, and from the Tafel slope
the jump distance a was determined. Recently film growth on platinum [90], tantalum [89], and iron [83, 90, 159, 160] has been discussed in a similar light.)

In the model of Mott and Cabrera [155-158] the crossing of the
first barrier, which metal ions have to surmount to enter the oxide,
is considered to be the rate-determining step; that is, the current is
under "interfacial control" when the film is very thin:

$$i = \text{const } n' \exp(-\frac{U_1 - \beta \, zea'E}{kT}) ,   \tag{25a}$$

where n' is the concentration of atoms on the metal surface and a'
is the distance from an equilibrium site in the metal to an equilibrium
site in the oxide film.

Dewald [161] and Young [162] have extended the above models
to include the effects of space charge due to ions in transit: both
the number of mobile species and the field are treated as a function
of distance. The system is under "internal-interface" control; that
is, the rate-determining step is either in the bulk or at the metal
surface, depending on conditions, giving rather complicated significance to the Tafel slope. Young [162] has suggested on empirical
grounds that the activation energy is a nonlinear function of the
electric field. He later justified the presence of a quadratic term
in the field on the basis of a postulated variation in activation
energy with the condenser pressure [163]. (The explanation of the
temperature independence of the Tafel slope on tantalum was tentatively given.)

Christov and Ikonopisov [164] have modified the model of Young
[162] and postulate that the activation energy, as a function of electric field potential, includes a nonlinear term:

$$E_{ac} = U - aE + bE^2,$$

which not only lowers the barrier but also shifts the equilibrium position of the barrier. The actual barrier parameters were calculated for a given field, and also for a field of zero, and compared with experimental data for the anodic oxidation of tantalum, niobium, and aluminum. The agreement obtained is significant.

Bean, Fisher, and Vermilyea [165] have postulated that the high field strength not only influences the migration of ions but is also the driving force for the production of Frenkel defects; that is, it lowers the activation energy needed to pull an ion out of a lattice into an interstitial position. The observed change in Tafel slope with field in the oxidation of tantalum is then explained by the field-dependent production of these defects.

Dignam [166] has pointed out that the condenser-pressure effect suggested by Young is probably too small to explain the observed field and temperature dependence of the Tafel slope for tantalum, niobium, and aluminum. Dignam suggested that the nonlinear relationship between the activation energy and the field can be accounted for by a model in which the field-independent component of the potential-energy function, for the displacement of a mobile charged species, is assumed to resemble a Morse function.

In an investigation of the conduction properties of valve-metal-oxide systems Dignam [167] developed a theory of cooperative ion transport. Thus the "amorphous" oxide is considered to be composed of very small crystallites, and ion transfer through such a "dielectric mosaic" is thought to be influenced by the effective, rather than the external, field. The cooperative-ion-transport phenomena arise then as a result of the time-dependent polarization properties of the "mosaic" oxide film. The comparison of these theories with experimental results gives a better agreement [168, 169] than the earlier theories [8, 161, 163, 165]. Furthermore, another advantage of Dignam's theories is that both anionic and cationic charge transport, in certain valve-metal oxides, are taken into account. Davies and colleagues [170, 171], in a radiotracer study with xenon-125 and radon-222 markers, were able to show that on tantalum and aluminum both metal and oxygen migration are contributing to film growth and that the oxide film on zirconium and niobium grows by oxygen migration alone.

Equations for high-field ionic conductivity were originally derived for the oxidation of valve metals, of which tantalum is a typical example. With these metals an oxide film is nearly always present, and the growth kinetics can be studied in sufficient detail to give a valid test of the growth theories.

The experimental data for the growth of thin films on other metals, formed either by low-temperature oxidation in an oxygen atmosphere or by anodic oxidation in solution, are very sparse and even more controversial.

For the growth of the thin, passive films on iron, for example, high-field-assisted ionic migration has been suggested [83, 90, 159, 160] as the growth-determining factor. Alternatively, field-assisted place exchange of the cations and anions in the film [53, 54] or some such similar mechanism has been proposed [76].

Direct kinetic data in the prepassive potential region are particularly scarce.

D. Kinetics and Mechanism of Film Growth; Quantitative Evaluation of the i-V Curve

1. Active Potential Region

With a view to understanding qualitatively and quantitatively the processes responsible for a typical i-V curve, two viewpoints will be brought into focus: (a) the work of Ebersbach, Schwabe, and Ritter [40, 42] containing the first theory that succeeded in calculating an i-V curve (for nickel) in good agreement with the experimental results, and (b) the work of Brusic [76] and Bockris, Genshaw, and Brusic [86], which gives the first direct evaluation of the growth mechanism of the film on iron in the active region. (Kinetic parameters were evaluated by a combination of electrochemical and ellipsometric measurements for iron in a slightly alkaline solution.)

a. The Model of Ebersbach, Schwabe, and Ritter [40]. It was assumed that the metal-dissolution reaction (rate i_1) and the film-forming reaction (i_2) at an initially bare electrode can proceed simultaneously with a possibility of the film's being dissolved by the action of the electrolyte, especially if some adsorbable anions are present. Thus

$$i = (i_1 + i_2)(1 - \theta) \,, \tag{28}$$

where θ, the electrode coverage, varies with time according to

$$\left. \frac{d\theta}{dt} \right|_{\Delta V, \, a_{A^-}, \, a_{H^+}} = Ci_2(1 - \theta) - K\theta - B\theta \,, \tag{29}$$

where B, the rate of oxide film degradation, is a function of c_{H^+} according to

$$MO + 2H^+ \rightleftarrows M^{2+} + H_2O \tag{30}$$

and K, the rate of oxide-film degradation, is a function of c_{A^-} according to

$$MO + 2A^- + H_2O \rightleftarrows MA_2 + 2OH^- \,. \tag{31}$$

Then

$$i = (i_1 + i_2)\left(1 - (1 + \frac{K+B}{Ci_2})^{-1} \{1 - \exp[-(Ci_2 + K + B)t]\}\right) \,, \tag{32}$$

where

$$i = i_1^0 \exp\left[\frac{\alpha_1 F}{RT}(\varepsilon - \varepsilon_1)\right], \quad i_1^0 = fc_{H+}, \tag{33}$$

$$i = i_2^0 \exp\left[\frac{\alpha_2 F}{RT}(\varepsilon - \varepsilon_2)\right], \quad i_2^0 = fc_{H+}, \tag{34}$$

with ε_1 and ε_2 being the equilibrium potentials for active dissolution and film formation.

The dissolution rate i_1 is determined by the reaction sequence given by Eqs. (16) through (18) (Bockris' dissolution mechanism) or by Eqs. (19) through (22) (Heusler's mechanism). The mechanism of film formation, resulting in i_2, is assumed to be as follows:

$$M + H_2O \rightleftharpoons MOH + H^+ + e^-, \tag{35}$$

$$MOH + H_2O \rightleftharpoons M(OH)_2 + H^+ + e^-, \tag{36}$$

$$M(OH)_2 \rightleftharpoons MO + H_2O, \tag{37}$$

with the second electron transfer being rate determining. The mechanism involves the direct formation of the passivating film at the electrode surface. This mechanism is supported by experimental evidence on a rotating nickel disk electrode [172], which indicates that the rate of the passivation process is independent of stirring and of the solubility of the nickel salt. Similar results were reported by Lovachev, Oshe, and Kabanov [173, 174] for nickel in acid and alkali, studied with a rotating–disk electrode. It was suggested that Ni_2O_3 forms in the acid, and $Ni(OH)_2$ and NiO, in alkaline solution, directly on the electrode, rather than through a dissolution-precipitation mechanism.

For the steady state, $t \to \infty$,

$$i = \frac{i_1^0 \exp\left[(\alpha_1 F/RT)(\varepsilon - \varepsilon_1)\right] + i_2^0 \exp\left[(\alpha_2 F/RT)(\varepsilon - \varepsilon_2)\right]}{1 + \left(\{Ci_2^0 \exp\left[(\alpha_2 F/RT)(\varepsilon - \varepsilon_2)\right]\}/(K + B)\right)} \tag{38}$$

and if no oxide-film degradation occurs,

$$i = \frac{i_1 + i_2}{\exp(i_2 Ct)} \tag{39}$$

According to Eq. (29), the constant C is the covered area per coulomb and was estimated on the basis of the area required for oxygen atoms as 10^4 cm^2/A-sec. The rate constants K and B at constant pH and anion concentration were given as first-order rate constants with estimated values of $10^4 > (K+B) > 10^{-8}$ sec^{-1}. These limits were chosen arbitrarily, assuming that film degradation is slower than film formation. Current-potential curves were calculated for large variation of parameters and did indeed show a resemblance to the experimental results (carried out for nickel in acid).

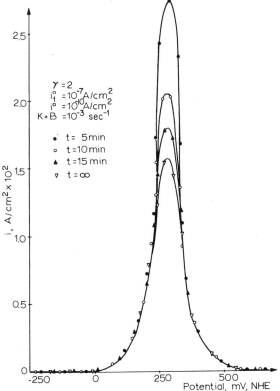

Fig. 14. Effect of time on the current density-potential curves by potentiostatic measurements (calculated for $K+B = 10^{-3}$ sec^{-1}). Reprinted from Ref. [40] by courtesy of Pergamon Press.

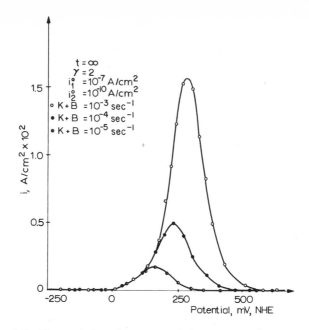

Fig. 15. Current density–potential curves calculated for t = ∞ showing dependence on the sum of the constants K and B. Reprinted from Ref. [40] by courtesy of Pergamon Press.

Examples are shown in Figs. 14 and 15 for $\gamma = \beta_2 z_2/\beta_1 z_1 = 2$ (or 1), $i_1^0 = 10^{-7}$ A/cm^2 (chosen arbitrarily from the region of values reported to be between 10^{-7} and 10^{-10} A/cm^2), $i_2^0 = 10^{-10}$ (again arbitrarily assumed to vary between 10^{-7} and 10^{-10} A/cm^2, that is, in the same region of values as the dissolution current i_{diss}). These figures show the variation in peak current or peak potential with both time (Fig. 14) and the velocity of the oxide–film degradation reactions (Fig. 15). Thus, according to this model, the total current will be zero when $\theta = 1$ (assumed to be reached at the bottom of the i–V curve, P in Fig. 1). Furthermore, according to Eq. (39), in the stationary state (when t → ∞) a finite current can flow only when $i_2 \to 0$. It is argued [35] that, in reality, a measurable current will always flow after passivation even if $i_2 > 0$ because the surface film is not perfect nor is the surface perfectly covered, and that the passive surface film is not absolutely impermeable.   Thus, as polarization increases (at $V > V_{pp}$, Fig. 1), ions increasingly penetrate into the film, increasing its thickness [40].   Thus this present theory satisfactorily describes the i-V curve in the active potential region, with a qualitative description of possibilities in the passive potential region.

b. The Model of Bockris, Genshaw, and Brusic [76, 86].   This model assumes, as does that of Schwabe [40], that in the active region both film formation and dissolution reactions proceed as simultaneous processes.   The advantage held here is that many of the unknown parameters concerning the mechanism of the film-forming reaction have been experimentally determined for iron in borate buffer.   The essential conclusions follow.

Thickness at constant potential is linearly proportional to the time of growth up to about one-third of a monolayer; thereafter it is proportional to log t.   Film growth commences less than 0.01 sec after switching on the current or potential.   Stirring does not affect the rate.   After the first several seconds, at any one time the growth rate is independent of the potential.

Comparison of the form of the galvanostatic transients in the prepassive region with that in the passive range indicates that the prepassive film is in the ferrous state, most likely $Fe(OH)_2$, which is at the passivation potential, one to two monolayers thick.   Above the passivation potential, the results suggest the formation of a higher oxide, probably Fe_2O_3.

The suggested mechanism, which yields good agreement (qualitative and quantitative) with the observed results, is as follows:

1. Discrete centers grow, initially two-dimensionally, up to 30% coverage.

2. At higher coverages the growth involves a rapid place-exchange

step with a rate-determining Temkin discharge of OH^- into sites where the metal is already attached to an OH group. This initial OH group is displaced into the first layer of metal atoms beneath the surface, forming a two-dimensional oxide lattice. Thus the processes that determine the i-V curve are

$$Fe + OH^- \rightleftarrows FeOH + e^- \qquad (40)$$

dissolution   film formation

$$FeOH \rightarrow FeOH^+ + e \quad (41) \qquad\qquad FeOH \rightleftarrows HOFe \qquad (42)$$

$$FeOH^+ \rightleftarrows Fe^{2+} + OH^- \quad (43) \qquad HOFe + OH^- \rightarrow Fe(OH)_2 + e^- \quad (44)$$

where Eq. (41) gives i_{diss} and Eq. (44) i_{ff}. Since, according to the determined mechanism of film growth $i_{ff} \neq f(V, L)$ at t = const and $i_{ff} \rightarrow 0$ at t large, the experimental i-V curve is largely determined by i_{diss}. The latter is calculated as a function of increasing $Fe(OH)_2$ coverage from 0 to about 0.8 at potentials in the prepeak region and in the regions of final covering of the electrode by the formation of Fe_2O_3 at potentials between V_p and V_{pp}. The theory was found to be in fair agreement with the observed values [76, 86].

Place exchange, in the formation of the first or second layer of the film, may be relatively rapid, helped by existing defects at the surface [175] or by the influence of metal image forces on the activation energy for the step [176]. At higher film thicknesses (thus at higher potentials) place exchange may become a rate-determining step, or, in general, the mechanism suggested for the prepassive potential region ceases to be valid.

A similar mechanism, where an oxygen-containing adsorbed intermediate, formed in an electrochemical step, reacts with the metal at the peripheries of oxide patches to form more oxide, was recently suggested by Dignam and Gibbs [177] for the anodic oxidation of

copper in alkaline solution. Both one- and two-dimensional lateral
growth mechanisms were considered, but one-dimensional growth was
suggested as more probable [177]. In alkaline solution copper shows
a typical i-V curve [178], and hence a typical active-passive transition has to be connected with the direct formation of the lowest oxide
(Cu_2O) and a decrease in the free surface area.

In summary, these theories have been able to construct the i-V
curve in the active potential region, as determined by the simultaneous dissolution and film-forming reactions with a constant decrease in
free electrode area.

2. Passive Potential Region

Phenomenologically, the current measured in the passive region
is independent of potential and decreases continuously with time, as
seen in the work of Nagayama and Cohen [51, 52] and many others.
When the potential is changed from one steady-state value to another,
at first the current suddenly increases and then slowly decreases with
time to very low values (10^{-8} to 10^{-10} A/cm^2), sometimes thought to
be limited by "chemical dissolution" of the film. Some experimental
evidence by Novakovsky and Likhachev [33] shows that chemical
dissolution (independent of potential and film thickness) accounts
for only a very small part of the total dissolution. Utilizing potentiostatic current-time diagrams together with radiometric control of the
true metal (iron) dissolution rate, it was shown that the changes in the
external current, after an abrupt potential change, take place largely
at the expense of the metal-dissolution rate. This is an anodic process, with the ionic movement through the film being regulated by the
film thickness and, in the stationary state, depending only on the ratio
of the anions and cations in the film [33]. These results have shown
that anodic dissolution in the passive region depends on the same

basic ionic movement as does film formation. Since $i = i_{ff} + i_{diss}$, once we understand the film-growth mechanism, we can also account for the current density in the "steady state" not being a function of potential.

The total current at any potential immediately after the primary act of passivation is limited by one of the following possible processes: (a) movement of the ions across the metal-oxide interface, (b) movement of ions through the oxide, or (c) movement of ions across the oxide-solution interface; each is influenced by the local electric field, which is in turn a function of the total potential drop.

Two basic mechanisms have been suggested for the rate-determining ionic movement: (a) place exchange and (b) high-field-assisted ionic migration.

a. Place-Exchange Mechanism. This mechanism was formulated by Sato and Cohen [53] following an earlier proposal by Lanyon and Trapnell [175]. The mechanism assumes that the propagation of ions through the film occurs by a simultaneous and independent exchange of all of the M-O in a given row (column upward from the metal). Thus all of the ions in a given M-O row must be in the activated state simultaneously. The probability of the ionic movement from one lattice position to another is assumed to be

$$P_m = \exp\left(-\frac{U - \beta zaF\varepsilon_m}{RT}\right) \tag{45}$$

and is thus similar to ionic movement in a high field, where the chemical activation energy U is diminished by the influence of the local field ε_m. The assumed simultaneous movement in the entire row will happen with a probability

$$P = P_1 P_2 P_3 \ldots P_x , \tag{46}$$

where x, the number of rotations, depends on the film thickness L

and the lattice constant a. Thus the expression for the reaction rate
will be of the form

$$\text{rate} = \text{const}\, \exp\left(\frac{\beta z a F \sum_{m=1}^{x} \varepsilon_m}{RT}\right)\, \exp\left(-\frac{LU}{aRT}\right)\,,\qquad (47)$$

or

$$\frac{dL}{dt} = \text{const'}\, \exp\left(\frac{\beta\, zFV_{MS}}{RT}\right)\, \exp\left(-\frac{LU}{aRT}\right)\,,\qquad (48)$$

giving

$$L = \text{const} + \frac{\beta\, zFaV_{MS}}{U} + \frac{aRT}{U}\, \ln t\,.\qquad (49)$$

Sato and Cohen [53] have observed that the experimentally found
rate of film growth on iron could be expressed by the overall equation

$$i\,\left(\approx \text{const}\, \frac{dL}{dt}\right) = k'\, \exp\left(\beta V_{MS} - \frac{L}{B'}\right)\,,\qquad (50)$$

where V_{MS} is the metal–solution potential difference and k', β,
and B' are constants, in excellent agreement with Eq. (48). A similar
mechanism was also suggested by Genshaw [179] for film growth on
platinum. Although this mechanism can explain the experimental
observations (such as linear increase in thickness with log t, linearity between thickness and potential, and the appearance of an activation energy that seems to increase with film thickness and decrease
with potential), this mechanism may still be criticized as highly
unlikely. This is because the activation energy needed for the direct
exchange of neighboring ions is very high, contrary to the low value
of 3. 5 kcal/mole obtained by comparison of the theory and experiment
in Sato and Cohen's work [53]. The decrease in the activation energy
due to the influence of metal image forces cannot be expected, because, in the passive potential region, the oxide film is several layers
thick. The alternate explanation of the observed linear increase in
the activation energy with film thickness can be sought in the mechanism which assumes the existence of a constant field through the
film (see Ref. [176]).

b. The Field-Assisted Ionic Migration Mechanism. The basic ionic movement is assumed to proceed with an energy of activation decreased by the applied field, the rate being given as Eq. (25), or

$$\frac{dL}{dt} = \text{const} \left[\exp\left(-\frac{U}{RT}\right) \exp\left(\frac{\beta\,zaFV}{LRT}\right) \right] , \tag{51}$$

with

$$\frac{1}{L} = A - \frac{RT}{aFa'V_{MS}} \ln t . \tag{52}$$

Such a mechanism explains film growth on tantalum and aluminum. Also, some results on iron, obtained by Kruger and Calvert [83], Ord and DeSmet [89], and Moshtev [159, 160] lend support to this mechanism. Ord and DeSmet [89] have also studied tantalum, and Ord and Ho [90] have examined film growth on platinum. Results on all three metals (i.e., Ta, Fe, and Pt) show remarkable similarities, suggesting that the anodic oxidation process may be independent of the metal involved even when metals are as dissimilar as these three [89]. This may be taken as one more indication that the passivation process depends on the film properties and that passivation theories that place the entire overpotential across the metal-electrical double layer (as, for example, the adsorption theory of Kolotyrkin [30]) are not tenable.

Referring to the i-V curve in the passive region, the constancy of the current may be explained by the interrelation of the film thickness and potential; that is, with increasing potential, the film thickens, and this reduces the field across the film to a value similar to that obtained for lower potentials and thinner films, thus giving approximately the same rate for ionic migration.

Some results on iron [76] show that, of the current that does flow in the passive region, i_{diss} may still be larger than i_{ff}. For passive iron in a borate electrolyte the direct recording of film-thickness change with time gives a rate of 10^{-3} Å/sec at t = 1000 sec. This rate is equivalent to $i_{ff} = 10^{-7}$ A/cm^2. The measured current density at this time was 10^{-6} A/cm^2. Assuming the absence of electronic current flow (a reasonable assumption when one considers the passive potential), this value of i_{diss} would correspond to 2×10^{12} Fe^{3+} ions entering the solution per second rather than reacting with OH at the outside interface to form the oxide film. This direct dissolution of iron into the solution through the film is consistent with the results reported by Novakovsky and Likhachev [32, 33].

Since the value of i_{diss} will be proportional to the ionic conductivity of the oxide films, in general, films with low ionic conductivity should be better passivating films.

IV. FINAL REMARKS

Modern passivation theory views the primary passivation act as a formation of a tightly held layer of monomolecular dimensions, containing oxide or hydroxide anions and metal cations; the process involves the formation of a new phase, an oxide, in steps that include the adsorption as an important intermediate stage. Thus the long-time controversial aspects of the adsorption and oxide theories are, in one sense, combined.

The observed current–potential behavior is a function of the simultaneous processes of film formation, its dissolution, and metal dissolution. The latter seems to be mostly responsible for the magnitude of the current at all potentials. In the active potential region dissolution is hindered by a decrease in the free electrode area, and in the passive region dissolution depends entirely on the properties of the passivating film.

APPENDIX

Tabular Summary of the Range of Views on Passivity

Experimental approach	Experimental evidence on thickness	Type of layer	Passivation model[a]		Refs.
			Thickness[b] at V_p or V_{pp}	Passivation process	
Electrochemical: potential sweep, potentiostatic oxidation, potential decay, galvanostatic oxidation, corrosion rate (Fe in pH 0-12)	Oxidizing capacity of passive film is 0.01 C/cm² of apparent surface [20, 25]	Chemisorbed oxygen, atomic and molecular $O \cdot O_2$	At V_{pp} film corresponds to atomic oxygen (thinner than $O \cdot O_2$; above $O \cdot O_2$; above V_{pp} multilayers exist of $O \cdot O_2$ or OH adsorbed	Primary protection mechanism satisfies surface affinities of metal through formation of chemical bonds between adsorbate and metal without metal atoms leaving their lattice. Only metals and alloys with unfilled d bond are expected to chemisorb oxygen (i.e., to chemisorb oxygen (i.e., to passivate). Chemi-	[20, 25-29, 69]

| Electrochemical: potentiostatic method with measured differential capacitance and ohmic resistance (Cr, Ni, stainless steel) | None | Adsorbed oxygen and oxygen-containing species (H_2O, OH^-) not oxide | Less than monolayer | Passivation process begins at potentials V_p; passivity is a specific case of decrease in kinetics of an electrodic reaction as a result of a specific adsorption of an anion | [30] |

sorbed oxygen decreases i, thereby increasing overvoltage for metal dissolution. Oxide films, if they form, may offer additional protection as diffusion-barrier layers.

Appendix — Continued

Experimental approach	Experimental evidence on thickness	Passivation model[a]			Refs.
		Type of layer	Thickness[b] at V_p or V_{pp}	Passivation process	
Electrochemical: galvanostatic, potentiostatic, electrode impedance at high frequencies (Fe in NaOH; Zn in alkaline solution)	0.3 mC/cm^2 of real surface at V_{pp}	Oxide (or salt) is formed in most cases. Pt, Fe (in alkali) adsorption of oxygen (O, OH, SO$_4$, I). Zn (in alkali) monolayer of adsorbed oxygen or some oxygen-containing species	Less than monolayer	Main reason in passivation process is not physical event of covering surface, but kinetic inhibition of anodic process; reaction (e.g.., formation of monolayer on Fe) occurs in solid phase, without metal ions passing into solution	[36–38]
Electrochemical, coulometry; electron	At potentials $V > V_{pp}$ film thick enough	Three-dimensional film, oxide		Three-dimensional film forms probably after chemisorption being	[18, 19]

diffraction (Fe)		to be stripped from electrode			first step
Slow galvanostatic transients with 2.5-7.5 A/cm² (Ni, alkaline solution)	Oxide	Measurement of Q at potential arrest gives 7.5-8.4 × 10¹⁵ atoms of oxygen per 7.4× 10¹⁵ atoms of Ni (true surface area); i.e., monolayer of NiO or Ni(OH)₂ at V = 0.11-0.059 pH	At V_p monomolecular Ni(OH)₂ starts to form	Passivation caused by formation of monomolecular Ni(OH)₂ starting at $\sim V_p$ and its transformation into Ni₂O₃ at high potential	[34]
Electrochemical: potentiostatic, galvanostatic (Ni in acid)	None	Oxide	Film starts to form below V_p, electrode not completely	At V_p, $i = i_{crit}$, i.e., film growth = dissolution rate; at higher potentials film-growth	[96]

Appendix — Continued

Experimental approach	Experimental evidence on thickness	Type of layer	Passivation model[a]		Refs.
			Thickness[b] at V_p or V_{pp}	Passivation process	
			covered till above V_{pp} (thickness 10–30 Å)	rate exceeds that of dissolution and surface becomes almost completely covered with an oxide with a consequent decrease in current (suggested discharge mechanism of film formation)	
Electrochemical: potentiostatic and galvanostatic, potential sweep oxidation (Ni in 1 N H_2SO_4)	At V_p = 340 mV thickness is zero (galvanostatically determined) but increases	Oxide (monolayer)	At V_p thickness is zero; at V_{pp} oxide is monomolecular	Formation of monomolecular passive layer begins at V_p and is completed at V_{pp}. Only after a pore-free mono-	[35]

Electrochemical: potentistatic and galvanostatic (Fe, acid)	with V to 30 mC/cm² at 1.8 V			molecular passive film is formed is electrode fully passive. Decrease in current at V_{pp} is connected to increasing coverage of electrode surface.	
	Above V_{pp} film thickness increases (coulometrically obtained) from 0 to 40 Å	Oxide: Fe_3O_4, Fe_2O_3	At V_p, Fe_3O_4 is there and Fe_2O_3 starts to form	First porous, nonconducting film is formed, which transforms irreversibly at V_p involving a change in the valency state of the oxide into electron-conducting passive film almost insoluble in media; passivation due to formation of oxide, which slowly corrodes in media [21].	[21, 23, 48–50]

Appendix — Continued

Experimental approach	Experimental evidence on thickness	Passivation model[a]			Refs.
		Type of layer	Thickness[b] at V_p or V_{pp}	Passivation process	
Electrochemical: galvanostatic and potentiostatic passivation and activation (Fe in neutral solution and weak acid)	4–4.5 mC/cm² at V 200 mV positive to V_{pp} [180]; at V_{pp} film is monomolecular [181]	Oxide (not related to any known oxide)	At V_{pp} electrode almost fully covered by three-dimensional film	Passivation due to formation of "primary layer" porous and hardly conducting in whose pores current density increases; by polarization at higher potential one obtains real passive layer, nonporous and electronically conducting	[46, 180, 181]
Electrochemical, electron diffraction, chemical analysis of stripped film (Fe)	None	Oxide, Fe_2O_3		Oxide film, nonstoichiometric, is formed. Film described in terms of n-p semiconducting Fe_2O_3. At oxide-	[47]

electrolyte boundary there is a slow reaction involving excess oxygen ions and bound positive holes $(O^{2-} + [^{+}_{+}] + 2H^{+} + 2e^{-} \rightarrow H_2O)$ which controls the Flade potential in the original sense (potential during decay after formation of film) and is responsible for the passive electrode behavior

Active-passive transition characterized by formation of higher oxide$^{\underline{c}}$ [51, 52]

10 Å at V_{pp}

Oxide (duplex)

Electrochemical (galvanostatic, potentiostatic); electron diffraction (Fe in buffer, pH 8.4)

1-2 mC at V_{pp} (up to 6.5-7.5 mC/cm² at high V) i.e., ~10 Å

Appendix — Continued

Experimental approach	Experimental evidence on thickness	Passivation model[a]			Refs.
		Type of layer	Thickness[b] at V_p or V_{pp}	Passivation	
Electrochemical: potential decay, potentiostatic and galvano-static oxidation (Ni in pH 0–14)	3.85 mC/cm² at V_p = 0.434 V (grows with potential about 12 mC/V); at V_p 10–15 Å (depending on roughness factor: 10 Å if R.f. = 2)	Oxide, NiO/Ni₃O₄		Prepassive film (formed by dis-solution–precipitation) trans-formed into higher valence oxide (higher valence oxide film theory)	[92, 93]
Electrochemical: potentiostatic oxidation, some experiments with rotating Ni electrode	None	Oxide (initially monolayer)	Film starts to form below V_p; at V_p half monolayer; at V_{pp} monolayer	Surface area blocked by com-petitive formation of MO (by discharge of H_2O or OH^-), which is presumed to become a monolayer at V_{pp}	[39–42]

Electrochemical (potentiostatic and galvanostatic measurement); ellipsometric (Ni in acid)	> 45 Å (< 80 Å by ellipsometry)	Oxide	Starts to form at $V < V_p$ with thickness 60 Å	Prepassive film formed by dissolution–precipitation and transformed into conducting film, nonstoichiometric oxide	[79]
Electrochemical: galvanostatic oxidation (Cd, Zn)	None. If, however, calculated from charge passed, neglecting diffusion, thickness would be 100 Å	Oxide, CdO	Monolayer	Film forms by dissolution–precipitation, but it is monolayer, which blocks surface and may grow in third dimension	[43]
Electrochemical: potentiostatic oxidation (Fe in NaOH)	None	Oxide film, single-phase Fe_2O_3 (monolayer at first)	At V_p film starts to grow; at V_{pp}, monolayer	Film starts to grow at V_p, by reaction of OH^- with M surface$\underline{{}^{d}}$	[44]

Appendix — Continued

Experimental approach	Experimental evidence on thickness	Passivation model[a]			Refs.
		Type of layer	Thickness[b] at V_p or V_{pp}	Passivation process	
Electrochemical: potential sweep and decay (Fe and Cr)	Galvanostatically at V_p less than monolayer; at V_{pp} one to three layers	Primary layer (adsorbed) and secondary oxide	Film starts to form before V_p; at V_{pp}, ~10 Å	Kink sites blocked by adsorbed oxygen; resulting film contains both oxygen and metal ions	[31, 63]
Electrochemical (potentiostatic and galvanostatic oxidation); chemical oxidation; ellipsometric; electron diffraction (Fe in alkaline solution)	Ellipsometrically at V_{pp} is 0–10 Å, growing with potential to ~50 Å at potential below O_2 evolution	Oxide: Fe_3O_4 + Fe_2O_3	At V_{pp}, 0–10 Å, growing with potential	Passive layer is examined mostly; passivity must require film with insulating outer layer (if it has more than one layer) that is not electron conducting	[61, 80–83]

Electrochemical (kinetics at constant potential); electron microscopy and diffraction (Zn, Cd)	Monolayer of ZnO on Zn	Ordered mono-molecular two-dimensional (at least) film of definite chemical phase	Monolayer (Zn); two to three layers (Cd)	Growth of layers as in column 4; reaction ceases either when electrode is "blocked" or when parent phase has been completely converted into new phase [45]
Electrochemical: potentiostatic current-time measurement with radiometric control of true dissolution rate (Fe)	No data	Adsorbed oxygen monolayer transformed into oxide by continuous process		Combination of adsorbed oxygen and oxide theory [32, 33]
Electrochemical: potentiostatic oxidation, potential decay, breakdown of passivity	No clear data	Oxide, initially monolayer	At V_{pp}, monolayer	Primary act of passivation is formation of tightly held monolayer containing oxide or hydroxide anions and metal cations (formed at V_p to V_{pp}) by discharge mechanism (combination of adsorption and monolayer-oxide view) [14, 98]

Appendix — Continued

Experimental approach	Experimental evidence on thickness	Passivation model [a]			Refs.
		Type of layer	Thickness [b] at V_p or V_{pp}	Passivation process	
Chronopotentiometry; differential capacity measurement after prior anodic oxidation (Fe-(2.7–19.1%)Cr)	0.7 mC/cm² (spent in reduction of inner film)	Monolayer of adsorbed oxygen and amorphous oxide	Slightly above one monolayer	Adsorption (depending on d character of alloy), electron transfer to adsorbate, and ultimately formation of bulk amorphous oxide (above adsorbed layer) via postulated cation migration into adsorbed array	[144, 182]
Electrochemical: alternating current (50 Hz), asymmetric square-wave current (10 Hz) (Ti in H₂SO₄)	Less than monolayer of oxygen (calculated from the shift of potential in anodic cycles)	Adsorbed oxygen + oxide	Less than monomolecular (degree of initial coverage not taken into consideration)	First stage associated with formation of adsorbed film; later formed oxide-type bonds are more stable and do not rupture.	[183, 184]

Electrochemical (anodic and cathodic polarization); ellipsometry (steel and Mo, Ni, Cr in 1 N Na$_2$SO$_4$ and 1 N H$_2$SO$_4$)	~10Å (at $V \ll V_p$) increasing in active potential region, decreasing to 10Å at V_{pp}	Adsorbed oxygen (from water)	Neither surface films nor adsorption layers predominate in passivation process, but both act simultaneously to provide metal protection	[84]
Electrochemical: anodic charging curves at continuously scoured surface (Ti, Cr, Ni, 1 N H$_2$SO$_4$)	None	Less than monomolecular at V_p; full monolayer at V_{pp}	Anodic dissolution hindered by oxygen adsorption, which diminishes active metal surface. At V_{pp}, electrode fully covered. At higher V Ti(III) reoxidizes in Ti(IV) and thickens (oxide?).	[74]

Appendix — Continued

Experimental approach	Experimental evidence on thickness	Passivation model[a]			Refs.
		Type of layer	Thickness[b] at V_p or V_{pp}	Passivation process	
Electrochemical: potential sweep and square-pulse potentiostatic measurements; rotating ring-disk electrode (Ti in H_2SO_4)	None	New phase containing Ti(III) or Ti(IV)		Active-passive transition shows no dependence on rotation speed and is explained by formation of passivating film directly at electrode, rather than by dissolution-precipitation in a competitive process with dissolution and hydrogen-evolution reaction	[185]
Electrochemical (Cu in alkaline solution)	None	Oxide		Both solution and film-forming reactions promoted by presence of OH^-. Formed new phase shields electrode, but passivation also explained	[186]

Electrochemical: galvanostatic i-V curves (Sn in NaOH)	None	$Sn_2O \rightarrow SnO$	by increasing concentration of highly charged O ions in surface layer. Passivity observed only in dilute NaOH (< 1N) and assumed to be due to formation of stannous oxide at rate exceeding rate of chemical dissolution	[187]
Electrochemical (galvanostatic oxidation, potential decay capacitance and resistance measurement); electron microscopy and diffraction (Pb in H_2SO_4)	None	$PbSO_4$, basic sulfates, PbO	Metal dissolution gives precipitate, $PbSO_4$; in its intercrystalline areas current density increases and passive film (basic sulfate and PbO) is formed	[94, 95]

Appendix—Continued

| Experimental approach | Experimental evidence on thickness | Passivation model[a] | | | Refs. |
		Type of layer	Thickness[b] at V_p or V_{pp}	Passivation process	
Electrochemical (anodic potential sweep) combined with in situ photomicrography (Zn in concentrated KOH)	None (but film must be "thick"; it is visible)	Oxide (ZnO?)		In quiescent solution a film—oxide is formed by dissolution—precipitation. In stirred solution formation of second, more compact oxide is visible; it is formed directly at electrode and seems to be responsible for appearance of passivity	[188, 189]
Electrochemical (linear sweep voltammetry at rotating-disk electrode (Zn in KOH)	None	ZnO	Initially mono-molecular	Two different surface films formed (probably directly on electrode rather than by dissolution-precipitation): first causes small retardation, but second, ZnO, causes passivity	[190]

Method	At V_p / V_{pp}	Oxide	Thickness	Remarks	Ref.
Electrochemical; ellipsometry (Cr K$_2$SO$_4$ in H$_2$SO$_4$, pH 1.7)	0Å at V_p; less than monolayer of Cr$_2$O$_3$ at V_{pp}	Oxide	Less than monolayer at V_{pp}	Simultaneous oxide formation and metal dissolution, with passivation resulting from formation of less than mono-layer of film	[85]
Electrochemical; ellipsometry (Fe in boric acid borate solution)	One to two monolayers of Fe(OH)$_2$ at V_p	Fe(OH)$_2$ at V_p; Fe(OH)$_2$ + Fe$_2$O$_3$ at V_{pp}	5–10Å (at V_p to V_{pp})	Dissolution hindered by direct formation of thin Fe(OH)$_2$, slightly above monomolecular at V_p; final drop in current determined by complete covering of free surface area with possible formation of Fe$_2$O$_3$	[76, 86]

[a] V_p = passivation potential; V_{pp} = potential of complete passivity.

[b] Author's viewpoint.

[c] Similar to that reported by Arnold and Vetter [35].

[d] Similar to process reported by Schwabe et al. [39–42].

REFERENCES

[1] G. Wranglen, Corrosion Sci., 10, 761 (1970).

[2] E. A. Gulbransen, Corrosion, 21, 76 (1966).

[3] M. V. Lomonosov, Collection, Vol. 1, Moscow, Izd. Akad. Nauk SSSR, 1950.

[4] J. Keir, Phil. Trans. Royal Soc. London, 80, 359 (1790).

[5] Ch. F. Schonbein, Pogg. Ann., 37, 490 (1836).

[6] Ch. F. Schonbein and M. Faraday, Phil. Mag., 9, 53, 57, 122, 153 (1836).

[7] M. Faraday, Experimental Researches in Electricity, Vol. 2, London, 1844, p. 244.

[8] T. P. Hoar, in Modern Aspects of Electrochemistry (J. O'M. Bockris, ed.), Vol. 2, Butterworths, London, 1959.

[9] D. A. Vermilyea, in Advances in Electrochemistry and Electrochemical Engineering (P. Delahay, ed.), Vol. 3, Interscience, New York, 1963.

[10] L. Young, Anodic Oxide Films, Academic Press, London and New York, 1961.

[11] O. Kubaschewski and B. E. Hopkins, Oxidation of Metals and Alloys, Academic Press, New York, and Butterworths, London, 1953.

[12] D. Gilroy and B. E. Conway, J. Phys. Chem., 69, 1259 (1965).

[13] S. Schuldiner, J. Electrochem. Soc., 115, 897 (1968).

[14] T. P. Hoar, Corrosion Sci., 7, 341 (1967).

[15] N. D. Tomashov and G. P. Chernova, Passivity and Protection of Metals Against Corrosion, Plenum Press, New York, 1967.

[16] J. M. Defranoux, Mem. Sci. Rev. Met., 66, 641 (1969).

[17] C. Wagner, Corrosion Sci., 5, 751 (1965).

[18] U. R. Evans, Metallic Corrosion, Passivity and Protection, Longmans, Green, New York, 1948.

[19] U. R. Evans, Z. Elektrochem. , 62, 619 (1958).

[20] H. H. Uhlig and P. F. King, J. Electrochem. Soc. , 106, 1 (1959).

[21] K. G. Weil, Z. Elektrochem. , 59, 711 (1955).

[22] U. F. Franck, Z. Naturforsch. , 4A, 378 (1949).

[23] K. J. Vetter, Electrochemical Kinetics Theoretical and Experimental Aspects (trans. by Scripta Technica), Academic Press, New York, 1967.

[24] J. M. West, Electrodeposition and Corrosion Processes, Van Nostrand, New York, 1965.

[25] H. H. Uhlig, Ann. N. Y. Acad. Sci. , 58, 843 (1954).

[26] H. H. Uhlig, Z. Elektrochem. , 62, 626 (1958).

[27] H. G. Feller and H. H. Uhlig, J. Electrochem. Soc. , 107, 864 (1960).

[28] Z. A. Foroulis and H. H. Uhlig, J. Electrochem. Soc. , 111, 13 (1964).

[29] F. Mansfeld and H. H. Uhlig, J. Electrochem. Soc. , 115, 900 (1968); Corrosion Sci. , 9, 377 (1969).

[30] Ya. M. Kolotyrkin, Z. Elektrochem. , 62, 664 (1958).

[31] R. P. Frankenthal, J. Electrochem. Soc. , 116, 580 (1969).

[32] V. M. Novakovsky, Electrochim. Acta, 10, 353 (1965).

[33] V. M. Novakovsky and M. A. Likhachev, Electrochim. Acta, 12, 267 (1967).

[34] D. E. Davies and W. Barker, Corrosion, 20, 47t (1964).

[35] K. Arnold and K. J. Vetter, Z. Elektrochem. , 64, 407 (1960).

[36] B. N. Kabanov and D. I. Leikis, Z. Elektrochem. , 62, 660 (1958).

[37] T. J. Popova, V. S. Bagotsky, and B.N. Kabanov, Zh. Fiz. Khim. , 36, 1432 (1962).

[38] B. N. Kabanov, in discussion on paper by J. O'M. Bockris, A. K. N. Reddy, and B. Rao, J. Electrochem. Soc. , 113, 1142 (1966).

[39] K. Schwabe, Electrochim. Acta, 3, 186 (1960).

[40] U. Ebersbach, K. Schwabe, and K. Ritter, Electrochim. Acta,
     12, 927 (1967).

[41] K. Schwabe, Proc. 3rd Intern. Congr. Metallic Corrosion, Mir,
     Moscow, 1968.

[42] K. Schwabe, Corros. Week, Manifestation Cent. Fed. Corros.
     41st (1968) (T. Farkas, ed.), Akad. Kiado Budapest, 1970,
     p. 739.

[43] M.A.V. Devanathan and S. Lakshmanan, Electrochim. Acta, 13,
     667 (1968).

[44] W.A. Mueller, J. Electrochem. Soc., 107, 157 (1960).

[45] M. Fleischmann and H.R. Thirsk, J. Electrochem. Soc., 110,
     688 (1963).

[46] K.F. Bonhoeffer and U.F. Franck, Z. Elektrochem., 55, 180
     (1951).

[47] M.J. Pryor, J. Electrochem. Soc., 106, 557 (1959).

[48] K.J. Vetter, Z. Elektrochem., 62, 642 (1958).

[49] K.J. Vetter, Z. Elektrochem., 66, 577 (1962).

[50] K.J. Vetter, J. Electrochem. Soc., 110, 597 (1963).

[51] N. Nagayama and M. Cohen, J. Electrochem. Soc., 109, 781
     (1962).

[52] N. Nagayama and M. Cohen, J. Electrochem. Soc., 110, 670
     (1963).

[53] N. Sato and M. Cohen, J. Electrochem. Soc., 111, 512 (1964).

[54] N. Sato and M. Cohen, J. Electrochem. Soc., 111, 519 (1964).

[55] A. Damjanovic, in Modern Aspects of Electrochemistry (J. O'M.
     Bockris, ed.), Vol. 5, Plenum Press, New York, 1969.

[56] N.D. Tomashov and V.N. Modestova, Tr. Inst. Fiz. Khim. Akad.
     Nauk SSSR, No. 5, 75, Moscow, 1955; Chem. Abstr., 50, 11138
     (1956).

[57] N.D. Tomashov and R.M. Al'tovskii, in Collection: Corrosion of
     Metals and Alloys, Metallurgizdat, Moscow, 1963, p. 141.

[58] W. Kossel, Naturwissenschaften, 18, 901 (1930).

[59] O. Knache and I. N. Stranski, Progr. Metal Phys., 6, 181 (1956).

[60] R. Piontelli, Electrochim. Metallorum, 1 (1966).

[61] C. L. Foley, J. Kruger, and C. J. Bechtoldt, J. Electrochem. Soc., 114, 994 (1967).

[62] A. R. Despic, R. Raicheff, and J. O'M. Bockris, J. Chem. Phys., 49, 926 (1968).

[63] H. W. Pickering and R. P. Frankenthal, J. Electrochem. Soc., 112, 761 (1965).

[64] A. R. Despic and J. O'M. Bockris, J. Chem. Phys., 32, 389 (1960).

[65] J. O'M. Bockris and A. Damjanovic, in Modern Aspects of Electrochemistry (J. O'M. Bockris and B. E. Conway, eds.), Vol. 3, Butterworths, Washington, 1964, p. 237.

[66] J. O'M. Bockris, A. Despic, and D. Drazic, Electrochim. Acta, 4, 325 (1961).

[67] K. E. Heusler, Z. Electrochem., 66, 177 (1962).

[68] G. Salie, Z. Physik. Chem., 239, 411 (1968).

[69] H. H. Uhlig, Corrosion Sci., 7, 325 (1967).

[70] A. N. Frumkin, V. S. Bagotskii, Z. A. Iofa, and B. V. Kabanov, Kinetics of Electrode Processes (English transl.), 1967, available from Clearinghouse for Federal Scientific and Technical Information, accession No. TT 7000987.

[71] L. Germer and A. MacRae, J. Appl. Phys., 33, 2923 (1962).

[72] A. MacRae, Science, 139, 379 (1963).

[73] A. MacRae, Surface Sci., 1, 319 (1964).

[74] N. D. Tomashov and Z. P. Vershinina, Electrochim. Acta, 15, 501 (1970).

[75] Ya. M. Kolotyrkin and G. G. Kossyi, Zashch. Metal., 1, 272 (1965).

[76] V. Brusic, Ph. D. thesis, University of Pennsylvania, 1971.

[77] K. Schwabe and W. Schmidt, Corros. Sci., 10, 143 (1970).

[78] L. Germer and A. MacRae, Proc. Natl. Acad. Sci. U.S., 48, 997 (1962).

[79] J. O'M. Bockris, A. K. N. Reddy, and B. Rao, J. Electrochem. Soc., 113, 1133 (1966).

[80] J. Kruger, J. Electrochem. Soc., 108, 504 (1961).

[81] J. Kruger, in Symposium on Ellipsometry and Its Use in the Measurement of Surfaces and Thin Films (E. Passaglia, R. R. Stromberg, and J. Kruger, eds.), National Bureau of Standards, Washington, D.C., 1964.

[82] J. Kruger, Corrosion, 22, 88 (1966).

[83] J. Kruger and J. P. Calvert, J. Electrochem. Soc., 114, 43 (1967).

[84] V. V. Andreeva and T. P. Stepanova, in Collection: Corrosion of Metals and Alloys, Metallurgizdat, Moscow, 1963, p. 44.

[85] M. A. Genshaw and R. S. Sirohi, University of Pennsylvania, private communication, 1969.

[86] J. O'M. Bockris, M. A. Genshaw, and V. Brusic, paper presented at Faraday Society Meeting, London, December 1970.

[87] J. L. Ord, J. Electrochem. Soc., 113, 213 (1966).

[88] J. L. Ord and D. J. DeSmet, J. Electrochem. Soc., 113, 1258, (1966).

[89] J. L. Ord and D. J. DeSmet, J. Electrochem. Soc., 116, 762 (1969).

[90] J. L. Ord and F. C. Ho, J. Electrochem. Soc., 118, 46 (1971).

[91] W. J. Muller, Z. Elektrochem., 33, 401 (1927).

[92] N. Sato and M. Okamoto, J. Electrochem. Soc., 110, 605 (1963).

[93] N. Sato and M. Okamoto, J. Electrochem. Soc., 111, 197 (1964).

[94] D. Pavlov, Z. Elektrochem., 71, 398 (1967).

[95] D. Pavlov and R. Popova, Electrochim. Acta, 15, 1483 (1970).

[96] T. S. de Gromoboy and L. L. Shreir, Electrochim. Acta, 11, 895 (1966).

[97] A. I. Krasil'shchikov, Elektrokhimiya, 6, 341 (1970).

[98] T. P. Hoar, J. Electrochem. Soc., 117, 17C (1970).

[99] A. J. Appleby, J. Electrochem. Soc., 117, 1373 (1970).

[100] H. Gerischer, Z. Elektrochem., 62, 256 (1958).

[101] E. Mattsson and J. O'M. Bockris, Trans. Faraday Soc., 55, 1586 (1959).

[102] H. Aida, I. Epelboin, and M. Garreau, J. Electrochem. Soc., 118, 243 (1971).

[103] V. V. Losev, Electrochim. Acta, 15, 1095 (1970).

[104] R. D. Armstrong, A. B. Suttie, and H. R. Thirsk, Electrochim. Acta, 13, 1 (1968).

[105] K. E. Heusler, Z. Elektrochem., 62, 582 (1958).

[106] T. P. Hoar and T. Hurlen, in Proc. 8th Meeting CITCE, Madrid 1956, Butterworths, London, 1958, p. 445.

[107] R. Piontelli, International Committee on Electrochemical Thermodynamics and Kinetics, Vol. 11, 1950, p. 136.

[108] Ch. Weissmantel, K. Schwabe, and G. Hecht, Werkstoffe Korros., 12, 353 (1961).

[109] K. E. Heusler, Elektrochemische Auflosung und Abscheidung von Metallen der Eisengrippe, inaugural dissertation, Stuttgart, 1966.

[110] I. K. Marshakov and V. K. Altukhov, Elektrokhimiya, 5, 658 (1963).

[111] K. Schwabe, J. Phys. Chem., 214, 6 (1960).

[112] K. Kunze and K. Schwabe, Corros. Sci., 4, 109 (1964).

[113] K. Schwabe, Werkstoffe Korros., 15, 70 (1964).

[114] R. D. Armstrong, J. A. Harrison, H. R. Thirsk, and R. Whitfield, J. Electrochem. Soc., 117, 1003 (1970).

[115] S. Asakura and K. Nobe, J. Electrochem. Soc., 118, 13 (1971).

[116] S. Asakura and K. Nobe, J. Electrochem. Soc., 118, 19 (1971).

[117] J. Yahalom and J. Zahair, Electrochim. Acta, 15, 1429 (1970).

[118] J. Kruger, J. Electrochem. Soc., 110, 654 (1963).

[119] Ya. M. Kolotyrkin, Corrosion, 19, 261t (1963).

[120] S. Brennert, J. Iron Steel Inst., 135, 101P (1937).

[121] N. Hackerman and O. B. Cecil, J. Electrochem. Soc., 101, 419 (1954).

[122] M. A. Streicher, J. Electrochem. Soc., 103, 375 (1956).

[123] C. Carius, Metaux, 33, 31 (1958).

[124] U. F. Franck, Werkstoffe Korros., 11, 401 (1960).

[125] I. L. Rosenfeld and V. P. Maximtschuk, Z. Phys. Chem., 215, 25 (1960).

[126] Ya. M. Kolotyrkin, G. B. Bolovina, and G. M. Florianovich, Dokl. Akad. Nauk SSSR, 198, 1106 (1963).

[127] N. D. Tomashov, G. P. Chernova, and O. N. Markova, Corrosion, 20, 166t (1964).

[128] W. Schwenk, Corrosion, 20, 129t (1964).

[129] B.·E. Wilde and E. Williams, J. Electrochem. Soc., 117, 775 (1970).

[130] H. J. Engel and N. D. Stolica, Z. Phys. Chem., 215, 167 (1960).

[131] H. Kaesche, Z. Phys. Chem., NF, 34, 87 (1962).

[132] Ya. M. Kolotyrkin and V. A. Gilman, Dokl. Akad. Nauk SSSR, 137, 642 (1961).

[133] D. V. Kokulina and B. N. Kabanov, Dokl. Akad. Nauk SSSR, 112, 692 (1957).

[134] T. P. Hoar, D. C. Mears, and G. P. Rothwell, Corrosion Sci., 5, 279 (1965).

[135] T. P. Hoar and D. C. Mears, Proc. Royal Soc. (London), A294, 486 (1966).

[136] M. W. Breiter, Electrochim. Acta, 15, 1195 (1970).

[137] M. Pourbaix, Atlas of Electrochemical Equilibria in Aqueous
      Solutions, Pergamon Press, New York, 1966.

[138] D. Brennan, D. O. Hayward, and B. M. W. Trapnell, Proc.
      Royal Society (London), A256, 81 (1960).

[139] H. Gohr and E. Lange, Naturwissenschaften, 43, 12 (1965).

[140] J. E. O. Mayne and M. J. Pryor, J. Chem. Soc., 1831 (1949).

[141] W. E. O'Grady, private communication, 1971.

[142] N. D. Tomashov, R. M. Al'tovskii, and M. Ya. Kushnerev,
      Dokl. Akad. Nauk SSSR, 141, 913 (1961).

[143] P. J. Gellings, Corrosion Sci., 6, 543 (1966).

[144] N. Hackerman, Z. Elektrochem., 62, 632 (1958).

[145] K. Siegbahn, C. Nordling, A. Fahlman, R. Nordberg, K. Hamrin,
      J. Hedman, G. Johansson, T. Bergmark, S. Karlsson, I. Lind-
      gren, and B. Lindberg, Nova Acta Regiae Soc. Sci. Uppsala
      Ser. IV, 20 (1967).

[146] E. K. Oshe and I. L. Rosenfel'd, Elektrokhimiya, 4, 1200 (1968).

[147] E. K. Oshe and I. L. Rosenfel'd, Zasch. Metall., 5, 5 (1969).

[148] E. K. Oshe, I. L. Rosenfel'd, and V. G. Doroshenko, Dokl.
      Akad. Nauk SSSR, 194, 614 (1970).

[149] K. Hauffe, Oxidation of Metals, Plenum Press, New York, 1965.

[150] K. J. Vetter, Z. Elektrochem., 55, 274 (1951).

[151] H. Takei and S. Chiba, J. Phys. Soc. Japan, 21, 1255 (1966).

[152] E. N. Paleolog, A. Z. Fedotova, and V. D. Fitiulina, Elektro-
      khimiya, 4, 700 (1968).

[153] S. Shibamori, S. Yoshizawa, and F. Hine, J. Electrochem.
      Soc. Japan, 35, 197 (1967).

[154] E. J. W. Verwey, Physica, 2, 1059 (1935).

[155] N. F. Mott, Trans. Faraday Soc., 43, 429 (1947).

[156] N. F. Mott, J. Chem. Phys., 44, 172 (1947).

[157] N. Cabrera and N. F. Mott, Rep. Progr. Phys., 12, 163 (1948-49).

[158] N. Cabrera, Phil. Mag., 40, 175 (1949).

[159] R. V. Moshtev, Z. Elektrochem., 71, 1079 (1967).

[160] R. V. Moshtev, Electrochim. Acta, 15, 657 (1970).

[161] J. F. Dewald, J. Electrochem. Soc., 102, 1 (1955).

[162] L. Young, Proc. Royal Soc. (London), A258, 496 (1960).

[163] L. Young, J. Electrochem. Soc., 110, 589 (1963).

[164] S. G. Christov and S. Ikonopisov, J. Electrochem. Soc., 116, 56 (1969).

[165] C. P. Bean, J. C. Fisher, and D. A. Vermilyea, Phys. Rev., 101, 551 (1956).

[166] M. J. Dignam, Can. J. Chem., 42, 1155 (1964).

[167] M. J. Dignam, J. Electrochem. Soc., 112, 722 (1965).

[168] M. J. Dignam, J. Electrochem. Soc., 112, 729 (1965).

[169] M. J. Dignam and D. Goad, J. Electrochem. Soc., 113, 381 (1966).

[170] J. A. Davies, J. P. S. Pringle, R. L. Graham, and F. Brown, J. Electrochem. Soc., 109, 999 (1962).

[171] J. A. Davies and B. Domeij, J. Electrochem. Soc., 110, 849 (1963).

[172] U. Ebersbach, K. Schwabe, and P. Konig, Electrochim. Acta, 14, 773 (1969).

[173] A. I. Oshe, V. A. Lovachev, and B. N. Kabanov, Elektrokhimiya, 5, 1383 (1969).

[174] V. A. Lovachev, A. I. Oshe, and B. N. Kabanov, Elektrokhimiya, 5, 958 (1969).

[175] A. M. Lanyon and B. M. W. Trapnell, Proc. Royal Soc. (London), A227, 387 (1955).

[176] F. P. Fehlner and N. F. Mott, Oxidation of Metals, 2, 59 (1970).

[177] M. J. Dignam and D. B. Gibbs, Can. J. Chem., 48, 1242 (1970).

[178] B. E. Wilde and G. A. Teterin, Br. Corros. J., 2, 125 (1967).

[179] M. A. Genshaw, Ph. D. thesis, University of Pennsylvania, 1965.

[180] K. G. Weil and K. F. Bonhoeffer, Z. Physik. Chem., NF, 4, 175 (1955).

[181] K. F. Bonhoeffer and H. Beinert, Z. Electrochem., 47, 147, 441, 536 (1964).

[182] G. Aronowitz and N. Hackerman, J. Electrochem. Soc., 110, 633 (1963).

[183] Yu. N. Mikhailovskii, Dokl. Akad. Nauk SSSR, 148, 617 (1963).

[184] Yu. N. Mikhailovskii, G. G. Lopovok, and N. D. Tomashov, Collection: Corrosion of Metals and Alloys, Metallurgizdat, Moscow, 1963, p. 267.

[185] R. D. Armstrong, J. A. Harrison, H. R. Thirsk, and R. Whitfield, J. Electrochem. Soc., 117, 1003 (1970).

[186] E. D. Kochman, Tr. Kasan. Khim. Tekhnol. Lust., No. 34, 104 (1965); through Ref. Zh. Khim., 1967, Pt. 11, Abstr. No. 5L269.

[187] S. A. Awad and A. Kassab, J. Electroanal. Chem. Interfacial Electrochem., 26, 127 (1970).

[188] R. W. Powers and M. W. Breiter, J. Electrochem. Soc., 116, 719 (1969).

[189] M. W. Breiter, Electrochim. Acta, 15, 1297 (1970).

[190] M. N. Hull, J. E. Ellison, and J. E. Toni, J. Electrochem. Soc., 117, 192 (1970).

Chapter 2

MECHANISMS OF IONIC TRANSPORT THROUGH OXIDE FILMS

M. J. Dignam

Department of Chemistry, University of Toronto

Toronto, Ontario, Canada

I. INTRODUCTION

The sustained growth of oxide films on metals, whether by the direct chemical attack of an oxidizing agent or through anodic oxidation, requires the transport of material through the phases involved and across the interfaces. In general the transport of metal (and charge) up to the metal-oxide interface, and of oxidizing agent (and charge) up to the film-environment interface will not be rate controlling once the metal surface is covered with a film. In the present

chapter this assumption is made explicitly, the problem being therefore reduced to a consideration of transport processes within the
oxide film and at its boundaries. A great deal of material has been
written on these topics, the publications of Mott and Gurney [1],
L. Young et al. [2, 3], and Vermilyea [4] having been particularly
influential in shaping the content of this chapter. In writing it, however, no attempt has been made to provide either a complete bibli-
ography or a review of the field, partly because excellent review
articles that are essentially up to date [2-7] are available, but
mainly in order to ease the problem of presenting a coherent treatment of the subject. It is therefore inevitable that the material
selected and its treatment reflect the author's views.

The kinetics of the growth of oxide films display a very wide
range of behavior. Thus the growth of porous or cracked (nonprotec-
tive) films is related to transport phenomena in a very much more
complicated fashion than is the growth of a continuous film of essentially uniform thickness. This chapter deals almost entirely with the
second class of systems, even though it is probable that no system
adheres strictly to this description. Thus excluded from consideration are such systems as those generated during the anodic formation
of platinum oxide on platinum [8] or of cuprous oxide on copper [9,10],
where only about one monolayer of oxide is formed before a secondary
reaction begins (oxygen evolution in the case of platinum, cupric
oxide formation followed by oxygen evolution in the case of copper).
For these systems, once the film is continuous, the initial reaction
ceases. The rate-controlling reactions presumably occur, therefore,
at the metal-electrolyte interface, with the only transport processes
involved being those within the electrolyte phase; that is, transport
of oxidizing species up to, and in some instances one or more of the
products away from, the electrode surface.

The remaining class of systems is usefully separated at this point into two divisions, one comprising the systems that behave in a manner characteristic of the valve metals, the other, those that do not. The same metal can behave under one set of conditions of electrolyte composition, etc., as a valve metal and under other conditions as a nonvalve metal, so that the above divisions are ones in which systems, not metals, are placed.

The two principal characteristics of valve-metal anodic oxidation behavior are the following:

1. The requirement of a very high overfield (overpotential divided by film thickness) for film formation, that is, on the order of 10^6 V/cm or larger.

2. The formation of an oxide film of essentially fixed stoichiometry, to the exclusion of other processes, such as metal dissolution and oxygen evolution. The metals that come the closest to exhibiting this behavior are tantalum, niobium, and aluminum in some electrolytes [7], with a large number of metals, semiconductors, and alloys approaching the behavior of these metals under certain conditions (e.g., W, Bi, Fe, Ti, Ge, Si, Zr, Sb, InSb [2]; alloys of Al with Mg, Zn, Cu, Si, Fe [11]; alloys of Nb with Zr, W, Ti, V, and Mo [12]; alloys of V with Nb, Ti, Ta, and Al [13]; and alloys of Ta with Nb, Ti, and Zr [13, 14].

No metal, when anodically oxidized, satisfies the second of the above two characteristics precisely, since film dissolution and/or oxygen evolution will inevitably occur, even if only to a small extent. For purposes of much of the discussion in this chapter, condition 2 relating to film dissolution and oxygen evolution may be relaxed, and all systems characterized by a high overfield and a film of essentially fixed stoichiometry may be included in the valve-metal category.

Sections II to VI are devoted to a development of equations describing the fundamental transport processes and accordingly contain little reference to specific systems or data. Sections VII and VIII deal primarily with the valve-metal-oxide systems, with the final section relating briefly to porous films and non-valve-metal-oxide systems.

II. DIFFERENTIAL EQUATIONS FOR IONIC TRANSPORT IN SOLIDS

A. Introduction

In this section we develop equations for the thermally activated transport of charged species through a solid under the combined action of concentration and electrostatic potential gradients, as such equations must clearly play a central role in the formulation of the kinetics of oxide-film growth. Any detailed consideration of the nature of the mobile defects will be deferred until Section VII. For the present they will be regarded simply as charged species moving on a three-dimensional, periodic, potential-energy surface, a consequence of the periodic arrangement of the atoms in the solid. This description will not be a satisfactory one for either vitreous or highly defective crystalline solids, a matter dealt with in Section II.C.

The factors affecting the transport of ionic species are their concentration, the local electrostatic field, and the spatial derivatives of these. Such quantities are not independent, but are coupled through the continuity condition on the flux density and through Poisson's equation. The driving force for the transport process is, of course, the electrochemical-potential gradient for the species. It is usual in treating equilibrium and transport processes in solids to employ the following approximate expression for the electrochemical potential $\bar{\mu}_i$ of the ith species:

$$\bar{\mu}_i = \mu_i^0 + kT \ln n_i + q_i V , \tag{1}$$

where μ_i^0, n_i, and q_i are the standard chemical potential, concentration, and charge, respectively, of species i and V is the local electrostatic potential. The approximate nature of Eq. (1) derives not only from the dilute-solution form chosen for the activity but also from the fact that a potential-energy term, which varies with the electrostatic field strength, has been excluded [15]. The excluded term arises from the variation with defect concentration in the local (static) dielectric constant, which in the presence of an electrostatic field introduces a field-strength-dependent potential energy of polarization. For conduction electrons or positive holes this energy term will likely be negligible. The effect of vacancies and interstitials on the local dielectric constant, though not totally insignificant, should introduce a negligible potential-energy term for all but perhaps extremely high field strengths, a topic that is treated in Section III. It is likely, therefore, that Eq. (1) represents a satisfactory approximation for all systems in which the defect concentration and field strength are not too high.

B. Derivation of Equations

In this section the transport equation is developed according to Dignam, D.J. Young, and Goad [16].

The system is taken to be a planar, parallel-sided film of thickness X, through which transport occurs in a direction normal to the film's surface. The origin is chosen to be at the interface away from which the defects move. The net activation energy for the migration of a charged defect over a diffusion barrier in the positive x direction and in the presence of an external applied field is given by $(Q + qV_x) - qV_{x-a}$, where the subscripts identify the position in the film where the subscripted variable is to be evaluated (see Fig. 1), Q is the

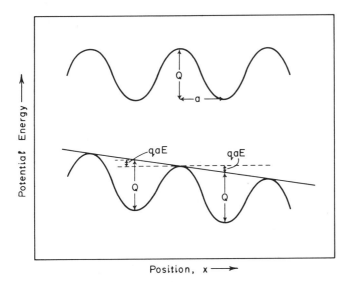

Fig. 1. Potential energy of a mobile defect as a function of position in the absence of an external electric field E.

height of the diffusion barrier, and a is the activation distance (or half the jump distance). If the applied field E is homogeneous (i.e., $\partial E/\partial x = 0$), the net activation energy can be written $Q - qaE$. The above form is retained, however, for greater generality. From elementary rate theory it now follows that the net particle flux J across a diffusion barrier located at position x is given by

$$J = 2a\nu \, \exp\left(-\frac{Q}{kT}\right) \left\{ n_{x-a} \, \exp\frac{q(V_{x-a} - V_x)}{kT} \right.$$

$$\left. - n_{x+a} \, \exp\left[\frac{q(V_{x+a} - V_x)}{kT}\right] \right\} , \tag{2}$$

where ν is the particle's attempt frequency. If the formalism of absolute-reaction-rate theory is employed, $\nu = \tau(kT/h) \exp(\Delta S^{0\neq}/k)$ and $Q = \Delta H^{0\neq}$, where τ is the transmission coefficient, usually set equal to unity, and $\Delta S^{0\neq}$ and $\Delta H^{0\neq}$ are the standard entropy and enthalpy of activation, respectively. Since these substitutions may be made at any time, the simpler formalism is used throughout this chapter.

In writing Eq. (2) it has been assumed for simplicity that only one kind of diffusion barrier is present and furthermore that it is symmetrical in the sense that the barrier maximum lies midway between adjacent minima. Two further, and perhaps more serious, approximations have been made in that the positions of the minima and maxima have been assumed to be unaltered by the applied field, and the three-dimensional potential-energy surface for diffusion has been replaced by a one-dimensional surface. The first of these approximations is treated in Section III, where it is shown to be valid for field strengths that are not too high, and the second is treated to a limited extent in Section V.

To simplify the algebra somewhat, the following parameters are defined:

$$c = \frac{n}{N} ; \quad s = \frac{x}{a} ; \quad p = \frac{qV}{kT} ; \quad \varepsilon = - \frac{\partial p}{\partial s} = \frac{qaE}{kT} ;$$

$$j = \frac{J}{2a\nu N \exp(-Q/kT)} ,$$

where N is the concentration of sites for the ionic defect in the crystal and $E = - \partial V/\partial x$ is the electric field strength. The dimensionless parameters c, s, p, ε, and j represent the defect concentration, position, electrostatic potential, electrostatic field strength, and the flux density, respectively. When we have made these substitutions, Eq. (2) becomes

$$j = - \exp(-p_s)[c_{s+1} \exp(p_{s+1}) - c_{s-1} \exp(p_{s-1})] . \tag{4}$$

In order to transform this difference equation into a differential equation, $c_{s+1} \exp(p_{s+1})$ and $c_{s-1} \exp(p_{s-1})$ are expanded as infinite Taylor series about s, with all terms being retained, to give

$$j = -2e^{-p}\{\sinh(\partial/\partial s)\} c \, e^{p} \tag{5}$$

where the term in braces is an operator. In examining the operation of the ℓth term in the series expansion of the operator on $c\,e^p$, it is seen that

$$\frac{1}{\ell!}\frac{\partial^\ell}{\partial s^\ell}\,c\,e^p = e^p\,\frac{1}{\ell!}\left\{\frac{\partial}{\partial s}+\frac{\partial p}{\partial s}\right\}^\ell\,c\,, \tag{6}$$

which with Eq. (5) and the identity $\varepsilon = -\partial p/\partial s$ leads to the result

$$j = 2\left\{\sinh\left(\varepsilon - \partial/\partial s\right)\right\}c\,. \tag{7}$$

The problem now is to obtain solutions of Eq. (7) that are valid under various limiting conditions and to ascertain the range of their validity. Fromhold [17] has shown, by numerical calculations, that for systems involving film growth in which the defect concentration is small compared with the lattice-ion concentration (i.e., $c << 1$) one may to a good approximation replace the continuity condition, div $\underset{\sim}{J} = -\partial n/\partial t$, by the steady-state approximation, div $\underset{\sim}{J} \simeq 0$ or $\partial j/\partial s \simeq 0$, even though J be time dependent. It seems that for $c << 1$ the fraction of the flux J involved in changing the defect-concentration profile, when E and J are changing, is very small and may be neglected.

The case in which the field strength is independent of position — that is, in which the space charge can be neglected — will be examined first. The problem is therefore to solve Eq. (7) for j and ε constant. The homogeneous part of the solution is found by setting $j = 0$. It is apparent that if $\{\varepsilon - \partial/\partial s\}\,c = 0$, the right-hand side of Eq. (7) will be zero. Solving $\{\varepsilon - \partial/\partial s\}\,c = 0$ gives $c = A\,e^{-p}$, where A is the integration constant. Since ε and j are constants, the particular solution is obtained on setting c equal to a constant. Thus the general solution becomes

$$c = c_0 - A(1 - e^{-p})\,, \tag{8}$$

where the boundary conditions $c = c_0$ and $p = 0$ at $s = 0$ have been used. Substituting Eq. (8) into Eq. (7) gives $j = 2(\sinh \varepsilon)(c_0 - A)$;

however, differentiation of Eq. (8) with respect to s leads to $\partial c/\partial s = A\epsilon e^{-p}$, which with Eq. (8) gives

$$c_0 - A = c - \frac{1}{\epsilon}\frac{\partial c}{\partial s}. \tag{9}$$

Thus the flux equation can be written

$$j = 2(\sinh \epsilon)(c - \frac{1}{\epsilon}\frac{\partial c}{\partial s}). \tag{10}$$

On converting back to the dimensioned variables and making use of Eq. (1), Eq. (10) can be written in the form

$$J = -\frac{Dn}{kT}\frac{\partial \bar{\mu}}{\partial x}\frac{\sinh(qaE/kT)}{qaE/kT} \tag{11}$$

where $D = 4a^2\nu \exp(-Q/kT)$. For $E = 0$, it follows that $[\sinh(qaE/kT)]/(qaE/kT) = 1$ and $\partial\bar{\mu}/\partial x = (kT/n)(\partial n/\partial x)$, so that Eq. (11) becomes $J = -D(\partial n/\partial x)$, from which it is apparent that D is simply the diffusion coefficient for this model. Equation (11) was first proposed by the author [18] on the grounds that it is the simplest equation that satisfies certain necessary conditions. These were that it reduce in the appropriate limiting cases to the linear transport equation

$$J = -D(\partial n/\partial x) + n\nu E \tag{12}$$

and the Mott-Cabrera conduction equation [19]

$$J = 4a\nu n[\exp(-Q/k\bar{T})]\sinh(qaE/kT) \tag{13}$$

as well as satisfy the condition that for $(\partial\bar{\mu}/\partial x) = 0$, $J = 0$. For field strengths that are low enough for the sinh function to be replaceable by its argument, Eq. (11) reduces to Eq. (12) on making use of the Einstein relation, $(\nu/D) = q/kT$, where ν is the mobility of the migrating species. Similarly for $(\partial n/\partial x) = 0$, Eq. (11) reduces to Eq. (13).

The question now arises as to whether or not Eq. (10) (or Eq. (11)) will represent a satisfactory approximation when the space charge is important and consequently the field strength is no longer independent of position. In general, ϵ may be related to the concentrations of the various charged species in the film through Poisson's equation. The problem can then be solved in principle, given both the stoichiometric equations that govern the relationships between the various flux densities and the equations relating the fluxes to the field strength, concentration, and their spatial derivatives. Of some importance is the case where the space charge is partially or entirely due to the moving ionic defects under consideration. For the present, however, the general case is considered.

The field derivative is related to the space-charge density ρ through Poisson's equation, which, expressed in terms of the dimensionless parameters, becomes

$$\partial\epsilon/\partial s = r , \tag{14}$$

where r, the dimensionless space-charge density, is given by

$$r = (4\pi qa^2/\epsilon kT)\rho \tag{15}$$

and ϵ is the dielectric constant of the medium.

Clearly Eq. (10) represents the solution of Eq. (7) in the limit as r approaches zero. The problem, however, is to determine how small r must be for Eq. (10) to represent a valid approximation. To ascertain this, the cases of low and of high field strengths will be considered separately.

For sufficiently small ϵ and r the following series expansion of Eq. (7) will converge rapidly:

$$j = 2 \sum_{\ell=1,3,5,\ldots}^{\infty} \{\epsilon - \partial/\partial s\}^\ell c/\ell ! \tag{16}$$

The approximate solution

$$j \approx 2\{\epsilon - \partial/\partial s\} c = 2(\epsilon c - \partial c/\partial s) \tag{17}$$

is now investigated. The procedure adopted is to employ Eq. (17) to obtain an estimate of the next higher order term in Eq. (16) and hence assess the conditions under which it may be neglected.

Equation (17) gives on rearrangement

$$\partial c/\partial s = \varepsilon c - j/2 \qquad\qquad (18)$$

from which expressions for $\partial^2 c/\partial s^2$ and $\partial^3 c/\partial s^3$ can be obtained relating them to c and the derivatives of ε. Evaluating the second term in the expansion of Eq. (16) and substituting for the derivatives of c as above and for $\partial\varepsilon/\partial s$ from Eq. (14) leads, after some algebra, to

$$j \approx 2(\varepsilon c - \partial c/\partial s)[1 - (r - \varepsilon^2)/6 + \cdots] . \qquad\qquad (19)$$

On the assumption that successive higher order terms are of diminishing importance, the condition that Eq. (17) lead to a fractional error in j which is $\leq \delta$ is that $|r - \varepsilon^2| \leq 6\delta$. Adopting the more stringent criterion

$$|r| + \varepsilon^2 \leq 6\delta \qquad\qquad (20)$$

and setting $\epsilon \approx 10$, $a \approx 2\,\text{Å}$, $q = 1e$, and $\delta = 0.1$, this places an upper limit on the space-charge concentration of about $9 \times 10^{19} e/\text{cm}^3$ and on the field strength of about 7×10^5 V/cm. For fields this low, sinh ε may be replaced by ε to a good approximation, so that the homogeneous-field solution, Eq. (10), reduces to Eq. (17). Thus Eq. (10) is a valid approximation for the space-charge case whenever Eq. (20) is satisfied.

In the case of large field strengths expansion (16) begins to converge only after a large number of terms. Accordingly, the summation is rearranged in the following manner: Equation (16) may be written

$$j = 2 \sum_{\ell=1,3,5,\ldots}^{\infty} \left(\sum_{i=0}^{\ell} T_{i\ell}/\ell! \right) , \qquad\qquad (21)$$

where $T_{i\ell}$ is the sum of all terms containing i differential operators in the total expansion of $\{\varepsilon - \partial/\partial s\}^\ell c$. As the binomial theorem cannot be used in this expansion (ε and $-\partial/\partial s$ do not commute), the evaluation of the terms $T_{i\ell}$ is rather involved and will not be repeated here (see Ref. [16]). Interchanging the order of summation in Eq. (21) gives

$$j = \sum_{i=0}^{\infty} Z_i , \qquad\qquad (22)$$

where

$$Z_i = 2 \sum_{\substack{\ell=m,\ m+2, \\ m+4,\ \ldots}}^{\infty} T_{i\ell}/\ell!$$

and m is the lowest odd integer $\geq i$.

In Ref. [16] the first two terms in this expansion are evaluated.

$$Z_0 = 2c \sinh \varepsilon , \qquad\qquad (23)$$

$$Z_1 = -2(\partial c/\partial s) \cosh \varepsilon - rc \sinh \varepsilon . \qquad\qquad (24)$$

As for the low-field case, the approximate equation

$$j \approx 2c \sinh \varepsilon \qquad\qquad (25)$$

is now investigated, using a similar procedure. Thus Z_1 can be estimated by using Eq. (25) to estimate c and $\partial c/\partial s$. The resulting expression on substitution into Eq. (22) leads to the following equation:

$$j = 2c \sinh \varepsilon \{1 + r[2 + (\sinh \varepsilon)^2]/4(\sinh \varepsilon)^2 + \cdots \} . \qquad\qquad (26)$$

Assuming again that the higher order terms continue to diminish in importance, the condition that Eq. (26) lead to a fractional error in j that is $\leq \delta$ is given by

$$|r|/(\sinh \varepsilon)^2 + |r|/2 \leq 2\delta . \qquad\qquad (27)$$

Using the same values for the parameters as before and setting $\delta = 0.1$, this condition places an upper limit on the space-charge concentration of about 6×10^{19} e/cm^3 for a field strength $\gtrsim 10^6$ V/cm. For

lower space-charge concentrations the approximation will of course
be valid for lower field strengths. Again it is easily shown that whenever condition (27) is satisfied, the homogeneous-field solution,
Eq. (10), reduces to a good approximation to Eq. (25). Thus the
homogeneous-field solution represents a satisfactory approximation
for the space-charge case over the entire range of field strength provided that the space-charge concentration does not exceed a value of
about 5×10^{19} e/cm^3.

C. Microcrystalline or Amorphous Films

The transport of ions through microcrystalline films under, say,
conditions of anodic oxidation will lead to films that are relatively
uniform in thickness only if the crystallites are small relative to the
film thickness or transport along grain boundaries is not favored over
bulk transport. If the latter is the case, the resulting transport properties will not be materially different from those for a crystalline film.
If, on the other hand, the former is the case, the complications that
arise will be very similar to those expected for an amorphous or vitreous film.

Thus for both microcrystalline and amorphous films there will
exist a range of values for the diffusion-barrier height, activation
distance, jump distance, attempt frequency, and local dielectric constant. The principal effect of local variations in the dielectric constant will be to introduce variations in the local electrostatic field
strength. This effect can accordingly be accommodated by defining
an effective activation distance that incorporates the influence of
local variations in the dielectric constant from the mean value.

In obtaining an equation for the transport rate for such a system
all possible combinations of sequential steps, parallel paths, branching paths, and the like must be considered — a formidable task.

Certain general properties of the resultant transport equation can be deduced, however. In the high-field limit, when Eq. (13) is valid for crystalline films, the transport equation for amorphous materials, if cast in the form of Eq. (13), will require that Q and a be functions of both field strength and temperature, the precise functional forms depending on the details of the transport network assumed.

If the transport paths are highly branched, with few regions that involve several sequential jumps without branching, Q would be expected to increase with increasing temperature and a to decrease with increasing field strength, since the higher the temperature, the greater will be the contribution to J of processes involving great barrier heights, with a similar argument applying to a. If, on the other hand, the transport paths are relatively unbranched, the reverse would be expected: Q should decrease with increasing temperature and a should increase with increasing field strength. The temperature variation of a and the field variation of Q will depend on the nature of the correlation between the local barrier heights and effective jump distances.

For low field strengths an equation formally identical with Eq. (12) will result since only first-order terms in E are retained. The activation energy determined from D or v will, however, be temperature dependent, the sign of the dependence being deduced as above.

To summarize, Eq. (10) (or Eq. (11)) can be used to describe ion transport through crystalline films subject to the conditions that the space charge does not exceed a value of about 5×10^{19} e/cm^3 and the field strength is not too high. In Section III the latter condition is shown to be satisfied for $|qaE/Q| << 3$. If the film is micro-crystalline or amorphous and possesses a range of values for the local transport parameters, this should be made readily apparent in the form of the temperature dependence of the flux density.

III. IONIC TRANSPORT ACROSS A BARRIER
IN THE HIGH-FIELD LIMIT

A. Introduction

In Section II the net activation energy for the migration of ionic species across a potential-energy barrier in the presence of a homogeneous field E was assumed to be given by

$$W(E) = Q - qaE , \tag{28}$$

where Q is the height of the barrier in the absence of the field, q is the charge on the ionic species, and a is the activation distance for the ion. As noted in Section II. B, Eq. (28) is obtained by assuming that the positions of minima and maxima in the potential-energy surface for the mobile ionic species are not displaced on applying an external field, an approximation that must break down for sufficiently high field strengths. This section is concerned with the behavior in the high-field limit.

Before proceeding with this, however, a somewhat more fundamental question — namely, that concerned with the nature of the quantities q and E appearing in Eq. (28) — will be examined first. It has from time to time been suggested that the polarization of the transport medium might make a contribution to the field assisting ion transport so that an effective field E_e should be used in place of the Maxwell field E. Furthermore, since bonding in most solids is by no means entirely ionic, the question has also been raised as to whether or not an effective ionic charge should also be used. The following discussion of these topics is taken, with minor changes, from a paper by the author [18].

B. Effective Charge and Effective Field Strength

In treating ionic transport in solids one must distinguish between two kinds of charge displacement. The first is associated with a more or less uniform polarization of the entire solid, a process that will

occur in the presence of an external field. The second is associated

with the displacement of the charged defect in question along the

"reaction coordinate." Both these charge displacements make a real

contribution to the kinetics of ion migration, the latter in a direct and

obvious manner, the former through the fact that a charge displace-

ment in the crystal will alter the field acting on the charged defect.

It is therefore difficult to make an unambiguous separation of charge

displacement into these two categories. Thus displacement of ions

that are nearest neighbors to the defect can be associated either with

the migration of the defect itself or with field-polarization effects, or

a combination of both of these.

To avoid this difficulty, two hypothetical systems are considered.

System 1 is a long, narrow crystalline solid aligned with a homoge-

neous external field and containing a single ionic defect. System 2

is as nearly identical with system 1 as possible except that it con-

tains no defect. In terms of these systems the defect-charge dis-

placement u is defined as the component in the field direction of the

first moment of the charge for system 1 less that for system 2, a cor-

responding origin being used for the two systems. Again the total

potential energy of the defect U is defined as the potential energy of

system 1 less that of system 2 and of necessity involves an arbitrary

constant since the composition of the two systems is not the same.

The effective field strength is now defined through the following

equation:

$$U = U_1 - \int_0^u E_e \, du , \qquad (29)$$

where U_1 is a function with the periodicity of the lattice and E_e is

the effective field strength, defined to go to zero for zero external

field. For zero field, then, U_1 becomes the unperturbed periodic

potential energy experienced by the defect on diffusing through the solid. The question now arises as to the properties of E_e as defined in Eq. (29). It is apparent that the displacement of the defect from one position to another, equivalent to the first by symmetry, is energetically equivalent to applying a particular symmetry transformation $\underset{\sim}{T}$ to the crystal as a whole. This follows from the fact that the external field is maintained homogeneous. On applying the symmetry transformation $\underset{\sim}{T}$ the following energy transformations therefore occur:

$$\underset{\sim}{T} U_1 \rightarrow U_1 , \tag{30}$$

$$\underset{\sim}{T} U \rightarrow U - q(\Delta x)_E E , \tag{31}$$

where q is the total charge carried by the defect (an integral multiple of the electron charge) and $(\Delta x)_E$ is the component of the displace-of the defect in the positive field direction brought about by the symmetry operation $\underset{\sim}{T}$. The point to note is that q and E in Eq. (31) are the valency charge and external field, respectively, and must be so whether or not bonding is covalent and/or extensive polarization of the medium occurs.

On comparing Eqs. (29) and (31) two alternatives present themselves. If E_e is assumed to be homogeneous (i. e., to possess no periodic variation), then the effective field E_e must be identical with the external field or Maxwell field E. On the other hand, if U_1 is defined as the periodic potential experienced by the defect in the absence of an external field, then, since polarization of the medium will, in general, contribute a field acting on the defect, E_e must contain a component that varies with the periodicity of the lattice. Furthermore, the mean of E_e with respect to u for any symmetry transformation $\underset{\sim}{T}$ is simply the external field. Thus one may either choose the effective field so that it is homogeneous (i. e., E_e = E), in which case U_1 must be a function of the field strength E, or one may choose U_1 to be independent of the external field, in which case the effective field E_e must contain a periodic component.

The former convention will be adopted. Thus setting $E_e = E$ in Eq. (29) and realizing that U_1 must be a function of E, we obtain

$$U = [U_1^0 + \Delta U_1(u, E)] - uE , \qquad (32)$$

where U_1^0 is the periodic potential function for the defect at zero field strength and $\Delta U_1(u, E)$ is a function of u and E that has the periodicity of the lattice and is zero at $E = 0$. If the adjacent positions of minimum and maximum potential energy for the migration of the defect in the direction assisted by the field are represented by u_0 and u_m, respectively, the net activation energy for migration in the field direction, $W(E)$, will be given by

$$W(E) = (U)_{u=u_m} - (U)_{u=u_0} , \qquad (33)$$

with u_0 and u_m given by the roots of the following equation:

$$\partial U / \partial u = 0 = dU_1^0/du + \partial[\Delta U_1(u, E)]/\partial u - E . \qquad (34)$$

The kinetic consequences of Eqs. (32) through (34) in the limit of very low field strengths are now considered. As before, the treatment is restricted to the one-dimensional case, in which U_1^0 contains only one kind of maximum and is symmetrical. Signifying the distance between consecutive minima or maxima in U_1^0 as $2a$, then for zero field it is clear from symmetry considerations that, on displacing the defect from a position of minimum potential energy to an adjacent position of maximum potential energy, the change in u will be simply qa, where once again q is the total charge carried by the defect regardless of whether bonding is largely covalent or not. This follows from the fact that for the assumptions made both the maximum and minimum potential-energy positions at zero field will represent positions of symmetry. From Eqs. (32) through (34) it now follows that

$$\lim_{E \text{ small}} W(E) = Q - [\Delta U_1(0, E) - \Delta U_1(qa, E)] - qaE , \qquad (35)$$

where for convenience u has been set equal to zero at a position corresponding to a minimum in U_1^0 and Q is the potential-energy difference between the positions of minimum and maximum potential energy for U_1^0.

The problem now is to determine the form of the field dependence of the term in brackets in Eq. (35). To do this, it is first noted that the positions u = 0 and u = qa represent positions of symmetry in the limit of small fields, in the sense that for every element of charge q' displaced from the defect at position u = 0 or u = qa by a vector displacement $\underset{\sim}{r}$ there will exist an identical element of charge displaced $-\underset{\sim}{r}$ from the defect. Considering a pair of such charges lying on the field axis, then, in the absence of an external field, the component of the potential energy of the defect arising from the interaction of the defect with this charge pair will be given by 2qq'/x, where x is the distance between the defect and either one of the charges. In the presence of the field, x will be altered to x + δx and x − δx, respectively, so that the change in the potential energy of the defect arising from these displacements is given by

$$\frac{qq'}{x+\delta x} + \frac{qq'}{x-\delta x} - \frac{2qq'}{x} \simeq \frac{2qq'}{x}\left(\frac{\delta x}{x}\right)^2 , \qquad (36)$$

where the approximation involves neglecting all terms in (δx/x) to the power 4 or higher. Setting δx proportional to E, it is seen that the potential-energy change for the defect, brought about by the field displacement of the charge pair, is proportional to E^2. As this argument can readily be extended to cover off-axes charge pairs, it may be concluded that

$$U_1 - U_1^0 = \Delta U_1(u, E) \propto E^2 \qquad (37)$$

for u = 0 or qa and E small.

It now follows from Eq. (35) and the above that

$$\lim_{E^2 \to 0} W(E) = Q - qaE . \tag{38}$$

The point to be made is simply that in the low-field limit, and hence for the linear transport equation, the quantities to be used are the Maxwell field E and the total charge carried by the defect, q. Thus corrections for effective field and effective charge are not required for low field strengths, regardless of whether or not covalent bonding is involved, at least for transport through a lattice exhibiting the symmetry assumed here. For high field strengths, on the other hand, some sort of correction is required. It is preferable, however, to include the correction terms explicitly rather than introduce an effective charge or effective field strength. The purpose of the following section is to obtain just such explicit equations for the higher order terms in the expansion of $W(E)$, the treatment being that due to Dignam and Gibbs [20].

C. Higher Order Field Terms in W(E)

1. First Approximation

In this section the model is generalized somewhat to include the case of ion transport across an activation barrier situated at an interface. The initial development is made on the assumption that

$$(\partial U_1/\partial E)_u = 0 = \Delta U_1(u, E) . \tag{39}$$

From the discussion in Section III. B it is clear that Eq. (39) is at best only approximately correct. The development will proceed nevertheless on the basis of Eq. (39), with the question of the field dependence of U_1 being dealt with later.

In this approximation Eq. (34) reduces to

$$dU_1^0/du = E \tag{40}$$

and the expression for the net activation energy, Eq. (33), becomes

$$W(E) = U_1^0(u_m) - U_1^0(u_0) - (u_m - u_0) E . \tag{41}$$

The form of $W(E)$ has been evaluated in the above manner by the author [21] for a number of assumed functional forms for $U_1^0(u)$. An example is given here for the case of a cosine potential-energy function, defined through the following equation:

$$U_1^0(u) = (Q/2)(1 - \cos \pi u/u_0^*) , \tag{42}$$

where Q is the height of the diffusion barrier, as before, and u_0^* is the charge-activation distance product, or activation dipole, in the limit of zero field strength. Thus for a symmetrical barrier $u_0^* = qa$. From Eqs. (40) and (42) the following equation is then obtained:

$$\sin(\pi u/u_0^*) = 2u_0^* E/\pi Q . \tag{43}$$

Solving Eq. (43) for u, the values of u_0 and u_m are given by the solutions that lie in the intervals $[0, u_0^*/2]$ and $[u_0^*/2, u_0^*]$, respectively. It therefore follows that

$$U_1^0(u_0) = \frac{Q}{2}\left\{ 1 - \left[1 - \frac{4}{\pi^2}\left(\frac{u_0^* E}{Q}\right)^2 \right]^{\frac{1}{2}} \right\} , \tag{44}$$

$$U_1^0(u_m) = \frac{Q}{2}\left\{ 1 + \left[1 - \frac{4}{\pi^2}\left(\frac{u_0^* E}{Q}\right)^2 \right]^{\frac{1}{2}} \right\} , \tag{45}$$

$$u_0 = \frac{u_0^*}{\pi} \arcsin \frac{2u_0^* E}{\pi Q} = u_0^* - u_m . \tag{46}$$

Substituting these values into Eq. (41) gives

$$W(E) = Q\left[1 - \left(\frac{2u_0^* E}{\pi Q}\right)^2 \right]^{\frac{1}{2}} - u_0^* E + \frac{2u_0^* E}{\pi} \arcsin \frac{2u_0^* E}{\pi Q} . \tag{47}$$

Expanding the square-root term and the arcsin function as a power series in $(2u_0^* E/\pi Q)$ gives

$$W(E) = Q - u_0^* E\left[1 - \frac{2}{\pi^2}\left(\frac{u_0^* E}{Q}\right) - \frac{4}{3\pi^4}\left(\frac{u_0^* E}{Q}\right)^3 + \cdots \right] . \tag{48}$$

Not all functions that might be used to represent U_1^0 will lead to a power-series form for $W(E)$. Thus a potential-energy function that possesses no true maximum, such as the Morse function or the Coulomb-interaction potential-energy function (U_1 linear in $1/u$), does not lead to a power-series form for $W(E)$ [20]. However, all functional forms for U_1 that can be expanded in a power series in u about their extrema, and this should include all functions that are reasonable representations of a diffusion barrier, do lead to a power-series form for $W(E)$, as shown in the following development. On choosing adjacent positions of minima and maxima in the potential-energy barrier $U_1^0(u)$ to coincide with $u = 0$ and $u = u_0^*$, respectively, u_0^* becomes the zero-field activation dipole. The zero-field activation energy is then defined through the equation

$$Q_0 = U_1^0(u_0^*) - U_1^0(0) . \tag{49}$$

The quantity $U_1^0(u)$ may now be represented over the range of interest (i.e., $u_0^* \geq u \geq 0$) in the following manner:

$$U_1^0(u) = Q_0 f_1(u/u_0^*) \quad (u_c \geq u \geq 0)$$
$$= Q_0[1 - f_2(1 - u/u_0^*)] \quad (u_0^* \geq u \geq u_c) , \tag{50}$$

where the functions $f_1(Z)$ and $f_2(Z)$ determine the "shape" of the function U_1^0 and Q_0 and u_0^* determine its dimensions. In order that U_1^0 be continuous and have a continuous first derivative $f_1(Z)$ and $1 - f_2(1 - Z)$ and their first derivatives, respectively, must be equal at $Z = u_c/u_0^*$. Since U_1^0 has a minimum at $u = 0$ and a maximum at $u = u_0^*$, it must contain at least one inflection point between these limits; u_c is chosen to coincide with the position at which dU_1^0/du is maximal. The reason for defining $U_1^0(u)$ in terms of two separate functions of (u/u_0^*) is one of mathematical convenience. Note that if the inflection point represents a center of symmetry for U_1^0, then

$f_1(Z) = f_2(Z)$, provided that the zero of potential energy is chosen so
that $U_1^0(0) = 0$. Adopting this latter convention, it is apparent that
$f_1(0) = 0 = f_2(0)$ in the general case. Furthermore, since both $f_1(Z)$
and $f_2(Z)$ have a minimum at $Z = 0$, $f_1'(0) = 0 = f_2'(0)$ in the general
case, where $f_j'(0)$ is the first derivative of $f_j(Z)$ evaluated at $Z = 0$.

Substituting Eq. (50) into Eq. (40) to obtain expressions for u_0
and u_m, and noting that u_0 must lie in the interval $(0, u_c)$ and u_m in
the interval (u_c, u_0^*), the following equations are obtained:

$$f_1'(u_0/u_0^*) = bE ; \qquad f_2'(1 - u_m/u_0^*) = bE , \tag{51}$$

where $b = u_0^*/Q_0$.

Representing the inverse of the functions f_j' by g_j, that is,

$$g_j[f_j'(Z)] \equiv Z \qquad (j = 1, 2) , \tag{52}$$

we obtain

$$u_0 = u_0^* g_1(bE) ; \qquad u_m = u_0^*[1 - g_2(bE)] . \tag{53}$$

By substituting Eq. (53) into Eq. (41), the following expression for the
net activation energy is obtained:

$$W(E) = Q_0[1 - bE + F_1(bE) + F_2(bE)] , \tag{54}$$

where

$$F_j(Z) \equiv Zg_j(Z) - f_j[g_j(Z)] \qquad (j = 1, 2) . \tag{55}$$

An alternative expression for $W(E)$ may be derived by noting that
the first derivative of $F_j(Z)$ is given by

$$F_j'(Z) = g_j(Z) . \tag{56}$$

To show this Eq. (55) is differentiated to obtain

$$F_j'(Z) = Zg_j'(Z) + g_j(Z) - f_j'[g_j(Z)] \cdot g_j'(Z) , \tag{57}$$

which reduces to Eq. (56) since $f'_j[g_j(Z)] \equiv Z$, f'_j and g_j being inverse functions. Integration of Eq. (56) for the boundary condition $F_j(0) = 0$ gives

$$F_j(bE) = \int_0^{bE} g_j(Z) \, dZ \, . \tag{58}$$

That this is the correct boundary condition follows from the conditions already stated, namely, $f_j(0) = 0 = f'_j(0)$, from which, by Eq. (52), $g_j(0) = 0$ and hence, from Eq. (55), $F_j(0) = 0$. By substituting Eq. (28) into Eq. (54), we obtain the following expression for W(E) in terms of the functions g_j:

$$W(E) = Q_0 \{ 1 - bE + \int_0^{bE} [g_1(Z) + g_2(Z)] \, dZ \} \, . \tag{59}$$

It will now be shown that expanding $f_1(Z)$ and $f_2(Z)$ about $Z = 0$ as a power series in Z permits Eq. (59) to be written in the form

$$W(E) = Q_0 [1 - (bE) + (bE)^2/C_2 + (bE)^3/C_3 + \cdots] \, , \tag{60}$$

where the dimensionless constants C_2, C_3, etc., are related to the coefficients in the expansions of $f_1(Z)$ and $f_2(Z)$. The relationships are developed by expressing $f_1(Z)$ and $f_2(Z)$ as follows:

$$f_j(Z) = \sum_{k=2}^{\infty} k^{-1} a_{jk} Z^k \, , \tag{61}$$

from which

$$f'_j(Z) = \sum_{k=2}^{\infty} a_{jk} Z^{k-1} \, . \tag{62}$$

That the first term in the expansion of $f_j(Z)$ is a term in Z^2 follows from the equations $f_j(0) = 0 = f'_j(0)$. Reversion of Eq. (62) gives

$$g_j(Z) = \sum_{k=2}^{\infty} A_{jk} Z^{k-1} \tag{63}$$

and hence

$$\int_0^{bE} g_j(Z)\, dZ = \sum_{k=2}^{\infty} k^{-1} A_{jk}(bE)^k ,$$ (64)

where the coefficients A_{jk} can be evaluated in terms of the coefficient a_{jk} by a successive-approximations procedure or from tables [22].

Comparing Eqs. (59) and (60), from Eq. (64) the coefficients C_K are given by

$$\frac{1}{C_K} = \frac{1}{K}(A_{1K} + A_{2K}) \quad (K = 2 \text{ to } \infty) .$$ (65)

It is clear from the form of Eq. (59) or from Eqs. (60) and (65) that, given U_1^0 as a function of u (and hence g_1 and g_2), W(E) is uniquely determined. On the other hand, if W(E) is known as a function of E (from empirical fit to data, say) it is impossible to determine the form of $U_1^0(u)$ uniquely since only the sum, $A_{1K} + A_{2K}$, is determined by specifying $1/C_K$, not the individual values, A_{1K} and A_{2K}.

The simplest form of barrier consistent with the limitations placed on $U_1^0(u)$ is one that can be represented by a pair of parabolas, one inverted with respect to the other and tangent to each other at $u = u_c$. Thus both $f_1(Z)$ and $f_2(Z)$ will be parabolas and can be represented as follows:

$$f_1(Z) = (u_c/u_0^*)\, f(Z\, u_0^*/u_c) ,$$
$$f_2(Z) = (1 - u_c/u_0^*)\, f[Z\, u_0^*/(u_0^* - u_c)] ,$$ (66)

where $f(Z) = Z^2$. It is easily shown that Eqs. (66) ensure that $f_1(Z)$ and $1 - f_2(1 - Z)$ join at $Z = u_c/u_0^*$ and have a common first derivative at this point, and further that $f_j(0) = 0 = f_j'(0)$. Any value between 0 and u_0^* may be chosen for u_c. Differentiating Eqs. (66) now gives

$$f_1'(Z) = f'(Z\, u_0^*/u_c) ; \qquad f_2'(Z) = f'[Z\, u_0^*/(u_0^* - u_c)]$$ (67)

and hence

$$g_1(Z) = (u_c/u_0^*)g(Z) \; ; \qquad g_2(Z) = [(u_0^* - u_c)/u_0^*]g(Z) \, , \qquad (68)$$

where $g(Z)$ and $f'(Z)$ are inverse functions.

The integral in Eq. (59) is therefore given by

$$\int_0^{bE} [g_1(Z) + g_2(Z)] \, dZ = \int_0^{bE} g(Z) \, dZ \, , \qquad (69)$$

which is seen to be independent of the parameter u_c. Since $f(Z) = Z^2$,

$f'(Z) = 2Z$, so that $g(Z) = \frac{1}{2}Z$ and $\int_0^{bE} g(Z) \, dZ = \frac{1}{4}(bE)^2$. Thus from

Eq. (59)

$$W(E) = Q_0 - u_0^* E(1 - u_0^* E/4Q) \, . \qquad (70)$$

Greater generality is achieved in the form of the barrier $U_1^0(u)$, and hence in the form of $W(E)$, by including cubic and higher order terms in the expression for $f(Z)$. As long as $f(Z)$ satisfies the conditions $f(1) = 1$, $f(0) = 0 = f'(0)$, Eq. (69) will still be valid. Thus setting $f(Z) = \omega Z^2 + (1 - \omega)Z^3$, $f'(Z) = 2\omega Z + 3(1 - \omega)Z^2$ so that $g(Z) = [1/2\omega]Z - [3(1 - \omega)/8\omega^3]Z^2$ correct to terms in Z^2. Evaluating the integral as before and comparing the resulting expression for $W(E)$ with Eq. (60) then gives

$$C_2 = 4\omega \; ; \qquad C_3 = -8\omega^3/(1 - \omega) \qquad (71)$$

and hence

$$C_3 = -C_2^3/2(4 - C_2) \, . \qquad (72)$$

The net activation energy can therefore be written

$$W(E) \simeq Q_0 - u_0^* E\left[1 - \left(\frac{u_0^* E}{C_2 Q_0}\right) + 2\left(\frac{4 - C_2}{C_2}\right)\left(\frac{u_0^* E}{C_2 Q_0}\right)^2\right]. \qquad (73)$$

Finally, from Eq. (50) it follows that

$$\left(\frac{d^2 U_1}{du^2}\right)_{u=0} = \left(\frac{Q_0}{u_0^{*2}}\right)f_1''(0) \, , \qquad (74)$$

$$\left(\frac{d^2 U_1}{du^2}\right)_{u=u_0^*} = -\left(\frac{Q_0}{u_0^{*2}}\right)f_2''(0) \, . \qquad (75)$$

However, from Eq. (61), $f_j''(0) = a_{j2}$ and hence $A_{j2} = 1/a_{j2} = 1/f_j''(0)$, so that, from Eq. (65),

$$\frac{1}{C_2} = \frac{1}{2}\left[\frac{1}{f_1''(0)} + \frac{1}{f_2''(0)}\right] \tag{76}$$

or on substituting from Eqs. (74) and (75),

$$\frac{1}{C_2} = \frac{Q_0}{2u_0^{*2}}\left[\left(\frac{d^2U_1}{du^2}\right)^{-1}_{u=0} - \left(\frac{d^2U_1}{du^2}\right)^{-1}_{u=u_0^*}\right] \tag{77}$$

from which it is seen that C_2 is determined by the radii of curvature of the potential-energy function at the extrema. A result essentially the same as Eq. (77) was first obtained by Christov and Ikonopisov [23] from a rather different approach.

It is clear from Eq. (60) that the condition that the quadratic term in the field be negligible compared with the first-order term is given by

$$u_0^* E/C_2Q_0 \ll 1 . \tag{78}$$

Thus setting $u_0^* \approx 2 \times 10^{-8}$ e/cm, $C_2 \approx 4$, and $Q_0 \approx 1$ eV, and requiring that the quadratic term be less than 5% of the first-order term, Eq. (78) leads to $E \lesssim 10^7$ V/cm, an extremely high field strength. Field strength exceeding this value would be expected to be very rare indeed for transport within thin films but might be ex-pected more commonly at interfaces (e.g., the electrolyte-electrode interface during hydrogen evolution from a metal like mercury, which exhibits an overpotential of about 1 V).

The foregoing development has been based on the approximation that U_1 is independent of field strength. The consequences of allowing for a field dependence of U_1 are examined next with the purpose of evaluating additional contributions to the quadratic term in the field expansion of $W(E)$.

2. Including the Field Dependence of U_1

Defining U_1 through Eqs. (50), (61), and (62), one may take account of the field dependence of U_1 by expanding Q, u^*, and the coefficients a_{jk} as power series in the field E. It is apparent that this procedure will still lead to a power-series form for $W(E)$. Only the terms that make a contribution to $W(E)$ and are of the second order or less in the field strength will be evaluated. Since the field-dependent terms in the expansions of the coefficients a_{jk} will contribute only terms of the third order and higher, only the parameters Q and u^* need be considered. The terms to be examined are accordingly the first three in the field expansion of Q and the first two in the expansion of u^*:

$$Q = Q_0 + Q_1 E + Q_2 E^2 , \tag{79}$$

$$u^* = u_0^* + u_1^* E . \tag{80}$$

The coefficient Q_1 is identical with zero if u_0^* is defined as follows:

$$u_0^* = \lim_{E \to 0} \frac{\partial[W(E)]}{\partial E} . \tag{81}$$

The question then arises, however, as to whether this definition of u_0^* is consistent with the original definition; that is, is u_0^* as defined by Eq. (81) necessarily the charge displacement required to go from the position of minimum to that of maximum potential energy in the absence of an applied field? If the potential barrier in question is symmetrical and separates positions of minimum potential energy that are also symmetrical (such as might be found for transport within a crystalline solid), then the analysis presented in Section III. B is applicable. Thus from Eqs. (81) and (38), $u_0^* = qa$, which is, of course, the charge displacement required to go from the position of minimum

potential energy to that of maximum potential energy in the absence
of an applied field.

It is apparent, however, that the above symmetry conditions will
not be satisfied for a potential barrier situated at an interface. In
that case $u_0^* \neq qa$ in general, and furthermore the two definitions of
u_0^* will not be equivalent. It seems probable, however, that u_0^* as
defined in Eq. (81) will not differ greatly from the charge displacement
required to go from the position of minimum to that of maximum potential energy in the absence of a field. In any event, for the following
development, u_0^* is defined through Eq. (81) so that $Q_1 = 0$.

As two different physical phenomena that will contribute to Q_2
can be identified, Q_2 is divided into two terms,

$$Q_2 = Q_{2e} + Q_{2d} , \tag{82}$$

where Q_{2e} arises from electrostriction effects and Q_{2d} arises from
the interaction of induced dipole moments with the electric field.
The term Q_{2e} will now be examined.

L. Young was the first to propose a quadratic field variation of
the activation energy Q arising from electrostriction [24]. The effect
may be regarded as arising from a volume change brought about by
electrostriction. As Q would be expected to change with volume, a
change in E will lead to a change in Q. Accordingly

$$\frac{Q_{2e}}{Q_0} \simeq - \frac{d(\ln Q)}{d(\ln \bar{v})} \cdot K_p \cdot \frac{\partial p}{\partial E^2} \tag{83}$$

where $\bar{v}$ is the specific volume of the relevant phase or interphase,
p is the hydrostatic pressure, and $K_p = -\partial(\ln \bar{v})/\partial p$ is the coefficient
of compressibility. Evaluating $\partial p/\partial E^2$ from the relationship $p = \epsilon E^2/8\pi$,
where ϵ is the dielectric constant, and assuming $Q \propto \bar{v}^{-\ell}$, Eq. (83)
becomes

$$Q_{2e}/Q_0 \approx K_p \epsilon \ell /8\pi . \tag{84}$$

Mott and Gurney [1] estimate that for crystals like rock salt, $\ell \approx 2$.
Although it is difficult to justify this particular estimate, it seems
unlikely that ℓ would be as large as 10, say.

Considering now the term Q_{2d}, on assigning a polarizability α
to the mobile defect, there will be a contribution $-\frac{1}{2}\alpha E_L^2$ to the
potential energy of the defect arising from the field-induced dipole
interaction, E_L being the local electrostatic field strength. In general both α and E_L will vary with position in the solid or interface,
so that one may write

$$Q_{2d}E^2 = -\frac{1}{2}\alpha_* E_{L*}^2 + \frac{1}{2}\alpha_0 E_{L_0}^2 , \tag{85}$$

where the subscripts $*$ and 0 refer to the positions $u = u_0^*$ and $u = 0$,
respectively. Setting $E_L = (1 + \frac{1}{3}\gamma\chi)E$, where χ is the electric
susceptibility of the medium and γ is a structural factor (equal to 1
if E_L is equal to the Lorentz field), Eq. (85) becomes

$$Q_{2d} = \frac{1}{2}\alpha_0 \left(1 + \frac{1}{3}\gamma_0\chi\right)^2 - \frac{1}{2}\alpha_*\left(1 + \frac{1}{3}\gamma_*\chi\right)^2 . \tag{86}$$

An equation similar to Eq. (86) was first proposed by Ibl [25] who employed a rather different approach to the problem.

An attempt is now made to estimate the terms in Eq. (86). If the
bonding in question can be regarded as essentially 100% ionic, then
α_0 and α_* should be nearly the same and approximately equal to the
atomic polarizability of the mobile ion. Inertial contributions will
presumably be small since the time available for relaxation will be
less than one-quarter of the period of vibration of the mobile defect
in the direction that corresponds to surmounting the potential barrier.

For crystalline solids γ_0 and γ_* will probably have opposite
signs. For simplicity, the case of a symmetrical barrier will be
examined. It follows from the symmetry considerations in Section III. B that

$$E = \frac{1}{2u_0^*} \int_{-\frac{1}{2}u_0^*}^{\frac{3}{2}u_0^*} E_L \, du = \frac{1}{2}\left(\frac{1}{u_0^*} \int_{-\frac{1}{2}u_0^*}^{+\frac{1}{2}u_0^*} E_L \, du + \frac{1}{u_0^*} \int_{\frac{1}{2}u_0^*}^{\frac{3}{2}u_0^*} E_L \, du \right). \quad (87)$$

Setting

$$E_{L_0} \approx \frac{1}{u_0^*} \int_{-\frac{1}{2}u_0^*}^{+\frac{1}{2}u_0^*} E_L \, du$$

and

$$E_{L*} \approx \frac{1}{u_0^*} \int_{+\frac{1}{2}u_0^*}^{+\frac{3}{2}u_0^*} E_L \, du ,$$

Eq. (87) becomes $(E_{L_0} + E_{L*})/2 \approx E$ or, on setting $E_L = (1 + \frac{1}{3} \gamma \chi)E$,

$$\gamma_0 + \gamma_* \approx 0 . \quad (88)$$

Provided that Eq. (87) can be regarded as defining the mean external field strength for a symmetrical barrier situated at an interface, Eq. (88) will apply to an interfacial potential barrier as well.

If the defects in question are vacancies, then for crystalline solids γ_0 should be on the order of, though perhaps smaller than, the value of γ (γ', say) used to obtain agreement between measured values of the high-frequency dielectric constant ϵ_0' and values calculated from ionic polarizabilities [1]. For interstitial ions the situation is less clear, but it seems likely that in that case $\gamma_0 \overset{\sim}{<} \gamma'$. Setting $-\gamma_* \approx \gamma_0 \approx \gamma$ and $\alpha_0 \approx \alpha_* \approx \alpha'$, the ionic polarizability of the mobile species, Eq. (85) becomes

$$Q_{2d} \overset{\sim}{\sim} \frac{1}{3} \gamma \chi \alpha' . \quad (89)$$

If the bonding in question involves an appreciable covalent component, then α_0 and α_* would no longer be expected to be the same. Thus the molecular orbitals centered on an interstitial ion situated at $u = 0$ will be localized roughly on this position. The same is true of the position $u = 2u_0^*$. In the transition state (i.e., $u = u_0^*$), however, the molecular orbitals would probably be delocalized over both regions, so that a higher polarizability is expected in this state. Setting $\frac{1}{2}\alpha_* \approx \alpha_0 \approx \chi_0/4\pi N$ and $-\gamma_* \approx \gamma_0 \approx \gamma$, where χ_0 is the electric susceptibility at optical frequencies and N is the lattice-ion concentration of ions of the type involved in the rate process, Eq. (86) becomes

$$Q_{2d} \sim (\chi_0/8\pi N)(2\gamma\chi - 1 - \gamma^2\chi^2/9) . \tag{90}$$

Since α', appearing in Eq. (89), will be on the order of, or less than, $\chi_0/4\pi N$ and $0 \leq \gamma \leq 1$, Eqs. (89) and (90) can be summarized as follows:

$$Q_{2d} \lesssim (\epsilon \ \epsilon_0'/6\pi N) , \tag{91}$$

where ϵ and ϵ_0' are the static and high (optical) frequency dielectric constants, respectively.

The term u_1^* is now considered. Presumably u_1^* could arise through a displacement by the field of positively charged species relative to negatively charged ones in the environment that constitutes the potential-energy barrier, that is, through an asymmetrical distortion of the barrier by the field. The magnitude of such an effect can be estimated for a barrier within a solid from a knowledge of the magnitude of the contribution of lattice-ion displacement to electrical polarization (i. e., to the dielectric constant). For this purpose a simple model consisting of alternately positioned ions of charge $\pm Ze$, separated from one another by a distance d_0, will be used. In the presence of a field E this distance will change to $d = d_0 + \Delta$ and $d = d_0 - \Delta$, resulting therefore in a displacement polarization per ion pair of magnitude ΔZe and hence to a displacement polarization per unit volume, P_d, given by

$$P_d = \Delta Ze/2d_0^3 . \tag{92}$$

However, P_d is also given by

$$P_d = (\epsilon - \epsilon_0')E/4\pi \ . \tag{93}$$

(This result may be obtained from the equations derived by Mott and Gurney [1] on setting their overlap parameter γ' equal to zero. It appears that $\gamma' \simeq 0$ for the alkali halides [1].)

Elimination of P_d therefore gives

$$|d_0 - d| = \Delta = (d_0^3/Ze)(\epsilon - \epsilon_0')E/2\pi \ . \tag{94}$$

Assuming, now, that $u^* \propto d$, then on comparing Eq. (94) with Eq. (80), the following is obtained:

$$|u_1^*/u_0^*| = (d_0^2/Ze)(\epsilon - \epsilon_0')/2\pi \ . \tag{95}$$

That u_1^*/u_0^* is positive can be seen by considering the case of a positively charged defect surmounting a potential-energy barrier. The position of maximum potential energy will reside at a region of local excess positive charge, with the adjacent positions of minimum potential energy residing at regions of local excess negative charge. Thus u^* for transport in the direction assisted by the field will increase with increasing field. The same conclusion is reached if a negatively charged defect is considered. Thus on setting $\Omega = 2d_0^3/Z$ and $(2/Z)^{\frac{1}{3}} \approx 1$, Eq. (95) becomes

$$u_1^*/u_0^* \sim (\Omega^{\frac{2}{3}}/e)(\epsilon - \epsilon_0')/4\pi \ , \tag{96}$$

where Ω is the volume occupied by an atom equivalent in the solid (i. e., Ω = equivalent weight/density times Avogadro's number).

On substituting for Q and u^* in Eq. (60) by using Eqs. (70), (80), and the estimates for Q_2 and u_1^* given by Eqs. (82), (84), (91), and (96), the following result is obtained:

$$W(E) = Q_0 - u_0^*E(1 - u_0^*E/C_2'Q_0 + \cdots) \ , \tag{97}$$

where

$$1/C_2' = 1/C_2 + T_e + T_d - T_a , \tag{98}$$

$$T_e \approx (K_p \epsilon \ell / 8\pi)(Q_0/u_0^*)^2 , \tag{99}$$

$$T_d \lesssim (\epsilon \; \epsilon_0'/6\pi N)(Q_0/u_0^{*2}) , \tag{100}$$

$$T_a \sim (\Omega^{\frac{2}{3}}/4\pi e)(\epsilon - \epsilon_0')(Q_0/u_0^*) . \tag{101}$$

The terms T_e, T_d, and T_a are the contributions to the quadratic field coefficient made by the electrostriction effect, the field-induced dipole interaction, and the asymmetrical barrier distortion, respectively. It appears likely that Eqs. (97) through (101) will represent the correct magnitude of the various contributions for a barrier situated at an interface. In that case, however, the system parameters K_p, ϵ, and ϵ_0' will take on values characteristic of the interphase. On setting $u_0^* \approx 2 \times 10^{-8} e$ cm and $Q_0 \approx 1$ eV as before, and $\epsilon \approx 10$, $\epsilon_0' \approx 3$, $N \approx 10^{23}$ cm^{-3}, $\Omega^{\frac{2}{3}} \approx 2 \times 10^{15}$ cm^{-2}, $\ell \approx 2$, and $K_p \approx 10^2$ cm^2/erg (a value typical of most solids), Eqs. (99) through (101) give

$$T_e \sim 0.02 ; \quad T_d \lesssim 0.24 ; \quad T_a \sim 0.02 . \tag{102}$$

Thus all three effects could conceivably make contributions to C_2' comparable to that arising from the curvature of the potential-energy barrier, since for a cosine potential-energy function, say, $1/C_2 = 2/\pi^2 \approx 0.2$ (see Eq. (48)). Since the contributions of the four effects to C_2' are fairly sensitive to the system parameters, no general statement about their relative importance is possible.

D. Summary

In summary, then, a quadratic term must be included in the field expansion of the net activation energy in the limit of very high field strength (qaE $\sim$ Q). At least four phenomena that could conceivably

make a significant contribution to the quadratic field term can be identified. In Section II it was pointed out that ionic conduction through
amorphous or microcrystalline films, if interpreted on the basis of the
equations developed for crystalline media, would give rise to a field
and temperature dependence of both Q and aq (i.e., of both Q_0 and
u_0^*). Effects due to the presence of a range of parameters for the diffusion barrier may be easily distinguished from those arising from the
presence of a very high field strength, since in the former case the
apparent or averaged Boltzmann factor for ion transport will take the
form $\exp[-W(T, E)/kT]$, whereas in the latter case the Boltzmann factor
will be $\exp[-W(E)/kT]$. It is primarily through the temperature
dependence, therefore, that the two cases can be distinguished. In
addition, however, the magnitude of the quadratic field coefficient
arising from the presence of a very high field strength can be estimated to within a factor of 2 or so, and its sign is positive.

Finally, it should be noted that although the equations in this
section were derived by assuming the field E to be homogeneous, it
was shown in Section II that for high field strengths ($qaE \gtrsim kT$) and
space-charge concentrations $\lesssim 5 \times 10^{19}\,e/cm^3$, the general transport
equation (Eq. (7)) reduces to the same form as that for ionic conduction in the absence of space charge (Eq. (25) or (13)). Thus the relevant equations of Section II.B can be generalized to include the very-
high-field effects simply by adding the higher order field terms. This
is done by the following substitution:

$$2 \sinh\left(\frac{qaE}{kT}\right) \;\rightarrow\; \left[\exp\left(\frac{Q}{kT}\right)\right]\left[\exp\left(-\frac{W(E)}{kT}\right) - \exp\left(-\frac{W(-E)}{kT}\right)\right] \quad (103)$$

with $W(E)$ given, from Eq. (97), by

$$W(E) = Q - qaE\left(1 - \frac{qaE}{C_2'Q} + \cdots\right). \quad (104)$$

The assumption of barrier symmetry has been introduced so that $u_0^* = aq$.

Furthermore, the subscript on Q has been dropped to be consistent with the equations of Section II. B.

IV. INTEGRATED TRANSPORT EQUATIONS

A. Introduction

In order to relate kinetic data on film growth to transport phenomena, it is necessary to have integrated forms of the transport equation which relate the flux density to the total voltage drop across the film and the concentration of the defects at the two boundaries. In deriving such equations it will be assumed that the field strengths are so low that only the linear term in the expansion of the net activation energy $W(E)$ need be retained, so that the equations of Section II can be used in their present form. Alternatively one may proceed as follows: the ion-flux density at high fields is given by

$$J = 2 a \nu n \exp[-W(E)/kT] \qquad (105)$$

which on substituting for $W(E)$ from Eq. (104) can be written as

$$J = (n/N) A_J e^{\beta E} , \qquad (106)$$

where N is the concentration of sites for the migrating defects,

$$A_J = 2 a \nu N \exp \left\{ - \frac{Q[1 - (aqE/Q)^2 /C_2']}{kT} \right\} \qquad (107)$$

and β, the differential field coefficient, is defined by

$$\beta = (\partial \ln J/\partial E)_{T,n} = -(1/kT)[\partial W(E)/\partial E]_T . \qquad (108)$$

Even under conditions in which the quadratic field term is important, A_J will vary much more slowly with E than will $\exp(\beta E)$. Over a moderate range of field strengths or flux densities it may therefore be treated as being essentially independent of field strength. Redefining the dimensionless position variable s and the dimensionless flux density j through the equations

$$s = \frac{x}{kT\beta/q} \, , \tag{109}$$

$$j = \frac{J}{A_J} \tag{110}$$

while maintaining the same definitions of the other dimensionless variables as set out in Section II (cf. Eq. (3)) leads to the following expressions for ε:

$$\varepsilon = -\partial p/\partial s = \beta E \, , \tag{111}$$

from which it follows that $j = c\,e^{\varepsilon}$, which is simply the high field form of Eq. (10). Equations (109) through (111) reduce to those of Section II under conditions in which the quadratic field term is negligible. Thus making the approximation that A_J and β are constant and substituting according to Eq. (5), the equations of Section II can be used, unaltered, to cover the case of very high field strengths.

It is not possible to develop a single integrated transport equation covering the entire range of field strengths and including space-charge effects. The situation in which the space charge may be neglected is therefore treated first, followed by several treatments that include the space charge. The latter cover cases in which the space charge is due solely to the presence of the migrating species and those for which there is in addition a constant background-space-charge density of opposite sign to the charge carried by the mobile defects.

To reduce algebraic complexity to a minimum, the dimensionless variables of Section II are used in this section.

B. Homogeneous Field

Equation (10), which is repeated here for convenience, represents the most general form of the ion-transport equation:

$$j = 2(\sinh \varepsilon)\,[c - (1/\varepsilon)(\partial c/\partial s)] \, . \tag{10}$$

Since both ϵ and j are assumed to be independent of position in the absence of space charge, Eq. (10) may be written in the form

$$(\partial/\partial s) \ln [(2c/j)(\sinh \epsilon) - 1] = \epsilon ,\qquad (112)$$

which on integrating from $s = 0$ to $s = S$, taking the antilog, and rearranging, becomes

$$j = 2(\sinh \epsilon)(c_0 e^{\epsilon S} - c_S)/(e^{\epsilon S} - 1) ,\qquad (113)$$

where c_0 and c_S are the values of c at $s = 0 = x$ and $s = S = X/a$, respectively, that is, the values of the concentration at the boundaries of the film. Equation (113), or rather its transformation on using Eq. (3), was first derived by Fromhold and Cook [26] directly from the difference equation, Eq. (2).

It is instructive to examine the terms in Eq. (113) for a case in which the free-energy driving force for ion transport is significant with respect to thermal energies (i.e., with respect to kT) and the voltage drop across the film is significant with respect to kT/q. More specifically, we examine the case in which

$$\bar{\mu}_0 - \bar{\mu}_X \geq 4kT ,\qquad (114)$$

$$XE \geq 4kT/q .\qquad (115)$$

Condition (114) will be satisfied for almost any system involving film growth in which transport through the film plays a significant role in controlling the rate. Condition (115) will be satisfied for a potential drop across the film of about 100 mV or more, which should almost invariably be the case during anodic film formation.

On substituting from Eqs. (1) and (3), Eq. (114) becomes $\ln (c_0/c_S) + \epsilon S \geq 4$, or alternatively $c_0 e^{\epsilon S} \geq c_S e^4$ so that c_S may be neglected in comparison with $c_0 e^{\epsilon S}$. Similarly, from Eq. (115), $\epsilon S \geq 4$ or $e^{\epsilon S} \geq e^4$, and hence unity may be neglected in comparison with $e^{\epsilon S}$. Thus Eq. (113) reduces to

$$j \approx 2c_0 \sinh \epsilon .\qquad (116)$$

Under these conditions, therefore, the transport rate depends only on the concentration of the defects at the boundary away from which the defects are moving. The validity of Eq. (116) implies that the mobile-ion concentration has the value c_0 throughout most of the film, a rapid adjustment to c_S occurring close the interface at $s = S$. For very thin films this means essentially within one jump distance, whereas for thicker films within a distance small compared with S. For all practical purposes, therefore, the concentration can be regarded as constant at c_0 throughout the film, changing discontinuously to c_S at $s = S$. It is easy to rationalize this behavior in the case of high field strengths, where motion against the field is totally negligible. For this case it is obvious that the concentration of mobile ions at the interface toward which the ions are moving can have no direct influence on the transport rate.

It is worth noting that neither of the inequalities leading to Eq. (116) requires that nonlinear transport conditions prevail; that is, neither requires high field strengths ($\varepsilon \overset{\sim}{>} 1$). Thus for systems in which the space charge can be neglected and ion transport is at least partially rate controlling Eq. (116) may be used in place of the much more complex Eq. (58) in the majority of cases.

C. Space Charge Due Solely to Mobile Defects

The case in which the space charge cannot be neglected and is due entirely to the mobile ions is now examined, first in the low-field limit and then for the case in which diffusion is unimportant.

1. Low-Field Approximation

In Section II it was shown that for space-charge densities less than about 5×10^{19} e/cm^3 the general transport equation, Eq. (10), remains a valid approximation. For low field strengths ($\varepsilon \ll 1$), sinh ε may be replaced by ε, so that the linear transport equation obtains:

$$j = 2(\varepsilon c - \partial c/\partial s) . \tag{17}$$

Furthermore, for the case under consideration, the dimensionless space-charge density r is equal to αc, with α given by

$$\alpha = 4\pi q^2 a^2 N/\epsilon kT , \tag{117}$$

so that Poisson's equation, Eq. (14), can be written

$$c = (1/\alpha) \partial \varepsilon/\partial s . \tag{118}$$

Substituting for c from Eq. (118) into Eq. (17) and integrating, we obtain

$$2(\partial \varepsilon/\partial s) - \varepsilon^2 = -\alpha js + 2\alpha c_0 - \varepsilon_0^2 , \tag{119}$$

where the boundary conditions $\varepsilon = \varepsilon_0$ and $c = c_0$ at $s = 0$ have been used. Equation (119) is a form of the Ricatti equation [27] and has been treated by Wright [28] in the application of a transport equation, formally identical with Eq. (17), to the case of space-charge-limited electronic currents. Applying the transformations

$$y = \exp\left(-\frac{1}{2}\int_0^s \varepsilon \, ds\right) = \exp(p/2) , \tag{120}$$

$$Z = \alpha js + \varepsilon_0^2 - 2\alpha c_0 \tag{121}$$

to Eq. (119) gives the equation

$$\partial^2 y/\partial Z^2 = Zy/(2\alpha j)^2 . \tag{122}$$

The solution of Eq. (122), on employing the steady-state approximation $\partial j/\partial s = 0$, may be expressed in terms of Bessel functions of one-third order, $I_{\frac{1}{3}}$. Thus the solution may be written

$$y = A Z^{\frac{1}{2}}\left[I_{\frac{1}{3}}(Z^{\frac{3}{2}}/3\alpha j) + B I_{-\frac{1}{3}}(Z^{\frac{3}{2}}/3\alpha j)\right], \tag{123}$$

where A and B are integration constants that may be evaluated from the conditions $y = 1$ and $Z = (\varepsilon_0^2 - 2\alpha c_0)$ when $s = 0$, whereas $y = \exp(\frac{1}{2}p_S)$ and $Z = (\alpha jS + \varepsilon_0^2 - 2\alpha c_0)$ when $s = S$. The quantity p_S is the dimensionless potential drop across the film, that is, the potential at $s = S$ relative to that at $s = 0$. From Eqs. (120) and (121), $\varepsilon = -2\alpha j \, \partial \ln y/\partial Z$, which with Eq. (123) gives

$$\varepsilon = -Z^{\frac{1}{2}} \left[\frac{I_{-\frac{2}{3}}(Z^{\frac{3}{2}}/3\alpha j) + B\,I_{\frac{2}{3}}(Z^{\frac{3}{2}}/3\alpha j)}{I_{\frac{1}{3}}(Z^{\frac{3}{2}}/3\alpha j) + B\,I_{-\frac{1}{3}}(Z^{\frac{3}{2}}/3\alpha j)} \right] , \qquad (124)$$

so that ε_0 may be evaluated in terms of the other boundary conditions. Thus an implicit equation for j is obtained in terms of the concentration and field strength at $s = 0$, the potential drop across the film, and the total film thickness. Alternatively, the concentration at the other boundary may be used in place of ε_0 by relating these through Eq. (119) and Poisson's equation, Eq. (118). Thus from Eqs. (118), (119), and (121) c is given by

$$c = \frac{1}{\alpha} \frac{\partial \varepsilon}{\partial s} = \frac{\varepsilon^2 - Z}{2\alpha} , \qquad (125)$$

which with Eq. (124) leads to the result

$$c = \frac{Z}{2\alpha} \left\{ \left[\frac{I_{-\frac{2}{3}}(Z^{\frac{3}{2}}/3\alpha j) + B\,I_{\frac{2}{3}}(Z^{\frac{3}{2}}/3\alpha j)}{I_{\frac{1}{3}}(Z^{\frac{3}{2}}/3\alpha j) + B\,I_{-\frac{1}{3}}(Z^{\frac{3}{2}}/3\alpha j)} \right]^2 - 1 \right\} , \qquad (126)$$

which may be evaluated at $s = S$ on substituting as above. The form of the resulting kinetic behavior is by no means immediately apparent on inspection and depends on the assumptions made with respect to the boundary concentrations. A brief discussion of the kinetics predicted by these equations will be deferred to Section VI, which deals also with the role of the boundaries in the overall transport process.

2. Diffusion Unimportant

Again in Section II it was shown that for a sufficiently low space-charge density the diffusion term in the general transport equation may be dropped, giving the conduction equation

$$j = 2c \sinh \varepsilon \, . \tag{25}$$

For $\tilde{\varepsilon} > 1$, the condition is satisfied for space-charge densities up to about 5×10^{19} e/cm^3, whereas for lower field strengths the condition becomes progressively more restrictive, being given approximately by $|\rho| \stackrel{\sim}{<} \varepsilon^2 \times 5 \times 10^{19}$ e/cm^3 (cf. Eq. (27)). Using Eq. (118), the ionic conduction equation, Eq. (25), may be written

$$j = (2/\alpha)(\partial \varepsilon / \partial s) \sinh \varepsilon \, , \tag{127}$$

which on integration gives

$$\cosh \varepsilon = \cosh \varepsilon_0 + \alpha j s / 2 \, . \tag{128}$$

Alternatively, eliminating s from Eq. (127) by using the identity $(\partial \varepsilon / \partial s) = -\varepsilon (\partial \varepsilon / \partial p)$ and integrating, the following equation is obtained:

$$-p = (2/\alpha j)[\varepsilon (\cosh \varepsilon) - (\sinh \varepsilon) - \varepsilon_0 (\cosh \varepsilon_0) + (\sinh \varepsilon_0)] \, . \tag{129}$$

To determine the potential p as a function of position s, ε must be eliminated from Eq. (129) by using Eq. (128). This is easily achieved with the aid of the identities

$$\varepsilon = \ln [\cosh \varepsilon + (\cosh^2 \varepsilon - 1)^{\frac{1}{2}}] \, , \tag{130a}$$

$$\sinh \varepsilon = (\cosh^2 \varepsilon - 1)^{\frac{1}{2}} \, , \tag{130b}$$

the result, however, being rather unwieldy. Furthermore, using Eq. (25) in the form

$$\sinh \varepsilon_0 = j/2c_0 \, , \tag{131}$$

the boundary value of the field, ε_0, may be replaced in Eqs. (128) and (129) by functions of the boundary concentration c_0, using the identities

$$\varepsilon_0 = \ln [\sinh \varepsilon_0 + (\sinh^2 \varepsilon_0 + 1)^{\frac{1}{2}}] \, , \tag{132a}$$

$$\cosh \varepsilon_0 = (\sinh^2 \varepsilon_0 + 1)^{\frac{1}{2}} \, . \tag{132b}$$

Other functions of interest may be obtained directly from Eqs. (25)
and (127) through (132). Thus the defect concentration c may be
expressed as a function of position s by eliminating ε from Eq. (25)
by using Eqs. (130) and (128) to give

$$c = j/2 \left[(\cosh \varepsilon_0 + \alpha js/2)^2 - 1 \right]^{\frac{1}{2}} \tag{133}$$

or alternatively, eliminating ε_0 from Eq. (133) by using Eqs. (132) and
(131) to give

$$c = j/2 \left\{ \left[(j^2/4c_0^2 + 1)^{\frac{1}{2}} + \alpha js/2 \right]^2 - 1 \right\}^{\frac{1}{2}}. \tag{134}$$

It can be seen that the defect concentration is a maximum at the
boundary at which the defects originate (i. e., at s = 0) and decreases
monotonically with increasing displacement from this boundary.

Again, an expression for the defect-concentration gradient may
be obtained as a function of s by differentiating Eq. (25) with
respect to s , eliminating $\partial \varepsilon/\partial s$ from the result by using Eq. (127),
and then substituting for ε by using Eqs. (130) and (128), the result
being

$$-\frac{\partial c}{\partial s} = \frac{\alpha j^2 (\cosh \varepsilon_0 + js/2)}{4[(\cosh \varepsilon_0 + js/2)^2 - 1]^{\frac{3}{2}}}. \tag{135}$$

Thus $-\partial c/\partial s$ is also a maximum at s = 0 , decreasing monotonically
with increasing s .

Since the field strength itself is not amenable to direct experimental determination during the course of anodic film growth, it is
useful to define related quantities that are. Two such quantities are
the mean field strength $\overline{E} = -V_X/X$ and the differential field strength
$\widetilde{E} = -(\partial V_X/\partial X)_{J,T}$. In terms of the dimensionless parameters these
become

$$qa\overline{E}/kT = \overline{\varepsilon} = -p_S/S, \tag{136}$$

$$qa\widetilde{E}/kT = \widetilde{\varepsilon} = -(\partial p_S/\partial S)_j. \tag{137}$$

An expression for $\bar{\varepsilon}$ as a function of S is obtained directly from Eq. (136) on substituting for p_S from Eq. (129) and for ε_S from Eq. (128); the result is, however, unwieldy. The differential field strength may be evaluated from Eqs. (136), (137), and (129) to give

$$\tilde{\varepsilon} = \frac{2}{\alpha j}\left[\varepsilon_S\left(\frac{\partial \cosh \varepsilon_S}{\partial S}\right)_j - \varepsilon_0\left(\frac{\partial \varepsilon_0}{\partial S}\right)_j \sinh \varepsilon_0\right]. \tag{138}$$

Evaluating $(\partial \cosh \varepsilon_S/\partial S)_j$ from Eq. (128) then gives

$$\tilde{\varepsilon} = \varepsilon_S + \frac{2}{\alpha j}(\varepsilon_S - \varepsilon_0)\left(\frac{\partial \varepsilon_0}{\partial S}\right)_j \sinh \varepsilon_0. \tag{139}$$

In Section VI it is shown that for every case in which Eq. (25) is a valid approximation of the transport equation (i. e. , for diffusion unimportant) ε_0 is a function of j only, so that $(\partial \varepsilon_0/\partial S)_j$ is zero and Eq. (139) becomes

$$\tilde{\varepsilon} = \varepsilon_S. \tag{140}$$

Thus the differential field strength is equal to the field strength at the boundary toward which the ionic defects are moving.

It is apparent from Eq. (128) that in the low–space–charge limit the mean and differential field strengths take on the value of the field strength at the boundary,

$$\bar{\varepsilon} = \tilde{\varepsilon} = \varepsilon_0 \quad \text{for} \quad \alpha c_0 S << (1 + 4c_0^2/j^2)^{\frac{1}{2}}. \tag{141}$$

On the other hand, in the high–space–charge limit it follows from Eqs. (128), (129), and (140) that

$$\bar{\varepsilon} = \ln \alpha j S/e \quad \text{for} \quad \alpha c_0 S >> (1 + 4c_0^2/j^2)^{\frac{1}{2}}, \tag{142}$$

$$\tilde{\varepsilon} = \ln \alpha j S \quad \text{for} \quad \alpha c_0 S >> (1 + 4c_0^2/j^2)^{\frac{1}{2}}, \tag{143}$$

where $e = \exp(1)$. Thus for sufficiently thin films the flux density j is a function only of the field strength and defect concentration at the boundary (for isothermal conditions) and hence is seen to be controlled by the kinetics of the interfacial processes. For sufficiently thick films, on the other hand, the flux density is not a function of the defect concentrations at the two boundaries, but it does depend on the total film thickness for fixed mean or differential field strength.

The equations derived in this section (Section IV. C. 2) are general in the sense that they apply to both low and high field strengths, provided that the condition on the space-charge density, Eq. (27), is satisfied. Such generality is useful in treating certain problems, one example being the kinetics of tarnishing reactions in the thin-film region, where the dimensionless field strength ε established by electron equilibrium is typically on the order of unity [16]. Under the conditions of anodic oxidation of many metals, and in particular of the valve metals, the field strength is high, so that $2 \sinh \varepsilon$ and $2 \cosh \varepsilon$ may be replaced by e^{ε}. On making these substitutions, a number of the equations are greatly simplified. Thus the expression for the dimensionless potential as a function of j, c_0, and s becomes

$$-p = s[\ln (j/c_0) + (1 + 1/\alpha c_0 s) \ln (1 + \alpha c_0 s) - 1] \qquad (144)$$

and that for the field strength is given by

$$\varepsilon = \ln (j/c_0) + \ln (1 + \alpha c_0 s) . \qquad (145)$$

The mean and differential field strengths are obtained directly from Eqs. (144) and (145), respectively, since $\bar{\varepsilon} = -p_S/S$ and $\tilde{\varepsilon} = \varepsilon_S$. In the low-space-charge limit (i. e., for $\alpha c_0 S \ll 1$) $\bar{\varepsilon}$ and $\tilde{\varepsilon}$ reduce to Eq. (141) as before. Furthermore, for $\alpha c_0 S \gg 1$ they reduce to Eqs. (142) and (143), respectively. Equations equivalent to these were first derived by Dewald [29]. They were then generalized by L. Young [30] to include the case of a constant background space charge due to impurities or nonstoichiometry. The extension is covered in the following section.

D. Migration in the Presence of Background Space Charge

The transport of ionic species in the presence of a constant background space charge can be solved analytically only in the case of migration (i. e., diffusion unimportant) in either the high- or the low-field limit. For a constant background charge density equal to $-qn_{\infty}$, Poisson's equation becomes

$$\partial \varepsilon / \partial s = \alpha(c - c_{\infty}) , \qquad (146)$$

where $c_{\infty} = n_{\infty}/N$. Eliminating c from Eq. (146) by using the conduction equation then gives

$$\partial \varepsilon / \partial s = \alpha j/2 \sinh \varepsilon - \alpha c_{\infty} . \qquad (147)$$

For low field strengths, replacing $\sinh \varepsilon$ by ε, Eq. (147) may be integrated to give

$$\varepsilon - \varepsilon_{\infty} = (\varepsilon_0 - \varepsilon_{\infty})\{\exp[-(\varepsilon - \varepsilon_0)/\varepsilon_{\infty}]\} \exp(-s/S_c) , \qquad (148)$$

where $\varepsilon_{\infty} = j/2c_{\infty}$ and is the field required to sustain a flux j when $c = c_{\infty}$ (i. e., in a region of zero space-charge density); and

$$S_c = j/2\alpha c_{\infty}^2 , \qquad (149)$$

S_c being a measure of the characteristic space-charge length in the limit of low space-charge density. Thus for $(\varepsilon - \varepsilon_{\infty})/\varepsilon_{\infty} << 1$, Eq. (148) becomes

$$\varepsilon = \varepsilon_{\infty} + (\varepsilon_0 - \varepsilon_{\infty})[\exp(\varepsilon_0/\varepsilon_{\infty} - 1)] \exp(-s/S_c) , \qquad (150)$$

so that the dimensionless space-charge density, $r = \alpha(c - c_{\infty}) = \partial \varepsilon/\partial s$ is proportional to $\exp(-s/S_c)$. The space-charge density therefore drops by a factor $1/e$ on moving a distance S_c from the boundary $s = 0$. The reason for the choice of the subscript ∞ is clear since for $s = \infty$, $\varepsilon = \varepsilon_{\infty}$ and $c = c_{\infty}$. Thus c_{∞} and ε_{∞} are the mobile-defect concentration and field strength in the space-charge-free region of the film.

Alternatively Eq. (147) can be solved in the low-field limit after replacing $\partial\varepsilon/\partial s$ by $-\varepsilon\,\partial\varepsilon/\partial p$ to give

$$\frac{\varepsilon^2 - \varepsilon_0^2}{2} + \varepsilon_\infty(\varepsilon - \varepsilon_0) + \varepsilon_\infty^2 \ln\frac{\varepsilon - \varepsilon_\infty}{\varepsilon_0 - \varepsilon_\infty} = \alpha c_\infty p \ . \tag{151}$$

Equations (148) and (151) can be solved by numerical methods to obtain p as a function of s and j, or j as a function of p_S and S. A more useful result is obtained by restricting the solution to the low-space-charge region, since the linear conduction equation is only a valid approximation for low space charge (see Eq. (27)). Thus integrating Eq. (150) and replacing ε_0 by $j/2c_0$ and ε_∞ by $j/2c_\infty$ gives

$$-p = \frac{j_s}{2c_\infty} - \alpha c_\infty \left(\frac{c_\infty}{c_0} - 1\right) \exp\left(\frac{c_\infty}{c_0} - 1\right) \tag{152}$$

$$\text{for } |c - c_\infty| \ll c_\infty,$$

from which j as a function of $\bar{\varepsilon}$ and s may be determined. Similarly, from Eq. (150), an expression for $\tilde{\varepsilon} = \varepsilon_S$ is obtained:

$$\tilde{\varepsilon} = \frac{j}{2c_\infty} + \frac{j}{2c_\infty}\left(\frac{c_\infty}{c_0} - 1\right) \exp\left(\frac{c_\infty}{c_0} - 1\right) \exp\left(-\frac{S}{S_c}\right) \tag{153}$$

$$\text{for } |c_X - c_\infty| \ll c_\infty$$

so that for j constant, $\ln|\tilde{\varepsilon} - j/2c_\infty|$ is a linear function of film thickness.

For high field strengths, replacing $2\sinh\varepsilon$ by e^ε, Eq. (147) may be integrated to give

$$\exp(\varepsilon - \varepsilon_\infty) - 1 = [\exp(\varepsilon_0 - \varepsilon_\infty) - 1]\exp(-\alpha c_\infty s) \ , \tag{154}$$

where $\varepsilon_\infty = \ln(j/c_\infty)$. In the low-space-charge region where $\varepsilon - \varepsilon_\infty \ll 1$, Eq. (154) reduces to $\varepsilon - \varepsilon_\infty \propto \exp(-\alpha c_\infty s)$. Thus in this limit the characteristic space-charge length is $1/\alpha c_\infty$. Solving Eq. (154) for ε and integrating again gives

$$p = \varepsilon_\infty s + \int_0^s \ln\{1 + [\exp(\varepsilon_0 - \varepsilon_\infty) - 1]\exp(-\alpha c_\infty s)\}\,ds\,. \qquad (155)$$

The integral cannot be expressed in terms of a finite number of simple functions except in the two limiting cases $\alpha c_\infty s \gg 1$ and $\alpha c_\infty s \ll 1$. Thus for $\alpha c_\infty s \gg 1$, the logarithmic function of Eq. (155) can be expanded to give on integrating

$$-p = \varepsilon_\infty s - [\exp(-\alpha c_\infty s)][\exp(\varepsilon_0 - \varepsilon_\infty) - 1]/\alpha c_\infty + \cdots + \text{constant} \qquad (156)$$

for $\alpha c_\infty s \gg 1$.

On replacing ε_0 by $\ln(j/c_0)$ and ε_∞ by $\ln(j/c_\infty)$, Eq. (154) can be written

$$\varepsilon = \ln(j/c_0) + \ln\{\exp(-\alpha c_\infty s) + (c_0/c_\infty)[1 - \exp(-\alpha c_\infty s)]\}\,, \qquad (157)$$

which for $\alpha c_\infty s \ll 1$ reduces to

$$\varepsilon = \ln(j/c_0) + \ln[1 + \alpha(c_0 - c_\infty)s + \cdots]\,. \qquad (158)$$

Integrating this then gives

$$-p = s\{\ln(j/c_0) + [1 + 1/\alpha(c_0 - c_\infty)s]\ln[1 + \alpha(c_0 - c_\infty)s] - 1\}\,, \qquad (159)$$

which reduces precisely to Eq. (144) on setting $c_\infty = 0$, as of course it must. It is apparent that in the thin-film limit both the mean and differential field strengths are equal to $\varepsilon_0 = \ln(j/c_0)$, whereas in the thick-film limit they are equal to $\varepsilon_\infty = \ln(j/c_\infty)$, a result that is hardly surprising.

A general expression for the differential field strength can be obtained from Eq. (157) by noting again that $\tilde{\varepsilon} = \varepsilon_s$. For $\alpha c_\infty S \ll 1$, equations for $\tilde{\varepsilon}$ and $\bar{\varepsilon}$ can be obtained directly from Eqs. (155) and (156), respectively. They differ from the corresponding ones derived for the case of no background space charge only in that $\alpha c_0 S$ is replaced by $\alpha(c_0 - c_\infty)S$.

The results of Sections IV.C.2 and IV.D can be summarized in part by noting that for a sufficiently low total space charge, which

means sufficiently thin films, $j = 2c_0 \sinh \bar{\varepsilon}$, whereas for a suffi -
ciently high total space charge, which means sufficiently thick films,
j is independent of the boundary concentrations of the defects and
hence independent of interfacial kinetic processes. The conditions
under which this high-space-charge limit can be achieved are examined in Section V, and the low-space-charge limit is examined further
in Section VI.

V. IONIC MIGRATION WITH CONTROL
ENTIRELY WITHIN THE FILM

A. Introduction

It is clear from the results of Section IV that in general ionic
transport will not be controlled entirely within the film. Thus for the
case in which the space charge can be neglected and diffusion is
unimportant, the concentration of defects is determined at the boundary away from which the defects move (see Eq. (116) and related discussion). In the most general case one requires the concentration
of defects at both boundaries in order to calculate the defect flux
density. If the boundary concentrations are essentially under equilibrium control, transport through the film can be regarded as the
sole rate-controlling process.

It does not necessarily follow, however, that the boundary concentration n_0 for this case will be independent of X and V_X, since
it might depend on the potential drop across the electrolyte-film
double layer, a point covered in Section VI. If n_0 is not under equilibrium control, the flux density will clearly be influenced by the
interfacial kinetics. In any event the flux density will in general
depend on n_0 and hence on the interfacial processes occurring near
the boundary defined by $x = 0$ (i.e., the boundary away from which
the defects move).

In Section IV it was shown that for films thick enough for the total space charge in the film to be large (i. e. , $\alpha c_0 S \gg 1$ or $\alpha c_\infty S \gg 1$) ionic conduction through the film is independent of the defect concentration at the boundaries. For such cases the space charge near the boundary $s = 0 = x$ acts to decouple the transport rate from the kinetic process occurring at this interface.

Thus for a film possessing a constant background-space-charge density $-q c_\infty$ the conduction equation becomes, in the thick-film limit, $j = 2c_\infty \sinh \bar{\epsilon}$. For zero background-space-charge density, on the other hand, it becomes $j = (e/\alpha S)\, e^{\bar{\epsilon}}$, where $\bar{\epsilon}$ is the mean dimensionless field strength in the film. The dependence of j on film thickness for fixed $\bar{\epsilon}$ in the latter case can be regarded as the vestiges of the dependence of j on the interfacial kinetic processes, though none of the interfacial parameters actually appears.

The case in which the space charge is due solely to the migrating ions is therefore somewhat arbitrarily excluded from consideration in this section, being treated instead in Section VI. With this in mind, the simplest systems for which the ionic flux density can be truly independent of the interfacial processes are those for which the film possesses, in addition to the mobile defects, point defects bearing a charge of opposite sign to that of the mobile defects. For such a system, and for sufficiently thick films, most of the film will be electrically neutral, the mobile-defect concentration being therefore determined by the neutrality condition. These immobile charge compensators may take the form of impurity centers, in which case the ionic conductivity of the film may be said to be impurity induced, or they may arise from lattice dissociation (e. g. , formation of a Frenkel defect from a lattice cation), in which case the ionic conductivity may be said to be intrinsic.

In this section these two cases are considered from the point of view of determining the conditions required for ionic conduction to be

independent of interfacial kinetic processes. As the impurity-induced
case has been largely dealt with in Section IV, the present section is
concerned primarily with the case of intrinsic ionic conduction. The
treatment given is that due to Dignam and Taylor [31]. It does not
include the contribution of diffusion to ionic transport, an approximation that will be valid provided that condition (27) is satisfied. This
will be the case for high field strengths ($qaE \stackrel{\sim}{>} kT$) and a space-
charge density of less than about 5×10^{19} e/cm^3. For lower field
strengths the condition on the space-charge density is more restrictive (see Eq. (27)).

B. Intrinsic Ionic Conduction

In 1956 Bean, Fisher, and Vermilyea [32] proposed a model to
account for the complex kinetics of the formation of anodic oxide
films on valve metals. The model was subsequently clarified in
certain details, first by Dewald [33] in 1957 and then by L. Young [30]
in 1959. It was L. Young who named it the high-field Frenkel-defect
model. The essential features of this model are summarized here.

The applied electric field is assumed to act on lattice cations in
the oxide film, reducing the activation energy for the formation of
Frenkel defects (interstitial cations and cation vacancies). The interstitial cations so generated are assumed to move in the direction of
the applied field, which acts to reduce the activation energy for the
transport motion as well as that for defect formation. The vacancies,
assumed to be immobile, function as traps for the mobile interstitials
and are assigned a characteristic capture cross section for this recombination process. Under conditions of constant field strength and
temperature a steady-state concentration of defect pairs is ultimately
established in which the rate of field-assisted generation of defect
pairs is equal to the rate of their recombination. Under these steady-

state conditions the ion-current density is a function of the field strength alone for a given temperature. On changing the current density or field strength suddenly, however, the concentration of defect pairs adjusts slowly, and the approach toward a new steady-state condition produces field or ion-current transients.

The purpose of this section is first to derive ionic conduction equations from the basic model proposed by Bean et al. [32] and then to ascertain under what conditions they may be applied. In order that low field strengths not be excluded, the derivation must be carried out for a three-dimensional model rather than for a one-dimensional model as in Section II. B. The reason for this is that consideration of the processes of lattice dissociation and recombination cannot be restricted to the direction of the applied field, since processes occurring in a direction normal to the field will contribute to the overall kinetics. In the derivation the particular case of the formation of Frenkel defects in which only the interstitial cations are mobile will be considered, although it can be generalized without difficulty to include other defect pairs and mobility for both defects.

1. Field-Assisted Generation of Defect Pairs

In order to simplify the problem it is assumed at the outset that the positions occupied by the lattice cations (from which the Frenkel defects are generated) represent centers of symmetry in the solid, at least with respect to the potential-energy surface for displacement of these lattice cations. From elementary rate theory, the probability that a given lattice cation will enter an interstitial position along a path within the infinitesimal solid angle $\sin \theta \, d\theta \, d\phi$, where θ is the polar angle and ϕ is the longitudinal angle (normal spherical polar coordinates), is given by

$$\nu^f_{\theta,\phi} \, \exp[-W^f_{\theta,\phi}(E_{\theta,\phi})/kT] \, \sin \theta \, d\theta \, d\phi \,, \qquad (160)$$

where $E_{\theta,\phi}^{f}$ is the component of the electric field in the (θ, ϕ) direction, $W_{\theta,\phi}^{f}(E_{\theta,\phi})$ is the net activation energy in this direction, and $v_{\theta,\phi}^{f}$ is the corresponding frequency factor. The direct functional dependence of $W_{\theta,\phi}^{f}$ on (θ, ϕ) and the functional form of $v_{\theta,\phi}^{f}$ will reflect the structural symmetry. Thus in this case, where the origin is assumed to be a center of symmetry,

$$v_{\theta,\phi}^{f} = v_{\pi-\theta,\,\phi+\pi}^{f} \tag{161}$$

and

$$W_{\theta,\phi}^{f}(E_{\theta,\phi}) = W_{\pi-\theta,\,\phi+\pi}^{f}(-E_{\pi-\theta,\,\phi+\pi}) . \tag{162}$$

Taking $\theta = 0$ as the direction of the applied field, the total probability of lattice dissociation in any direction, P_{f}, is given by

$$P_{f} = \int_{-1}^{1} \int_{0}^{2\pi} v_{z,\phi}^{f} \exp\left[-\frac{W_{z,\phi}^{f}(Ez)}{kT}\right] d\phi \; dz , \tag{163}$$

where E is the electric field strength and $z \equiv \cos\theta$. This integral can be expressed as the sum of two integrals, the first over the range $-1 \leq z \leq 0$, the second over $0 \leq z \leq 1$. If the limits of integration on the first integral are changed to $(0, 1)$ by substituting $-z$ for z, then, on applying the symmetry conditions in the form $v_{-z,\,\phi}^{f} = v_{z,\,\phi+\pi}^{f}$ and $W_{-z,\,\phi}^{f}(-Ez) = W_{z,\,\phi+\pi}^{f}(-Ez)$, Eq. (163) becomes

$$P_{f} = \int_{0}^{1} \int_{0}^{2\pi} v_{z,\phi}^{f} \exp\left[-\frac{W_{z,\phi}^{f}(Ez)}{kT}\right] d\phi \; dz$$

$$+ \int_{0}^{1} \int_{0}^{2\pi} v_{z,\,\phi+\pi}^{f} \exp\left[-\frac{W_{z,\,\phi+\pi}^{f}(-Ez)}{kT}\right] d\phi \; dz . \tag{164}$$

Since $v_{z,\phi}^{f}$ and $W_{z,\phi}^{f}(Ez)$ must be periodic functions of ϕ of period 2π or a submultiple of 2π, the integrals in Eq. (164) differ only in the sign of the parameter E. Making the substitution

$$\nu_f \exp\left[-\frac{W_f(E)}{kT}\right] = \int_0^1 \int_0^{2\pi} \nu_{z,\phi}^f \exp\left[-\frac{W_{z,\phi}^f(Ez)}{kT}\right] d\phi \, dz \,, \qquad (165)$$

where ν^f is the weighted mean value of $\nu_{z,\phi}^f$, the weighting factor being $\exp[-W_{z,\phi}^f(Ez)/kT]$, whose mean value is $\exp[-W_f(E)/kT]$, Eq. (164) becomes

$$P_f = \nu_f\{\exp[-W_f(E)/kT] + \exp[-W_f(-E)/kT]\} \,. \qquad (166)$$

For any given E, $W_{z,\phi}^f(Ez)$ would be expected to display one or more pronounced minima on variation of (z, ϕ), the number depending on the rotational symmetry order of the structure about the $z = 1$ ($\theta = 0$) axis (specified by the direction of the field). In this case W_f would reflect very closely the function $W_{z,\phi}^f$ at the minima. Setting W_f equal to $W_{z,\phi}^f$ at the minimum, or saddle, point, and ν_f equal to $\nu_{z,\phi}^f$ at the same point, corresponds to the usual approximation made in treating thermally activated kinetic processes.

The rate of formation of the current carriers per unit volume, $(\partial n/\partial t)_f$, can now be written in the form

$$(\partial n/\partial t)_f = (N - n^*)\nu_f\{\exp[-W_f(E)/kT] + \exp[-W_f(-E)/kT]\} \,, \qquad (167)$$

where N is the number of cation lattice positions per unit volume and n^* is the number of cation vacancies.

2. Ion Flux and Defect Recombination

By proceeding in a manner exactly analogous to that of the preceding section, the flux density of ions in a direction within the solid angle $\sin\theta \, d\theta \, d\phi$ is shown to be

$$d^2J_{\theta,\phi} = n d_{\theta,\phi} \nu_{\theta,\phi} \exp[-W_{\theta,\phi}(E\cos\theta)/kT]\sin\theta \, d\theta \, d\phi, \qquad (168)$$

where $d_{\theta, \phi}$ is the mean distance traveled in the (θ, ϕ) direction per activation. Assuming as before that the interstitial positions are centers of symmetry, the net flux in the direction of the field, J, is given by

$$J = n \int_0^1 \int_0^{2\pi} z\, d_{z, \phi}\, v_{z, \phi} \left\{ \exp\left[-\frac{W_{z, \phi}(Ez)}{kT} \right] - \exp\left[-\frac{W_{z, \phi}(-Ez)}{kT} \right] \right\} d\phi\, dz .$$
(169)

On making the substitution,

$$v d \exp\left[-\frac{W(E)}{kT} \right] = \int_0^1 \int_0^{2\pi} z\, d_{z, \phi}\, v_{z, \phi} \exp\left[-\frac{W_{z, \phi}(Ez)}{kT} \right] d\phi\, dz ,$$
(170)

Eq. (169) becomes

$$J = n\, dv\, \{ \exp[-W(E)/kT] - \exp[-W(-E)/kT] \} ,$$
(171)

where dv is the weighted mean value of $d_{z, \phi}\, v_{z, \phi}$, the weighting factor being $z \exp[-W_{z, \phi}(Ez)/kT]$, whose mean value is $\exp[-W(E)/kT]$.

Defining $\sigma_{z, \phi}(E)$ as the capture cross section of a cation vacancy for an interstitial moving in the (z, ϕ) direction, the rate of recombination of defect pairs per unit volume within the solid angle $\sin\theta\, d\theta\, d\phi$ is given by $n^* \sigma_{z, \phi}(E)\, d^2 J_{z, \phi}$ provided the vacancies are immobile. On substituting from Eq. (168), integrating over all values of (z, ϕ), and applying the symmetry condition, the total rate of recombination of defect pairs per unit volume is obtained as

$$\left(\frac{\partial n}{\partial t} \right)_r = n^* n \int_0^1 \int_0^{2\pi} \sigma_{z, \phi}(E)\, d_{z, \phi}\, v_{z, \phi} \left\{ \exp\left[-\frac{W_{z, \phi}(Ez)}{kT} \right] \right.$$
$$\left. + \exp\left[-\frac{W_{z, \phi}(-Ez)}{kT} \right] \right\} d\phi\, dz .$$
(172)

Defining $\bar{\sigma}$ through

$$\bar{\sigma}d\nu \, \exp\left[-\frac{W(E)}{kT}\right] =$$

$$\int_0^1 \int_0^{2\pi} \sigma_{z,\phi}(E) \, d_{z,\phi} \nu_{z,\phi} \, \exp\left[-\frac{W_{z,\phi}(Ez)}{kT}\right] d\phi \, dz , \qquad (173)$$

Eq. (172) becomes

$$(\partial n/\partial t)_r = n^* n\bar{\sigma} \, d\nu \, \{\exp[-W(E)/kT] + \exp[-W(-E)/kT]\} . \qquad (174)$$

By solving for $\bar{\sigma}$ from Eqs. (173) and (170), it can be seen that this parameter is the weighted mean value of $\sigma_{z,\phi}(E)/|z|$, the weighting function being $|z| d_{z,\phi} \nu_{z,\phi} \exp[-W_{z,\phi}(E|z|)/kT]$. For values of E that are not too small the weighting function will be a maximum for $|z|$ close to unity and will fall off rapidly with decreasing $|z|$. Under these conditions $\bar{\sigma}$ will be very nearly the capture cross section for motion in the field direction. As a direct result of the Coulombic attraction between the defect pairs, $\sigma_{z,\phi}(E)$ would be expected to depend somewhat on both field strength and temperature. Even in the absence of such dependence, however, $\bar{\sigma}$ should show a field and temperature dependence introduced through the averaging procedure. This effect should be small compared with the temperature and field variation inherent in $\sigma_{z,\phi}(E)$.

3. Steady-State Ionic Conduction

For specimens of such thickness that space-charge effects may be neglected, under steady-state conditions $n = n^*$ and $(\partial n/\partial t)_f = (\partial n/\partial t)_r$, so that from Eqs. (167) and (174)

$$n^2 = \frac{(N-n)\nu_f}{\bar{\sigma}\nu d} \left\{\frac{\exp[-W_f(E)/kT] + \exp[-W_f(-E)/kT]}{\exp[-W(E)/kT] + \exp[-W(-E)/kT]}\right\} . \qquad (175)$$

The quantities $W(E)$ and $W_f(E)$ may be represented by the following expressions, obtained from Eq. (104) on replacing qa by u^*:

$$W(E) = Q - u^*E(1 - u^*E/C_2'Q + \cdots),\qquad (176)$$

$$W_f(E) = Q_f - u_f^*E(1 - u_f^*E/C_{2f}'Q_f + \cdots).\qquad (177)$$

For high field strengths, (u_f^*E/kT) and $(u^*E/kT) \gg 1$, Eq. (175) reduces to

$$n^2 = \frac{(N-n)\nu_f}{\bar\sigma\nu d} \ \exp\left[\frac{W(E)}{kT} - \frac{W_f(E)}{kT}\right]\qquad (178)$$

and from Eq. (171)

$$J = n\, d\nu \ \exp\left[-W(E)/kT\right].\qquad (179)$$

Elimination of n from Eqs. (178) and (179), and setting $N \gg n$, leads to the following steady-state equation for high-field ionic conduction:

$$J_{ss} = (N\nu_f\nu\, d/\bar\sigma)^{\frac{1}{2}} \exp\left[W(E)/2kT + W_f(E)/2kT\right].\qquad (180)$$

Note that the subscript "ss" refers to the steady state and bears no relation to the dimensionless position coordinate defined in Eqs. (3).

If terms of order 3 or higher in the field strength are neglected, Eq. (180) may be written as

$$J_{ss} = \left(\frac{N\nu_f\nu d}{\bar\sigma}\right)^{\frac{1}{2}} \exp\left(-\frac{Q_{ss}}{kT}\right) \exp\left[\frac{u_{ss}^* E}{kT}\left(1 - \frac{u_{ss}^* E}{C_{ss}Q_{ss}}\right)\right],\qquad (181)$$

where $u_{ss}^* = (u_f^* + u^*)/2$, $Q_{ss} = (Q_f + Q)/2$, and

$$\frac{1}{C_{ss}} = \frac{1}{2C_{2f}'}\frac{(u_f^*/u_{ss}^*)}{(Q_f/Q_{ss})} + \frac{1}{2C_2'}\frac{(u^*/u_{ss}^*)}{(Q/Q_{ss})}.$$

Equation (181) is essentially the same as that proposed by Bean et al. [32], except for the inclusion of the quadratic field term.

For intermediate field strengths, defined by $Q_f \gg u_f^*E \gtrsim kT$ and $Q \gg u^*E \gtrsim kT$, the quadratic field term is negligible and Eqs. (175),

(176), and (177) reduce to

$$n^2 = \frac{(N-n)}{\bar{\sigma}vd} \exp\left(\frac{Q - Q_f}{kT}\right) \frac{\cosh(u_f^* E/kT)}{\cosh(u^* E/kT)} \qquad (182)$$

and Eqs. (171) and (176) and (177) can be combined to give

$$J = 2ndv \exp\left(-\frac{Q}{kT}\right) \sinh\frac{u^* E}{kT} \qquad (183)$$

so that when n is eliminated (assuming $N \gg n$)

$$J_{ss} = 2\left(\frac{Nv_f vd}{\bar{\sigma}}\right)^{\frac{1}{2}} \exp\left(-\frac{Q}{kT}\right) \left[\frac{\cosh(u_f^* E/kT)}{\cosh(u^* E/kT)}\right]^{\frac{1}{2}} \sinh\frac{u^* E}{kT} \quad . \ (184)$$

For low field strengths, defined by $u_f^* E/kT$ and $u^* E/kT \ll 1$, Eqs.
(182) and (184) reduce to

$$n^2 = \frac{(N-n)v_f}{\bar{\sigma}vd} \exp\left(\frac{Q - Q_f}{kT}\right) \qquad , \qquad (185)$$

and

$$J = 2ndv \frac{u^* E}{kT} \exp(-Q/kT). \qquad (186)$$

Hence

$$J_{ss} = 2\left(\frac{Nv_f vd}{\bar{\sigma}}\right)^{\frac{1}{2}} \frac{u^* E}{kT} \exp\left(-\frac{Q_{ss}}{kT}\right) \cdot \qquad (187)$$

Note that for low field strengths n is independent of field
strength, and hence the defects must be generated by thermal activation alone. Thus n takes on its equilibrium value, so that J and
J_{ss} are not really distinguishable. Furthermore,

$$K_d = \frac{n^2}{N(N-n)} = \frac{v_f}{\bar{\sigma}vdN} \exp\left(\frac{Q - Q_f}{kT}\right) \qquad , \qquad (188)$$

where K_d is the equilibrium constant for the formation of defect pairs.
Since

$$\left[\frac{\partial (k \ln K_d)}{\partial (1/T)}\right]_{\bar{V}} = -\Delta U_d^0 , \qquad (189)$$

where ΔU_d^0 is the standard internal energy of formation of a defect pair, and since $\bar{\sigma}$, ν_f, ν, d, Q_f, and Q may be considered to be essentially temperature independent under conditions of constant volume,

$$Q_f - Q = \Delta U_d^0 . \qquad (190)$$

Noting that $kT \ln K_d = -\Delta A_d^0 = -\Delta U_d^0 + T\Delta S_d^0$, and that $u^* = qa = qd/2$ for a symmetrical barrier, a combination of Eqs. (187), (188), and (190) leads to

$$J = \left(\frac{\nu Nqa}{kT}\right) E \exp\left(\frac{\Delta S_d^0}{2k}\right) \exp\left[-\frac{(\Delta U_d^0/2) + Q}{kT}\right] . \qquad (191)$$

This expression is the same as that obtained by Mott and Gurney [1] for low-field ionic conduction in the alkali halides.

Turning briefly once again to ionic conduction under conditions of very high field strengths, the equations derived in the preceding sections are applicable not only for steady-state but also for transient conditions. Thus making the assumption that the film is electrically neutral ($n^* = n$), the ion-flux density is given from Eq. (171) as

$$J = nd\nu \exp[-W(E)/kT] \qquad (192)$$

and an expression for the net rate of formation of defect pairs per unit volume can be obtained from Eqs. (167), (171), and (174):

$$\partial n/\partial t = (N - n)\nu_f \exp[-W_f(E)/kT] - n\bar{\sigma}J . \qquad (193)$$

Except for the allowance for higher order field terms in $W_f(E)$ and $W(E)$, Eqs. (192) and (193) are identical with the equations given by Dewald [33] to describe the transient behavior of valve-metal anodic oxide systems.

For future convenience expressions for the various differential field coefficients (cf. Eq. (166)) for this model are given below:

$$\beta = (-1/kT)[\partial W(E)/\partial E]_T , \tag{194}$$

$$\beta_{ss} = (-1/2kT)\{\partial[W_f(E) + W(E)]/\partial E\}_T , \tag{195}$$

$$\beta_f = 2\beta_{ss} - \beta = (-1/kT)[dW_f(E)/dE]_T . \tag{196}$$

For specimens of such thickness that space-charge effects may be neglected, and for high field strengths, it follows from Eqs. (179) and (180) that

$$\beta = [\partial \ln(J)/\partial E]_{T,n} , \tag{197}$$

$$\beta_{ss} = [\partial \ln(J_{ss})/\partial E]_T . \tag{198}$$

Thus Eq. (180) can be written

$$J_{ss} = A_{J_{ss}} \exp(\beta_{ss}E) , \tag{199}$$

where $A_{J_{ss}}$ is given by

$$A_{J_{ss}} = \left(\frac{N\nu_f vd}{\bar{\sigma}}\right)^{\frac{1}{2}} \exp\left(\frac{Q_{ss}}{kT}\right) \exp\left[\left(\frac{u_f^{*2}}{C_{2f}'Q_f} + \frac{u^{*2}}{C_2'Q}\right)\frac{E^2}{2kT}\right] . \tag{200}$$

The factor $A_{J_{ss}}$ is a slowly changing function of J_{ss} and may therefore be treated as being essentially constant over a moderate range of flux densities for a given temperature.

4. Nature of Space Charge

Before leaving the case of intrinsic ionic conduction, we investigate the nature of the space-charge distribution near the interface $x = 0$. That such a space-charge region must in general exist can be seen from the following. For steady-state conditions the ion-flux

density across the interface $x = 0$ must equal that within the film
some distance from the interface, where the space-charge density
will be essentially zero. Since the migration of ions across the
interface and their migration within a space-charge-free region of the
film are independent kinetic processes, their rate can in general
only be equal if the electrostatic field strengths are different. This
difference in field strength can only be sustained if the intervening
space contains a net charge. Thus for intrinsic ionic conduction as
well as for impurity-induced conduction (i. e., constant background-
space-charge density) a space-charge region is required to decouple
the bulk and interfacial transport processes.

a. Low Field Strengths. The case of ionic conduction at low field
strengths is treated first. In the absence of the assumption of charge
neutrality, but retaining the steady-state condition, Eq. (175) is
altered only in that n^2 is replaced by nn^* and n by n^*. In the low-
field limit, on setting $N \gg n^*$, Eq. (188) then becomes

$$nn^* \approx N^2 K_d \ . \tag{201}$$

Furthermore Eq. (186) may be written

$$J = nvE \ , \tag{202}$$

where v is the mobility of the mobile defect. In writing Eq. (202)
diffusion has been ignored, an approximation that will be valid only
for a sufficiently low space charge. Thus from Eq. (27) the condition
$|n - n^*| \leq \delta(\epsilon/4\pi kT)E^2$ must be satisfied. Poisson's equation may be
written

$$dE/dx = 4\pi(n - n^*)q/\epsilon \ . \tag{203}$$

Elimination of n and n^* from Eqs. (201) through (203) leads to the
differential equation

$$\frac{dE^2}{dx} + \left(\frac{8\pi N^2 K_d vq}{\epsilon J} \right) E^2 = \frac{8\pi Jq}{\epsilon v} \ , \tag{204}$$

which on using the boundary condition $E = E_0$ at $x = 0$ can be solved to obtain

$$E^2 - E_\infty^2 = (E_0^2 - E_\infty^2) \exp(x/X_c) , \qquad (205)$$

where

$$E_\infty = J/NvK_d^{\frac{1}{2}} , \qquad (206)$$

$$X_c = \epsilon J/8\pi N^2 K_d qv = X_c^0 \exp(Q_f/kT) , \qquad (207)$$

$$X_c^0 = (\epsilon k\bar{T\sigma}J/8\pi Naq^2 v_f) . \qquad (208)$$

From Eq. (205) it follows that E_∞ is the field strength in the space-charge-free region of the film, where $n = n^*$. For $|E - E_\infty| \ll E_\infty$, Eq. (205) reduces to $E - E_\infty \dot{\propto} \exp(x/X_c)$, and hence the space-charge density, which is proportional to $\partial E/\partial X$, is also proportional to $\exp(x/X_c)$. Thus X_c is the characteristic space-charge length in the limit as $E \to E_\infty$ (i. e., the length, or distance, in the film over which the space-charge density drops by a factor $1/e$). The second expression for the characteristic space-charge length in Eq. (207) was obtained from the former by substituting $v = (4a^2 qv/kT) \exp(-Q/kT)$ and $K_d = (v_f/2\bar{\sigma} v_f aN) \exp[(Q - Q_f)/kT]$.

The condition that the space charge within a film of thickness X will not influence the conduction properties of the film is that $X_c \ll X$, which with Eq. (207) becomes

$$X/X_c^0 \gg \exp(Q_f/kT) . \qquad (209)$$

On setting $X \approx 10^4 \text{Å}$, $\epsilon = 10$, $T = 298°K$, $a = 2\text{Å}$, $q = 1e$, $v_f = 10^{12} \text{ sec}^{-1}$, $N \approx 10^{22} \text{ cm}^{-3}$, $\bar{\sigma} \approx N^{-\frac{2}{3}}$, and $J \approx 10^{16} \text{ cm}^{-2}$ (corresponding to a current density of 1.6 mA/cm^2), condition (209) requires that Q_f be no greater than about 0.8 eV. Thus low-field ionic conduction through thin films at room temperature will be independent of the interfacial transport process only if the activation energy for the

formation of a defect pair is significantly less than 0.8 eV. Very few
film materials are likely to satisfy this condition, except possibly
certain of the silver salts. Thus the enthalpy of formation of Frenkel-
defect pairs in silver bromide is about 20 kcal, whereas the activation energy for the migration of an interstitial cation in silver bromide
is about 8 kcal [1], leading to $Q_f \approx 1.2$ eV, which is still too large
for Eq. (209) to be satisfied. In general, therefore, and certainly for
most metal oxides, Q_f will be much larger than 0.8 eV.

b. High Field Strengths. An estimate of the characteristic space-
charge length for the high-field case can also be obtained. Following the derivation of Dignam and Ryan [34], Eq. (171) can be written
in the approximate form

$$\dot{J} \; \dot{\alpha} \; n\, e^{\beta E} , \tag{210}$$

where an approximation analogous to that used in obtaining Eq. (199)
has been employed. Similarly Eqs. (167), (174), and the steady-
state condition $(\partial n/\partial t) = 0 = (\partial n^*/\partial t)$ give

$$n^* n \, \exp(2\,\beta\, E) \; \dot{\alpha} \; \exp(2\,\beta_{ss} E) . \tag{211}$$

Denoting the field strength in the space-charge-free region of the
film by E_∞ and the corresponding defect concentration by n_∞, an
expression for the variation in n with field strength within the
space-charge zone is obtained from Eq. (210):

$$n \approx n_\infty \, \exp[\beta(E_\infty - E)] . \tag{212}$$

The variation in n^* with field strength within the space-charge zone
is obtained from Eqs. (210), (211), and (196):

$$n^* \approx n_\infty \, \exp[\beta_f(E - E_\infty)] . \tag{213}$$

Substituting for n and n^* from Eqs. (212) and (213) into Poisson's equation, Eq. (203), the resulting equation can be written

$$dy/ds' = (1 - y^{1+\gamma})/(1 + \gamma) \, , \qquad (214)$$

where y, s', and γ are dimensionless parameters defined as follows:

$$y = \exp\left[-\left|(E - E_\infty)2\beta_{ss}\big/(\gamma+1)\right|\right] \, , \qquad (215)$$

$$s' = [8\pi q\beta_{ss} n_\infty/\epsilon\,]x = x/X'_c \geq 0 \qquad (216)$$

for $qE_0 < qE_\infty$, $\gamma = \beta_f/\beta$, and for $qE_0 > qE_\infty$, $\gamma = \beta/\beta_f$, E_0 being the value of the field strength at $x = 0 = s'$ and being determined by interfacial-transport processes.

Equation (214) can be solved in terms of simple functions for certain values of γ only. Asymptotic solutions for any value of γ, however, can be found. The asymptotic solution in the limit of low space-charge — that is, in the limit as y approaches unity — is considered in the following development.

The general solution of Eq. (214) can be expressed as follows:

$$s' + s'_0 = f(\gamma, y) - \ln(1 - y) \, , \qquad (217)$$

where s'_0 is the integration constant and

$$f(\gamma, y) = \int_0^y \left(\frac{1+\gamma}{1-y^{1+\gamma}} - \frac{1}{1-y}\right) dy \, .$$

In Ref. [34], $f(\gamma, y)$ was evaluated for various values of γ lying between 0 and 6, and for y covering its full range of 0 to 1. Neglecting the slight curvature of the graphs of $f(\gamma, y)$ versus y, the function $f(\gamma, y)$ may be represented approximately in the range $y = 0$ to 1 by the equation $f(\gamma, y) \simeq 4\gamma y/5$. The point is that $f(\gamma, y)$ is a slowly varying function of y that is continuous and finite in the region $y = 0$ to 1. Equation (217) can be rearranged to give

$$\ln y = \ln\{1 - \exp[f(\gamma, y) - (s' + s'_0)]\} \, . \qquad (218)$$

On substituting for y from Eq. (215), and noting that for $1 - y \ll 1$, $1 \gg \exp[f(\gamma, y) - (s' + s_0')]$ and $f(\gamma, y) \simeq f(\gamma, 1) \equiv f(\gamma)$, Eq. (218) reduces to the following approximate relation, valid in the limit of low space charge:

$$\left| E - E_\infty \right| \stackrel{\cdot}{=} A_{\Delta E} \exp(-x/X_c') , \qquad (219)$$

where $A_{\Delta E} = \left| \gamma/2\beta_{ss} \right| \exp[f(\gamma) + s_0']$.

The condition that the space charge within a film of thickness X will not influence the conduction properties of the film is again given by $X_c' \ll X$, which with Eq. (216) becomes

$$n_\infty \gg (\epsilon/8\pi q\beta_{ss}X) . \qquad (220)$$

Setting $\beta_{ss} \approx aq/kT$, $X \approx 10^4 \,\text{Å}$, and using the same values for the parameters as before leads to the result $n_\infty \gg 3 \times 10^{16}$ cm^{-3}, or $n_\infty \gtrsim 10^{18}$ cm^{-3}. From Eqs. (179) and (180) respectively we have that $J \propto n_\infty [\exp(-Q/kT)] \exp(\beta E_\infty)$ and $J \propto [\exp(-Q_{ss}/kT)] \exp(\beta_{ss}E_\infty)$, which, on eliminating E_∞, give

$$n_\infty \propto \{\exp[(\beta_f Q - \beta Q_f)/(\beta_f + \beta)kT]\} J^{(\beta_f - \beta)/(\beta_f + \beta)} . \qquad (221)$$

It can be seen that n_∞ is a sensitive function of the system parameters, particularly of Q_f and Q. Since n_∞ cannot be much greater than 10^{21} cm^{-3}, this allows only a three-decade range for n_∞ in which the space charge, and hence interfacial effects, is unimportant. It is improbable that n_∞ will happen to lie in this narrow range for room-temperature conditions and accessible current densities. Thus if by chance at a given temperature and current density n_∞ for some system should fortuitously lie in the range 10^{18} to 10^{21} cm^{-3}, altering the current density or temperature by a relatively small amount would almost certainly alter n_∞ to lie outside this range.

The argument presented here is a fairly general one and should apply to any ionic conduction mechanism involving the homogeneous generation of defect pairs within the film under high-field conditions in which the steady-state concentration of defect pairs is field dependent. Such mechanisms are therefore improbable in the case of thin films and high fields, though they cannot be ruled out for specimens of macroscopic dimensions.

In the analysis of the high-field Frenkel-defect model given by Dignam and Taylor [31] essentially the same conclusion as above was reached based on an analysis of a quantity called the mean free path λ_s, defined as the mean distance traveled in the field direction and in bulk material by a mobile defect from its formation to its annihilation. We asserted, incorrectly, that the condition $\lambda_s \ll X$ must be satisfied in order that the interfacial kinetic processes not influence the conduction properties of the film. The error in the assertion has been pointed out by L. Young [35]. The proper condition is that the characteristic space-charge length X_c', not λ_s, be much smaller than X. That the above analysis leads to essentially the same result as obtained by Dignam and Taylor arises from the fact that λ_s and X_c' are closely related, since both are inversely proportional to n_∞, an expected result.

C. Impurity-Induced Ionic Conduction

The case of ionic conduction in the presence of a uniform background-space-charge density $-qn_\infty$, generated by impurities, has already been treated in Section IV.D. For a low space-charge density and low field strengths the characteristic space-charge length is given by Eq. (149). The condition that the space-charge region not influence the film conductance is given by $S \gg S_c$, which on transforming to dimensioned variables by using Eqs. (3) and (117) becomes

$$X/X_C^{0\prime\prime} \gg (N/n_\infty)^2 \exp(Q/kT) ,\tag{222}$$

where

$$X_C^{0\prime\prime} = (\epsilon kTJ/16\pi N^2 a^2 q^2 v) .\tag{223}$$

Using the same values for the parameters as before, Eq. (222) requires
that

$$Q < [0.8 - 0.12 \log (N/n_\infty)] \text{ eV} ,\tag{224}$$

or, since n_∞/N is unlikely to be larger than about 10^{-5}, $Q \lesssim 0.5$ eV.
Once again, in general, this condition will be difficult to meet,
though not nearly so difficult as in the case of intrinsic ionic conduction at low field strengths, since activation energies for ionic
conduction as low as 0.5 eV are not that uncommon.

For high field strengths, again in the limit of low space-charge
density, the condition that the space charge not influence the film
conductance is, from Eq. (154), given by $S \gg 1/\alpha c_\infty$, which on
transforming to dimensioned variables as before becomes

$$n_\infty \gg \epsilon kT/4\pi q^2 a .\tag{225}$$

Again, using the same values for the parameters as before leads to
the result $n_\infty \gg 6 \times 10^{16}$ cm^{-3} , or $n_\infty \gtrsim 10^{18}$ cm^{-3}. Unlike the case
of high-field intrinsic ionic conduction, this condition could be satis-
fied for almost any system, the sole requirement being one on the
value of the impurity-induced background-space-charge concentration.

D. Summary

It has been shown that high-field ionic conduction in thin films
via an intrinsic mechanism is improbable. This conclusion was
reached from an analysis of the behavior of the interfacial space-
charge zone for this type of mechanism. The permitted range of the
defect concentration was shown to be very narrow, and the concen-

tration itself is a sensitive function of the system constants. Thus if conditions for intrinsic ionic conduction at high fields should fortuitously be met for some system at a given temperature and current density, a relatively small change in temperature or current density would cause the defect concentration to lie outside its permitted range.

The intrinsic ionic conduction mechanism for low fields, applied to thin films near room temperature, fares only slightly better than its high-field counterpart. The condition required in order that the space-charge region not dominate the transport kinetics is that the activation energy for the formation of the defect pairs be less than about 0.8 eV. Thus the sum of the energy of formation of a defect pair and the activation energy for defect migration must be less than about 0.8 eV, which is an extremely restrictive condition.

Ionic conduction via an impurity-induced mechanism fares considerably better on application to thin films near room temperature than does the intrinsic mechanism. Thus for high field strengths the only condition is that the impurity concentration be greater than about 10^{18} cm^{-3}, whereas for low field strengths, in addition, the activation energy for ion transport must be less than about 0.5 eV.

It is clear from the material presented in this section that ion transport through thin films near room temperature will commonly be controlled, at least in part, by interfacial-transport processes. Section VI is concerned with an analysis of these processes and their relationship to the bulk-transport process.

VI. THE ROLE OF INTERFACES IN IONIC CONDUCTION THROUGH THIN FILMS

A. Introduction

From the material presented in Sections IV and V it is clear that ionic transport through thin films must be controlled, if not predominantly, at least in part at one or both of the interfaces. In this section the role of the interfacial processes is examined principally for those cases in which ion transport through the film can be considered to be due to migration alone, with no significant contribution arising from diffusion. From Section II this condition will be satisfied for

$$|Eaq/kT| \overset{\sim}{>} 1 , \qquad \rho \overset{\sim}{<} 10^{19} \text{ e/cm}^3 ,$$

and at lower field strengths for lower space-charge concentrations (cf. Eq. (27)).

In treating transport processes at interfaces two new phenomena must be taken into account. The first arises whenever a surface charge σ exists at the interface, the second as a result of dielectric-constant variation in the interfacial region. A surface charge σ causes a discontinuity in the electric displacement, $D = \epsilon E$, at the charged plane equal to $4\pi\sigma$. In a region devoid of charged planes, or alternatively, treating a charged plane as a concentrated volume charge, D is continuous through any interphase, its variation with position being given by the following form of Poisson's equation:

$$\partial D/\partial x = 4\pi\rho . \tag{226}$$

Equation (226) is written in a form that remains valid in a region in which the dielectric constant is a function of x. Since the matter of prime interest is the distribution of the electrostatic potential V throughout the film and the interfacial region, it is useful to transform Eq. (226) by defining an effective charge density ρ^* and an effective distance parameter x^* through the equations

$$\rho^* = \frac{\epsilon'(x)}{\epsilon} \rho , \tag{227}$$

$$x^* = \int_0^x \left[\frac{\epsilon}{\epsilon'(x)} \right] dx , \tag{228}$$

where ϵ is the dielectric constant in the bulk of the oxide and $\epsilon'(x)$ is the dielectric constant at any position x. Substituting for D the expression $-\epsilon'(x)[\partial V/\partial x]$ and for ρ and dx according to Eqs. (227) and (228), respectively, Eq. (226) becomes

$$\partial^2 V/\partial x^{*2} = -4\pi\rho^*/\epsilon \, , \qquad (229)$$

which is of the same form as Poisson's equation for the bulk of the film. Because of the form of Eq. (229), x^* is a convenient choice for the distance parameter in analyzing rate processes at the oxide interfaces. In Sections VI. B and VI. C, rate processes at the metal-oxide and oxide-electrolyte interfaces are considered in turn.

B. Defect Injection at the Metal-Oxide Interface

The particular case in which the mobile ionic species are interstitial cations is dealt with first, and in some detail, the treatment being extended later to include other possible ionic defects. The present treatment follows closely that given by D. J. Young and Dignam for metal-tarnishing reactions [36].

Ion transport is considered to be achieved via interstitial cation movement through the film, and that process, coupled with the generation of interstitials at the metal-oxide interface, is taken to be rate controlling. Thus the process by which the interstitial ions, having surmounted the final diffusion barrier at the oxide-electrolyte interface, react electrochemically to form the oxide is assumed to proceed sufficiently rapidly so as to have no direct effect on the transport rate. It will, of course, affect the rate indirectly in that part of the total driving force for the electrochemical formation of the oxide film will be dissipated by this process.

The potential-energy profile for the mobile species near the metal-oxide interface, and under conditions in which the electrostatic

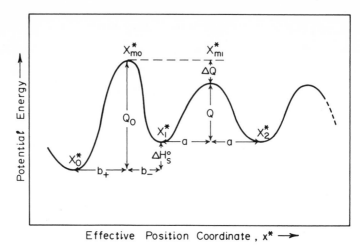

Effective Position Coordinate , x^* ⟶

Fig. 2. Potential energy of a mobile defect as a function of the effective position coordinate near the interface x = 0 .

field strength in the oxide at the interface is zero, is shown in Fig. 2. Position x_0^* represents that of a surface metal atom, with x_1^* and x_2^* representing the positions of interstitial sites; Q_0 and Q are the entrance- and diffusion-barrier heights, respectively, ΔH_s^0 the standard heat of solution of metal ions in the oxide, and $\Delta Q = Q_0 - Q - \Delta H_s^0$ is the amount by which the peak of the entrance barrier exceeds that of the diffusion barrier and may be negative. Under conditions in which the field strength at x_1^* is not zero but E_0, say, a net charge will reside on the metal near the metal–oxide interface. It is assumed that this charge resides entirely on the metal side of x_{mo}^*. It is not clear, however, that it will reside entirely on the metal side of x_0^*. The approximation that the surface charge resides precisely at x_0^* is made initially and examined later.

In the bulk of the film, under conditions in which Eq. (27) is satisfied and the field strength is not too high, Eq. (13) will be a valid approximation:

$$J \approx 4a\nu n \left[\exp(-Q/kT)\right] \sinh(qaE/kT) \ . \tag{230}$$

Furthermore, whenever Eq. (27) is satisfied, the percent change in both field strength and defect concentration over one jump distance (e.g., from x_1^* to x_2^*) is small, so that space–charge effects may be neglected to first order in the range x_0^* to x_2^*. Thus from Eq. (229), $-\partial V/\partial x^*$ is essentially constant in this region and equal to E_0, the boundary value for the field strength in the film. Similarly, the defect concentrations at x_1^* and x_2^* are essentially equal to one another and to n_0, the boundary value of the defect concentration.

Adopting the steady-state approximation, $(\partial J/\partial x)_t \simeq 0$, as before, the net ionic fluxes across the barriers at x_{m0}^* and x_{m1}^* may be equated. Thus across the interfacial barrier

$$J \approx M\nu_0 \exp[-(Q_0 - qb_+ E_0)/kT]$$

$$-(a + b'_-)\nu n_0 \exp[-(Q + \Delta Q + qb_- E_0)/kT] \ , \qquad (231)$$

where M is the surface concentration of metal ions at x_0^* in a position to enter the oxide and b'_- is the true distance corresponding to b_-. Across the first diffusion barrier

$$J \approx 4a\nu n_0 [\exp(-Q/kT)] \sinh(qaE_0/kT) \ . \qquad (232)$$

Equating the two expressions for J and making the approximation $(a + b'_-) \simeq 2a$ in the second preexponential factor of Eq. (231) leads to the following expression for n_0:

$$n_0 \approx \frac{[M(\nu_0/\nu)/2a] \exp[-(Q_0 - Q - qb_+ E_0)/kT]}{2 \sinh(qaE_0/kT) + \exp[-(\Delta Q + qb_- E_0)/kT]} \ . \qquad (233)$$

Substituting this value of n_0 into Eq. (232) then gives

$$J = \frac{M\nu_0 \exp[-Q_0 - qb_+ E_0)/kT]}{1 + \{\exp[-(\Delta Q + qb_- E_0)/kT]\}/2 \sinh(qaE_0)/kT)} \qquad (234)$$

If the space charge in the metal lies partially or entirely within the region (x_0^*, x_{m0}^*), then the potential drop from x_{m0}^* to x_0^* will not be given by b_+E_0. Over a fairly narrow range of field strengths, however, the potential drop will vary linearly with E_0, at least to a first approximation, so that Eq. (234) may still be used, the interpretation placed on b_+ being, however, in this case somewhat different from that illustrated in Fig. 2.

For high field strengths and $\Delta Q \overset{\sim}{>} 0$, Eq. (234) reduces to

$$J \approx M\nu_0 \exp\left[-(Q_0 - qb_+E_0)/kT\right] , \tag{235}$$

which is the Mott–Cabrera transport equation for "very thin films" (i.e., high fields (19)). Even for ΔQ negative, Eq. (235) remains valid for sufficiently high field strengths. For ΔQ sufficiently negative such that $-(\Delta Q + q(a + b_-)E_0) \overset{\sim}{>} 2kT$ and for $qaE_0 \overset{\sim}{>} 2kT$, Eq. (234) becomes

$$J \approx M\nu_0 \exp\left\{-[Q + \Delta H_s^0 - q(a + b_- + b_+)E_0]/kT\right\} . \tag{236}$$

Again, this has the form of the Mott–Cabrera high–field equation, the effective interfacial barrier in this case being situated at x_{m1}^* and the effective activation distance being the effective distance between x^* and x_{m1}^*.

For low field strengths Eq. (233) reduces to

$$n_0 \approx \frac{[M(\nu_0/\nu)/2a]\exp(-\Delta H_s^0/kT)}{1 + 2(qaE_0/kT)\exp(\Delta Q/kT)} \tag{237}$$

and Eq. (234) becomes

$$J \approx \frac{2M\nu_0(qaE_0/kT)\exp[-(Q + \Delta H_s^0/kT)]}{1 + 2(qaE_0/kT)\exp(\Delta Q/kT)} . \tag{238}$$

For sufficiently low fields, or for $\Delta Q \overset{\sim}{<} 1$, Eq. (237) reduces to the equilibrium expression for n_0 :

$$n_0 \approx [M(\nu_0/\nu)/2a] \exp(-\Delta H_s^0/kT) \,, \tag{239}$$

whereas Eq. (238) reduces to the Mott-Cabrera "thin-film" (i. e., low-field) equation

$$J \approx 2M\nu_0(qaE_0/kT) \exp[-(Q + \Delta H_s^0)/kT] \,. \tag{240}$$

Note that for $\Delta Q > 0$ and $2(qaE_0/kT) \exp(\Delta Q/kT) \gg 1$, J becomes independent of field strength:

$$J \approx M\nu_0 \exp(-Q_0/kT) \,. \tag{241}$$

Equation (241) corresponds to rate control by the thermally activated passage of ions across the metal-oxide interface.

In general, n_0 differs from its equilibrium value, Eq. (239), and is a function of the boundary value for the field strength. Under certain conditions, however, n_0 takes on its equilibrium value over the full range of field strengths. Thus, on making the assumption that $\Delta Q \approx 0$ and $b_- \approx a$, Eqs. (233) and (234) become

$$n_0 \approx [M(\nu_0/\nu)/2a] \exp\{-[\Delta H_s^0 - q(b_+ - a)E_0]/kT \,, \tag{242}$$

$$J \approx 2M\nu_0[\sinh(qaE_0/kT)] \exp\{-[Q_0 - q(b_+ - a)E_0]/kT\} \,. \tag{243}$$

Equation (242) now reduces to the equilibrium expression provided that $a \approx b_+$, and Eq. (243) reduces to

$$J \approx 2M\nu_0[\sinh(qb_+ E_0/kT)] \exp(-Q_0/kT) \,. \tag{244}$$

The conditions required for the validity of Eq. (244) appear at first glance to be rather specialized and hence highly improbable. However, to the extent that a metal ion at x_0^* can be thought of as situated on a "surface interstitial site" (i. e., a site at the surface that transforms to an interstitial site on extending the oxide lattice), the effective activation distance and peak energy level of the entrance barrier should approximate those of the diffusion barriers. The further assumption must be made, however, that the surface charge on the metal resides on a plane situated close to x_0^*.

Equation (244) represents a convenient summary of the various limiting transport equations even for cases in which ΔQ is not close to zero and $b_+ \not= a$, since it reduces precisely to Eq. (235) in the high-field case and is of the same form as Eq. (240) for low fields, becoming identical with it on making the substitutions $Q_0 \rightarrow \Delta H_s^0 + Q$, $b_+ \rightarrow a$. It is a particularly useful approximation for cases in which the electrostatic field strength is on the order of kT/qa, so that neither the low-field nor the high-field approximations can be used. Such a case is encountered in the kinetics of the gaseous oxidation of aluminum at temperatures below about 450°C, where a kinetic equation derived from Eq. (244) has been shown to fit the data with no deviation beyond experimental scatter (37).

If, instead of interstitial cations, anion vacancies are generated at the metal-oxide interface, Eqs. (230) through (244) will still apply under the various conditions, with n_0 representing the boundary concentration of anion vacancies and M the surface concentration of anions suitably located for the creation of an anion vacancy.

Two different mechanisms for the injection of either interstitial cations or anion vacancies into the oxide at the metal-oxide interface can be described. These will be referred to as the vacancy-interstitial mechanism and the dislocation-climb mechanism.

Interstitial cations may be formed at the interface by a process in which a metal ion at the surface of the metal phase enters an interstitial position in the oxide, creating in the process a vacancy in the metal lattice at the surface. Film growth via this vacancy-interstitial mechanism can lead to the creation of voids at the metal-oxide interface as a result of condensation of these vacancies, a behavior frequently observed during the high-temperature oxidation of metals.

Alternatively, metal ions may leave only from lattice step sites on the metal surface to enter interstitial positions in the oxide. Thus the metal atoms are removed in such a manner as to bring about the motion of metal-lattice steps in the interfacial plane, the mechanism being virtually identical with that of dislocation climb. Screw dislocations in the metal will of course guarantee an inexhaustible supply of lattice steps at the interface. The dislocation-climb mechanism should preserve contact between the phases (as indeed it does in the case of dislocation climb within a crystal) and should not lead to void formation, since no vacancies are generated.

Similarly, anion vacancies can be injected into the oxide at the metal-oxide interface, either by a process in which oxygen ions enter the interstices of the metal lattice or by one in which oxygen ions, adjacent to surface-lattice steps in the oxide lattice, enter step sites. In the latter process the oxide film grows at the metal-oxide interface via the motion of these steps in the interfacial plane, again in a manner virtually identical with dislocation climb. The former (vacancy-interstitial) mechanism is unlikely to be operative unless the lattice spacing of the metal atoms in the metal is on the order of, or greater than, the metal-ion spacing in the oxide. If the metal spacing is smaller than the oxide spacing, the oxide will form under a lateral compressional stress, whereas if it is greater, under a lateral tensile stress, which may cause film rupture. The dislocation-climb mechanism for anion-vacancy injection, on the other hand, will not, in any direct way, introduce stress into the growing film, regardless of the extent of mismatch between the lattices. This is also the case for both mechanisms of interstitial cation injection.

C. Defect Injection at the Electrolyte-Oxide Interface

Before proceeding with a development of equations describing the kinetic processes occurring at the electrolyte-oxide interface, a

particular model for these processes is required. The following model
will be assumed and is consistent with that arising out of studies of
the double layer at oxide-solution interfaces [38-41]. It is fair to
say, however, that at present no generally accepted model of the
oxide-solution double layer exists. This then clearly constitutes the
major obstacle in developing kinetic equations applicable to rate control at the oxide-electrolyte interface. The details of the following
development are accordingly in considerable doubt.

Equilibrium is assumed with respect to reactions of the following
type, written for the particular case of Al_2O_3:

$$
\left(
\begin{array}{c}
\diagdown \\
-Al \\
\diagdown \\
\diagup O \\
-Al \\
\diagup
\end{array}
\right)_{surface}
+ H_2O \; \rightleftarrows \;
\left(
\begin{array}{c}
\diagdown \\
-Al-OH \\
\\
-Al-OH \\
\diagup
\end{array}
\right)_{surface}
\tag{A}
$$

For a number of oxides, it appears that at the zero point of charge
(zpc) approximately half the surface OH groups are acidic in character, the rest being basic [40]. There is some evidence from the form
of the variation of the surface-charge density with pH that the surface
OH groups are largely autodissociated at the zpc [41], so that one
should perhaps replace reaction (A) by

$$
\left(
\begin{array}{c}
\diagdown \\
-Al \\
\diagdown \\
\diagup O \\
-Al \\
\diagup
\end{array}
\right)_{surface}
+ H_2O \; \rightleftarrows \;
\left(
\begin{array}{c}
\diagdown \\
-Al-O^{-} \\
\\
-Al-O^{H^+}_{H} \\
\diagup
\end{array}
\right)_{surface}
\tag{B}
$$

In either case, under sufficiently anodic conditions (or high pH) and
for equilibrium conditions, a significant fraction of the surface protons will be missing. For each acidic proton abstracted from the
surface there will remain on the surface a local region bearing a net

charge of $-e$, which can be represented conveniently by $-O^-$. Thus the reaction of importance can be represented by

$$(>Al-OH)_{surface} \rightarrow (>Al-O^-)_{surface} + H^+_{aq} \qquad (C)$$

For sufficiently cathodic conditions (or sufficiently low pH) proton addition can lead to a net excess of positively charged surface species, $-OH_2^+$. In addition, anions may replace surface hydroxyl groups through a basic dissociation to become chemisorbed on the metal atoms of the surface. It is convenient to regard all of these surface species as lying in the plane that defines the surface of the oxide film. Next to this surface there will be a layer of oriented water molecules, some of which will probably form part of the hydration shell of aqueous ions situated at the outer Helmholtz plane. Following the usual description of the double layer, certain aqueous ions may under appropriate conditions be specifically adsorbed, that is, be situated at the inner Helmholtz plane, in direct contact with the surface of the oxide.

With this model in mind, one may expect any one or more of the following interfacial reactions to occur under anodic conditions:

1. Injection of conduction electrons into the oxide accompanied by the discharge of hydroxyl ions or oxidation of some other aqueous species.

2. Injection of cation vacancies into the oxide accompanied by the transport of metal ions from the oxide surface into the electrolyte.

3. Injection of cation vacancies into the oxide accompanied by the transport of molecular ionic species (e.g., $Al(OH)_2^+$ or $AlSO_4^+$) from the oxide surface into the electrolyte.

4. Injection of cation vacancies into the oxide accompanied by the abstraction of protons from surface hydroxyl groups.

5. Injection of oxygen ions or specifically adsorbed anions into the oxide accompanied by proton abstraction, as in process 4.

Process 1 corresponds to the anodic oxidation of species in the electrolyte, which most commonly will result simply in oxygen evolution. Process 2 amounts to metal dissolution, whereas process 3 involves simultaneous metal and oxide dissolution. The final two processes correspond to oxide-film growth, with process 4 being perhaps the more probable, although oxygen-ion migration along grain boundaries seems not at all unlikely in the case of microcrystalline films. The outcome of the competition between these reactions will determine the extent to which oxygen evolution, metal dissolution, film dissolution, or film formation dominates the electrode process. At present only the processes leading to film formation will be examined.

The electrolyte-oxide interface is taken as origin, to conform with the convention established earlier, the flux therefore being that of negatively charged defects (presumably cation vacancies or sites of excess oxygen; however, see Section VII). The electrochemical steps at the oxide-electrolyte interface are given by reaction (C) and its reverse, according to this model.

The forward direction of this reaction corresponds to a positive flux J_e of species with charge q_e equal to that on the electron. With this convention, elementary kinetic theory leads to the following equation for the net flux across the double layer:

$$J_e = (M_p - m)\nu_+ \exp[-(Q_+ - q_e d_+ E_d)/kT]$$

$$- mX(H^+)\nu_- \exp[-(Q_- + q_e d_- E_d)/kT] , \qquad (245)$$

where M_p is the total number of surface hydroxyl groups and $-O^-$ groups per unit area at the film-double layer interface; m is the number of surface $-O^-$ groups per unit area at this interface arising from proton abstraction; $X(H^+)$ is the mole fraction (or corresponding activity) of hydrogen ions at the outer Helmholtz plane; Q_+ and Q_- are the activation energies for $E_d = 0$ for the forward and reverse transport processes, respectively; d_+ and d_- are the corresponding activation distances; E_d is the field strength in the double layer; and v_+ and v_- are the attempt frequencies. It is entirely possible that the field strength in the double layer could be sufficiently large to require the inclusion of the higher order terms in the expressions for the net activation energies. They have, nevertheless, been omitted in order to minimize mathematical complexity.

The simplest assumptions to make with regard to the process of defect injection into the oxide is that it follows an equation of the form of Eq. (244), with each $-O^-$ group at the double layer–oxide interface acting as a site for the injection of an ionic defect into the oxide (i.e., $M = m$). Thus in the oxide

$$J \approx 2mv_0 \left(\sinh \frac{qb_+ E_0}{kT} \right) \exp\left(-\frac{Q_0}{kT} \right) . \tag{246}$$

Again, when we adopt the steady-state approximation, $J_e q_e = Jq$, Eqs. (245) and (246) lead to the following expression for m:

$$\frac{M_p - m}{m} = X(H^+) \frac{v_-}{v_+} \exp\left[-\frac{Q_- - Q_+ + q_e(d_+ + d_-)E_d}{kT} \right]$$

$$+ 2 \left(\frac{q v_0}{q_e v_+} \right) \left(\sinh \frac{qb_+ E_0}{kT} \right) \exp\left(-\frac{Q_0 - Q_+ + q_e d_+ E_d}{kT} \right) . \tag{247}$$

The field strengths E_0 and E_d and the boundary charge $(m - m^*)q_e$ are related through the equation

$$\epsilon E_0 = \epsilon_d E_d + 4\pi(m - m^*)q_e , \tag{248}$$

where ϵ_d is the effective dielectric constant of the double layer and m^* is the concentration of compensating positive charge, the presence of which cannot be ruled out. Thus in principle Eqs. (246) through (248) can be solved to obtain J as a function of E_0. For sufficiently low field strengths the first term on the right-hand side of Eq. (247) will be the dominant one, giving

$$\left(\frac{m}{M_p - m}\right) X(H^+) = \frac{\nu_+}{\nu_-} \exp\left(-\frac{Q_+ - Q_-}{kT}\right) = \exp\left(-\frac{\Delta G_{pa}^0}{kT}\right), \quad (249)$$

where ΔG_{pa}^0 is the standard free-energy change for proton abstraction. Equation (249) is an equilibrium expression analogous to that for the dissociation of water. If the first term continues to dominate at high field strengths, m is given by

$$m = (M_p - m)\left[\frac{1}{X(H^+)}\right] \exp\left(-\frac{\Delta G_{pa}^0 - q_e dE_d}{kT}\right), \quad (250)$$

where $d = d_+ + d_-$. If, on the other hand, the second term of Eq. (247) dominates, one obtains

$$m = (M_p - m)\left(\frac{q_e \nu_+}{q\nu_0}\right) \frac{\exp\{-[(Q_+ - Q_0) - q_e d_+ E_d]/kT\}}{2 \sinh(qb_+ E_0/kT)}. \quad (251)$$

Thus combining Eqs. (249), (250), and (251) in turn with (246) and making the approximation $M_p \gg m$ leads to the following expressions for the ion-flux density:

$$J \approx 2M_p \nu_0 \left[\frac{1}{X(H^+)}\right]\left(\frac{qaE_0}{kT}\right) \exp\left(-\frac{\Delta H_s^0 + Q + \Delta G_{pa}^0}{kT}\right) \quad (252)$$

for low field strengths;

$$J = M_p \nu_0 \left[\frac{1}{X(H^+)}\right] \exp\left(-\frac{\Delta G_{pa}^0 + Q_0 - q_e dE_d - qb_+ E_0}{kT}\right) \quad (253)$$

for high field strengths and defect injection being rate controlling;

$$J \approx M_{p^+} \nu \left(\frac{q_e}{q}\right) \exp\left(-\frac{Q_+ - q_e d_+ E_d}{kT}\right) \qquad (254)$$

for high field strengths and proton abstraction being rate controlling. (In Eq. (252) the substitutions $Q_0 \rightarrow \Delta H_s^0 + Q$, $b_+ \rightarrow a$ have been made according to the discussion following Eq. (244).)

Thus for sufficiently low field strengths the flux density is controlled by the migration of ionic defects within the film, the boundary value for the defect concentration being given by its equilibrium value, whereas for sufficiently high field strengths the rate is controlled by one of the interfacial processes. The above statement does not depend for its validity on the particular mechanism chosen, but only on the assumption that field-assisted charge-transport processes occur across the double layer, the double layer-oxide interface, and within the oxide film. Even in the high-field limit ion transport within the film can influence the overall kinetic behavior of the system if the total ionic space charge within the film is significant, a matter that was first noted in Section IV. C. 2 and will be discussed further in Section VI. D.

It is clear that one could proceed for high-field conditions on the basis of either Eq. (253) or Eq. (254), the choice perhaps varying with the system. Vermilyea [42] has demonstrated that rapid exchange of protons takes place between the surface region of anodically formed Ta_2O_5 and the electrolyte. This observation is consistent with a mechanism in which essentially equilibrium conditions prevail with respect to the proton-exchange reaction across the double layer. Equation (253) is of course based on just such an equilibrium approximation, since Eq. (250) is simply the equilibrium condition in the presence of a field E_d in the Helmholtz double layer.

The term involving the boundary charge, $(m - m^*)q_e$, in Eq. (248) can be neglected provided that $|\epsilon_d E_d| \gg |4\pi(m - m^*)q_e|$. This will be satisfied if some mechanism exists for adjusting m^* so that $m^* \simeq m$. If, on the other hand, $m^* \simeq 0$, on substituting for m from Eq. (250), the inequality becomes

$$\Delta G^0_{pa} > q_e dE_d + kT \ln [4\pi q_e M_p / \epsilon_d E_d X(H^+)] .\qquad (255)$$

Neglecting the logarithmic term, the condition reduces to $\Delta G^0_{pa} > q_e dE_d$. Using the estimates for the double-layer parameters for the tantalum oxide system given in Section VIII. C. 3. b and setting $E \sim 6 \times 10^6$ V/cm, Eq. (255) requires that $\Delta G^0_{pa} \overset{\sim}{>} 0.8$ eV. Since the corresponding standard-free-energy change for proton abstraction from water is 1.03 eV at 25°C (i. e., $[H^+_{aq}][OH^-_{aq}] = [H_2O]^2 \exp[-1.03/kT(eV)])$, it is not unreasonable to assume that Eq. (255) is satisfied. If neither this condition nor the condition $m^* \simeq m$ is satisfied, complex steady-state kinetic behavior that is not in accord with that observed for the valve metals is predicted. Assuming, therefore, that this boundary charge may be neglected for steady-state conditions, Eq. (248) reduces to $E_d \simeq (\epsilon/\epsilon_d)E_0$, so that the flux equation, Eq. (253), becomes for high field strengths

$$J = M_p \nu_p \exp[-(Q_e - u^*_e E_0)/kT] ,\qquad (256)$$

where

$$\nu_p = \frac{\nu_0 [\exp(\Delta S^0_{pa}/k)]}{X(H^+)} ,$$

$$Q_e = \Delta H^0_{pa} + Q_0 ,\qquad (257)$$

$$u^*_e = (\epsilon/\epsilon_d)q_e d + qb_+ .$$

It is apparent that other equally plausible mechanisms for the interfacial reactions may be proposed. The particular mechanism presented here should therefore be considered only as an example that points the way to other approaches. Thus rather than postulating the formation of sites of negative charge at the oxide-double layer interface by proton abstraction, these sites then giving rise to defect formation in the oxide, one might instead postulate the formation of sites of positive charge by defect injection, these sites then giving rise to proton abstraction. Or one could assume defect injection to be completely unrelated to the surface hydroxyl groups. In all cases, however, an equation of the form of Eq. (256) is expected, though the interpretation of the terms will differ.

D. Composite Conduction Equations

1. High-Field Limit

In Sections VI. B and VI. C it is shown that in the high-field limit the transport rate near the appropriate interface is determined by the defect-injection process. Taking the example of defect injection at the electrolyte-oxide interface, the ion-flux density is assumed to be given by Eq. (256). However, within the film, at the boundary, the ion flux is given by Eq. (230), with n and E replaced by n_0 and E_0, respectively. Equating the two expressions for the flux gives, in the high-field limit,

$$n_0 = \left(\frac{M_p \nu_p}{2a\nu}\right) \exp\left[-\left(\frac{Q_e - Q}{kT}\right) + \left(\frac{u_e^* - u^*}{kT}\right) E_0\right], \qquad (258)$$

where qa has been replaced by u^*. Alternatively, eliminating E_0 from Eqs. (256) and (258) leads to the following expression for n_0:

$$n_0 = [(M_p \nu_p)^{\gamma_e}/2a\nu][\exp\{(Q - \gamma_e Q_e)/kT\}] J^{(1-\gamma_e)}, \qquad (259)$$

where $\gamma_e = u^*/u_e^*$ and is the ratio of the activation dipole in the film to the effective activation dipole for the interfacial reactions.

In general the ionic species will produce a space charge in the film, so that the mean field strength $\bar{E} = -V_X/X$ will not be equal to E_0. An equation relating $\bar{E}$ to n_0 was derived in Section IV.C.2 for precisely the case under consideration. Thus Eq. (144), on transforming the dimensionless field strengths and potentials according to Eqs. (3) and evaluating at $x = X$ (or $s = S$), becomes

$$\bar{E} = E_0 + (kT/qa)[(1 + 1/\alpha c_0 S) \ln (1 + \alpha c_0 S) - 1] , \qquad (260)$$

where the dimensionless quantity $\alpha c_0 S$ may be related to the dimensioned variables n_0 and X by using Eqs. (3) and (117) to obtain

$$\alpha c_0 S = (4\pi q^2 a/\epsilon kT)n_0 X . \qquad (261)$$

Expressions analogous to Eqs. (259) through (261) were first derived by Dewald [29] and applied to the kinetics of the anodic oxidation of tantalum.

Differentiating Eq. (260) with respect to $\ln j$ gives the following expression for the differential field coefficient, defined in terms of the mean field strength:

$$\frac{1}{\bar{\beta}} = \left(\frac{\partial \bar{E}}{\partial \ln J}\right)_{T,X} = \frac{kT}{u_e^*}\left[\frac{\ln (1 + \alpha c_0 S)}{\alpha c_0 S}\right] + \frac{kT}{u^*}\left[1 - \frac{\ln (1 + \alpha c_0 S)}{\alpha c_0 S}\right]. \quad (262)$$

In deriving Eq. (262) use was made of the result $(\partial c_0/\partial \ln j)_S = c_0(1 - \partial \epsilon_0/\partial \ln j)$, which in turn follows from $j = c_0 e^{\epsilon_0}$. Furthermore, $(\partial E_0/\partial \ln J)_{T,X}$ was evaluated from Eq. (256) as kT/u_e^* and qa was replaced by u^*. In precisely the same way an expression for the variation in $\bar{E}$ with total film thickness X can be evaluated to give

$$\left(\frac{\partial \bar{E}}{\partial \ln X}\right)_{T,J} = \frac{kT}{u^*}\left[1 - \frac{\ln (1 + \alpha c_0 S)}{\alpha c_0 S}\right] . \qquad (263)$$

A quantity that is in some respects more amenable to experimental evaluation than the mean field strength is the differential field strength, $\tilde{E} = -\partial V_X/\partial X$. Under galvanostatic conditions, and for a system in which the current efficiency for film formation is close to 100% and the film density is independent of thickness, $\tilde{E}$ is given by

$$\tilde{E} = (F/i\Omega')(\partial V_X/\partial t)_{i,T} , \tag{264}$$

where i is the ion-current density, F is Faraday's constant, and Ω' is the volume per electrochemical equivalent weight of oxide. For the present model the double-layer potential will be constant for J or i constant, so that $(\partial V_X/\partial t)_{i,T}$ will be equal to the time derivative of the electrode potential measured with respect to a reference electrode.

Proceeding as for the case of the mean field strength, Eq. (145) gives for the differential field strength

$$\tilde{E} = E_0 + (kT/qa) \ln(1 + \alpha c_0 S) , \tag{265}$$

from which expressions analogous to Eqs. (262) and (263) can be obtained:

$$\frac{1}{\tilde{\beta}} = \left(\frac{\partial \tilde{E}}{\partial \ln J}\right)_{T,X} = \frac{kT}{u_e^*}\left(\frac{1}{1+\alpha c_0 S}\right) + \frac{kT}{u^*}\left(\frac{\alpha c_0 S}{1+\alpha c_0 S}\right) , \tag{266}$$

$$\left(\frac{\partial \tilde{E}}{\partial \ln X}\right)_{T,J} = \frac{kT}{u_e^*}\left(\frac{\alpha c_0 S}{1+\alpha c_0 S}\right) . \tag{267}$$

The above treatment for the differential field strength may be readily extended to include the case in which a constant background-space-charge density $-qn_\infty$ is present in the film. Equations (265) through (267) then become (cf. Eq. (157))

$$\tilde{E} = E_0 + \left(\frac{kT}{qa}\right) \ln\left\{\exp(-\alpha c_\infty S) + \left(\frac{c_0}{c_\infty}\right)[1 - \exp(-\alpha c_\infty S)]\right\} , \tag{268}$$

$$\frac{1}{\tilde{\beta}} = \left(\frac{\partial \tilde{E}}{\partial \ln J}\right)_{T,X} = \frac{kT}{u_e^*}\left\{\frac{\exp(\alpha c_\infty S)}{1 + (c_0/c_\infty)[\exp(\alpha c_\infty S) - 1]}\right\}$$

$$+ \frac{kT}{u^*}\left\{\frac{(c_0/c_\infty)[\exp(\alpha c_\infty S) - 1]}{1 + (c_0/c_\infty)[\exp(\alpha c_\infty S) - 1]}\right\}, \tag{269}$$

$$\left(\frac{\partial \tilde{E}}{\partial \ln X}\right)_{T,J} = \frac{kT}{u^*}\left\{\frac{\alpha(c_0 - c_\infty)S}{1 + (c_0/c_\infty)[\exp(\alpha c_\infty S) - 1]}\right\}, \tag{270}$$

where $c_\infty = n_\infty/N$. Equations analogous to these were first derived by L. Young [30].

From the above it is clear that for sufficiently thin films or low space-charge density (i.e., for $\alpha|c_0 - c_\infty|S \ll 1$) $\bar{E} = \tilde{E} = E_0$ whether or not the background space charge is included, and furthermore

$$\bar{\beta} = \tilde{\beta} = u_e^*/kT, \tag{271}$$

$$(\partial \bar{E}/\partial \ln X)_{T,J} = 0 = (\partial \tilde{E}/\partial \ln X)_{T,J} \text{ for } \alpha|c_0 - c_\infty|S \ll 1; \tag{272}$$

that is, for $\alpha|c_0 - c_\infty|S \ll 1$, Eq. (256), with $\tilde{E}$ or $\bar{E}$ replacing E_0, becomes a valid approximation for the composite conduction equation. On the other hand, for $\alpha c_0 S \gg 1$ and $\alpha c_\infty S \ll 1$, from Eq. (260), $\tilde{E} = E_0 + (kT/qa)[\ln(\alpha c_0 S) - 1]$, or on substituting for c_0 according to $j = c_0 e^\varepsilon$,

$$\bar{E} = (kT/u^*)[\ln(\alpha jS) - 1] \tag{273}$$

for $\alpha c_0 S \gg 1$, $\alpha c_\infty S \ll 1$.

Similarly

$$\tilde{E} = (kT/u^*)\ln(\alpha jS) \tag{274}$$

for $\alpha c_0 S \gg 1$, $\alpha c_\infty S \ll 1$, and therefore

$$\bar{\beta} = \tilde{\beta} = u^*/kT \text{ for } \alpha c_0 S \gg 1, \alpha c_\infty S \ll 1. \tag{275}$$

Furthermore, from Eqs. (263) and (267)

$$(\partial \bar{E}/\partial \ln X)_{T,J} = (\partial \tilde{E}/\partial \ln X)_{T,J} = 1/\bar{\beta} = 1/\tilde{\beta} = u^*/kT \qquad (276)$$

for $\alpha c_0 S \gg 1$, $\alpha c_\infty S \ll 1$.

Thus in this limit the mean and differential field strengths vary in an identical fashion with $\ln J$ and $\ln X$, a prediction that should make system behavior in this limit easy to identify. Transforming to dimensioned variables, Eq. (273) can be written

$$JX = (\epsilon kTev/2\pi q^2) \exp[-Q - u^*\bar{E})/kT] \qquad (\alpha c_0 S \gg 1, \ \alpha c_\infty S \ll 1), \qquad (277)$$

where $e = \exp(1)$.

Essentially the same expression applies if the differential field strength is employed, the only difference being that e in Eq. (277) is replaced by unity.

Finally, for $\alpha c_\infty S \gg 1$, Eq. (268) leads to

$$\tilde{E} = (kT/qa) \ln(j/c_\infty) = \bar{E} \qquad (\alpha c_\infty S \gg 1), \qquad (278)$$

a result that can be obtained more directly from Eq. (153). In this thick-film limit, once again $\tilde{\beta}$ and $\bar{\beta}$ are both given by u^*/kT; the derivatives $(\partial \bar{E}/\partial \ln X)_{T,J}$ and $(\partial E/\partial \ln X)_{T,J}$ are, however, both zero. Transforming to dimensioned variables, Eq. (278) becomes simply

$$J = 2av n_\infty \exp[-(Q - u^*\bar{E})/kT] \qquad (\alpha c_\infty S \gg 1). \qquad (279)$$

In Section V. C it was shown that for $X = 10^4 \text{Å}$ the condition $\alpha c_\infty S \gg 1$ will be satisfied for $n_\infty \gtrsim 10^{18} \text{ cm}^{-3}$. For $X = 100 \text{Å}$ the condition becomes $n_\infty \gtrsim 10^{20} \text{ cm}^{-3}$. Thus for thin films an extremely high background-space-charge density is required for Eq. (279) to be a valid approximation. An essentially identical condition on n_0 rather than n_∞ is required for Eq. (277) to be a valid approximation.

On the other hand, for Eq. (271) to be valid, so that the transport rate is controlled entirely at the interface $x = 0$, the condition $\alpha |c_0 - c_\infty| S \ll 1$ must be satisfied. Setting $\epsilon \approx 10$, $T = 273°$ K, $q = 1e$, $a = 2\mathring{A}$, and $X = 10^4 \mathring{A}$, this becomes $|n_0 - n_\infty| \ll 0.6 \times 10^{17}$ cm^{-3}. Alternatively, on assuming $n_0 \gg n_\infty$ and substituting for n_0 from Eq. (232), the condition reduces in the high-field case to

$$qaE_0/kT > Q/kT + \ln(i/A_i') , \tag{280}$$

where $A_i' = (\nu_\epsilon kT/2\pi qX)$. Using the above values and setting $\nu \approx 10^{12}$ sec^{-1}, $i \approx 1$ mA/cm^3, Eq. (280) becomes

$$qaE_0/kT \overset{\sim}{>} (Q/kT) - 13 . \tag{281}$$

Thus for $Q \approx 0.6$ eV and $E_0 \approx 6 \times 10^6$ V/cm (typical of valve-metal anodic oxidation) qa must be on the order of 5 e$\mathring{A}$, a not unreasonable value.

A further possibility exists, however, whereby the condition $\alpha c_0 S \ll 1$ can be met even if Q is too large for Eq. (280) to be satisfied. On surmounting the first barrier at the interface, a mobile defect has sufficient potential energy to move directly, in a time on the order of $\frac{1}{2}\nu$, to the top of the next diffusion barrier. In so doing it would gain energy from the applied electric field equal to $2qaE_0$. Thus if the defect were to transfer its excess energy to the surrounding lattice at a rate that did not exceed $4qaE_0$ per vibration, it would arrive at the top of the second diffusion barrier with positive momentum and so proceed in a like manner through the entire film. That is to say, the defect, once put in motion, would retain its momentum and move through the film with an effective activation energy of zero:

$$J \overset{\sim}{>} 4a\nu n_0 . \tag{282}$$

This sort of behavior is only possible at sufficiently high field strengths, so that $4qaE_0$ exceeds the energy loss to the lattice per vibration. If this condition were achieved, however, it is apparent that ac_0S would be $<<< 1$ for any realizable film thickness, so that ion transport would be controlled entirely at the interface at which the defects were formed. Although there is no direct experimental or theoretical support for such a "momentum-retention" ion-conduction mechanism, a similar phenomenon has been observed experimentally in connection with sputtering processes induced by positive-ion bombardment [43]. Ion bombardment of thin foils can lead to sputtering from the far surface, the process evidently requiring a focused transfer of momentum without serious loss of energy to the lattice.

The equations derived in this section have been for the case of defect generation at the electrolyte-film interface. However, the equations may readily be transformed to apply to the metal-film interface. Thus on replacing Eq. (256) with Eq. (235) in the derivations, all of the equations of this section remain valid simply on substituting qb_+ for u_e^*, Q_+ for Q_e, and $M\nu_+$ for $M_p\nu_p$.

2. Relationship Between Overpotential and Capacitance in the High-Field Limit

Before leaving the case of ionic conduction in the high-field limit, where both ionic space charge within the film and surface charge at the oxide-double layer interface may be neglected, we investigate the relationship between the electrode overpotential η for anodic oxide-film formation and the electrode capacitance c predicted by this model.

The overall electrode reaction may be broken into the following four steps:

$$\left(\begin{array}{c}\diagdown\\\diagup\end{array}M\text{—OH}\right)_{\text{surface}} \quad\rightarrow\quad \left(\begin{array}{c}\diagdown\\\diagup\end{array}M\text{—O}^{-}\right)_{\text{surface}} + H^{+}_{aq}$$

$$\left(\begin{array}{c}\diagdown\\\diagup\end{array}M\text{—O}^{-}\right)_{\text{surface}} \quad\rightarrow\quad \left(\begin{array}{c}\diagdown\\\diagup\end{array}M\text{—O}^{-}\right)_{x \simeq 0}$$

$$\left(\begin{array}{c}\diagdown\\\diagup\end{array}M\text{—O}^{-}\right)_{x \simeq 0} \quad\rightarrow\quad \left(\begin{array}{c}\diagdown\\\diagup\end{array}M\text{—O}^{-}\right)_{x \simeq X}$$

$$\left(\begin{array}{c}\diagdown\\\diagup\end{array}M\text{—O}^{-}\right)_{x \simeq X} + (M)_{\text{metal}} \quad\rightarrow\quad \left(\begin{array}{c}\diagdown\\\diagup\end{array}M\text{—O—M}\begin{array}{c}\diagup\\\diagdown\end{array}\right) + e_{\text{metal}}$$

where it has been assumed that the mobile defects are singly charged, singly bonded oxygen atoms in the vitreous oxide, which move via a partner-exchange mechanism (for further details see Section VII). Much the same final expression is obtained, however, if the mobile defects are assumed to take a different form. By postulate, the first step occurs under equilibrium conditions and hence makes no contribution to the free energy change for the reaction. Breaking the total free-energy change for the final three steps into a chemical and an electrical portion, in the usual way, leads to the following expression for the free-energy change ΔG:

$$\Delta G = kT\left[\ln\left(\frac{n_0}{n_0^0}\right) - \ln\left(\frac{m}{m^0}\right) + \ln\left(\frac{n_X}{n_0}\right) - \ln\left(\frac{n_X}{n_X^0}\right)\right] + q_e(V_X - V_0) ,$$

$$(283)$$

where n_0 and n_X are the defect concentrations at $x = 0$ and $x = X$, respectively (i.e., at the electrolyte and metal boundaries, respectively) and the zero superscript refers to the value for the quantity concerned under equilibrium conditions ($\Delta G = 0 = \eta$). The first two logarithm terms arise from the second step, the final two from the third and fourth steps, respectively. The electrical term represents the change in potential energy on transporting a charge q_e through the entire film and comes mainly from the third step, with small contributions coming from the second and fourth steps. Substitution for m and m^0 according to Eq. (251) and setting $M_p - m \approx M_p$ gives for the overpotential $\eta = \Delta G / q_e$ the following relationship:

$$\eta = (V_X - V_0) - d(E_d - E_d^0) + (kT/q_e) \ln (n_X^0/n_0^0) , \qquad (284)$$

where $-d(E_d - E_d^0)$ is simply the displacement of the double-layer potential from its value at the equilibrium electrode potential. At the equilibrium potential, $V_X = V_X^0$ and $V_0 = V_0^0$, so that

$$(kT/q_e) \ln (n_X^0/n_0^0) = -(V_X^0 - V_0^0) \qquad (285)$$

and hence

$$\eta = (V_X - V_{aq}) - (V_X - V_{aq})^0 , \qquad (286)$$

where the substitution $E_d = -(V_0 - V_{aq})/d$ has been made. (The diffuse double-layer potential is neglected in this development, as it was in the derivation of Eq. (250).)

Taking the origin at the double layer-oxide interface, then, since it has been assumed that negligible space charge exists within the region bounded by the surface charge on the metal and that on the electrolyte, it follows that $(\partial D/\partial x) = 0$ in this region so that $D = 4\pi\sigma_{aq}$ in $(-d, X)$, where σ_{aq} is the charge per unit area on the electrolyte phase. Integration of this equation gives, with Eq. (286),

$$\eta = -4\pi\sigma_{aq} \int_{-d}^{X} \frac{dx}{\epsilon} - \Delta V^0 , \qquad (287)$$

where $\Delta V^0 = (V_X - V_{aq})^0$.

The electrode capacitance per unit area in the limit as the measuring frequency goes to zero, $\underline{c}$, may be obtained from the relation

$$1/\underline{c} = -\partial\eta/\partial\sigma_{aq} , \qquad (288)$$

which with Eq. (287) gives

$$\frac{1}{\underline{c}} = 4\Pi \int_{-d}^{X} \frac{dx}{\epsilon} \ . \tag{289}$$

If the capacitance is measured at a frequency that is high enough for the slow polarization processes in the film to make no contribution to the polarization changes, but low enough for equilibrium conditions to be maintained across the oxide-electrolyte double layer, the resulting capacitance per unit area, $\underline{c}_t$, will be given by Eq. (289), with ϵ replaced by the high-frequency dielectric constant ϵ^t.

Elimination of $\int_{-d}^{X} dx/\epsilon$ from Eqs. (287) and (289) gives

$$\eta = -(\sigma_{aq}/\underline{c}) - \Delta V^0 \ . \tag{290}$$

A similar expression may be obtained in terms of $\underline{c}_t$ only if ϵ^t/ϵ is independent of position in the film, for which case

$$\eta = -(\epsilon^t/\epsilon)(\sigma_{aq}/\underline{c}_t) - \Delta V^0 \ . \tag{291}$$

If the electrode kinetics are assumed to be in accord with Eqs. (256) and (257), for a fixed temperature the ion-current density will be a function of E_0 only. However, $\sigma_{aq} = \epsilon_d E_d/4\pi = \epsilon_0 E_0/4\pi$, where ϵ_0 is the value of ϵ at $x = 0$, so that for a given ion-current density, σ_{aq} is a constant. Thus

$$\left[\frac{\partial \eta}{\partial(1/\underline{c}_t)}\right]_{i,T} = -\frac{\epsilon_d E_d}{4\pi} = \frac{\epsilon_0^t}{4\pi u_e^*}\left[kT \ln\left(\frac{i}{M_p^\nu p^q}\right) + Q_e\right]$$

$$= \frac{\epsilon_d/4\pi}{q_e d + b_+ q(\epsilon_d^t \epsilon_0^t)}\left[kT \ln\left(\frac{i}{M_p^\nu p^q}\right) + Q_e\right], \tag{292}$$

with the corresponding expression for $[\partial \eta/\partial(1/\underline{c})]_{i,T}$ obtained simply by replacing ϵ^t by ϵ. As this will be true for all the equations involving $\underline{c}$ and $\underline{c}_t$, only those for $\underline{c}_t$ will be derived from here on.

Note that no assumption was made with regard to the constancy of the dielectric constants, either with respect to position in the film or the total film thickness. Thus the dielectric constant, oxide composition, and density of the film may vary throughout the film, their mean values varying with the total film thickness, and η will still be a linear function of $1/\underline{c}$ for i and T constant, provided only that the double-layer properties remain essentially constant with increasing film thickness. On the other hand, neither the differential field strength $(\partial\eta/\partial X)_{i,T}$ nor the reciprocal capacitance will in general be linear functions of the film thickness if such variations in film properties occur. The same statements will apply to η versus $1/\underline{c}_t$ only if (ϵ/ϵ^t) is independent of position and total film thickness.

There exists the possibility that, adjacent to either the metal or the electrolyte interface, a thin film of constant thickness X' may reside — a thin film that under conditions of anodic polarization of the electrode will form a depletion layer of space-charge density ρ'. If the depletion layer were to reside at the metal-oxide interface, ρ' would be expected to be positive, and negative if it were to reside at the oxide-electrolyte interface.

Repeating the above analysis to include such a depletion layer at the metal-oxide interface, using Eq. (226), gives in place of Eq. (287)

$$\eta = -4\pi \left(\sigma_{aq} \int_{-d}^{X} \frac{dx}{\epsilon} + \frac{\rho'X'^2}{2\epsilon'} \right) - \Delta V^0 , \qquad (293)$$

whereas for a depletion layer at the oxide-electrolyte interface

$$\eta = -4\pi \left[(\sigma_{aq} + \rho'X') \int_{-d}^{X} \frac{dx}{\epsilon} - \frac{\rho'X'^2}{2\epsilon'} \right] - \Delta V^0 , \qquad (294)$$

where ϵ' is the value of ϵ within the depletion layer. By postulate, ρ' and X' are independent of field strength and hence of σ_{aq} in the range of interest. Thus both Eqs. (293) and (294) lead to the same expression for the capacitance as before, that is, to Eq. (289). The expressions for the overpotential are, however, different. Thus for the depletion layer at the metal-oxide interface the overpotential η'' is given by

$$\eta'' = -(\epsilon^t/\epsilon)(\sigma_{aq}/\underline{c}_t) - (2\pi\rho'X'^2/\epsilon') - \Delta V^0 . \tag{295}$$

The slope $[\partial\eta''/\partial(1/\underline{c}_t)]_{i,T}$ is therefore unchanged, being given by Eq. (292). The intercept, however, is decreased by $2\pi\rho'X'^2/\epsilon'$. Setting $\rho' \approx 5\times 10^{19}$ e/cm^3, $X' \approx 20$ Å, and $\epsilon' \approx 10$, $2\pi\rho'X'^2/\epsilon' \approx 0.2$ V.

If the depletion layer resides at the oxide-electrolyte interface, the situation is somewhat more complex. Thus the expression for the overpotential η' becomes

$$\eta' = -(\epsilon^t/\epsilon)(\sigma_{aq} + \rho'X')/\underline{c}_t + (2\pi\rho'X'^2/\epsilon') - \Delta V^0 . \tag{296}$$

Since for this case ρ' is expected to be negative and of course σ_{aq} is negative, the slope $[\partial\eta'/\partial(1/\underline{c}_t)]_{i,T}$ is increased over that given by Eq. (292) by a factor $1 + \rho'X'/\sigma_{aq}$, and once again the intercept is reduced. Setting $\rho' \approx -5\times 10^{19}$ e/cm^2, $X' \approx 20$Å, $\epsilon' \approx 10$, $\epsilon \approx 30$, $d \approx 2$Å, and $|E| \approx -4\pi\sigma_{aq}/\epsilon \approx 10^7$ V/cm, the result $(\rho'X'/\sigma_{aq}) \approx 0.1$ is obtained, so that for these values of the parameters the slope $[\partial\eta/\partial(1/\underline{c}_t)]_{i,T}$ would be increased by about 10% over that given by Eq. (292), and the intercept would be decreased by 0.2 V as before. Thus in both cases the presence of a depletion layer causes the intercept to shift to a lower value, whereas the slope is affected only for the case in which the depletion layer resides at the oxide-electrolyte interface. If the space-charge layers are not fully depleted, the resulting space-charge redistribution with potential will make a contribution to the capacitance, and the entire analysis becomes more complex.

If the rate of ion transport is controlled at the metal-oxide inter-face, an identical analytical procedure gives the following results:

$$\eta = (\epsilon^t/\epsilon)(\sigma_M/\underline{c}_t) - \Delta V^0 ,$$
(297)

$$\left[\frac{\partial \eta}{\partial(1/\underline{c}_t)}\right]_{i,T} = \frac{\epsilon_0^t}{4\pi q b_+} \left[kT \ln\left(\frac{1}{Mv_+ q}\right) + Q_+\right] ,$$
(298)

where σ_M is the charge per unit area on the metal and E_0 and ϵ_0 are the field in, and the dielectric constant of, the oxide at the metal-oxide interface. The quantity E_0 has been expressed in terms of the current density by using Eq. (235). For a depletion layer at the oxide-electrolyte interface, the slope is unchanged, whereas the intercept is decreased by $(-2\pi\rho'X'^2/\epsilon')$. For a depletion layer at the metal-oxide interface, on the other hand, both the slope and intercept are changed, the slope being increased by the factor $(1 +X'\rho'/\sigma_M)$, with the intercept again decreased by $2\pi\rho'X'^2/\epsilon'$. Thus, as for the case of control at the oxide-electrolyte interface, the effect of a depletion layer is to shift the intercept to a lower value. The situation is reversed, however, with regard to the effect of a depletion layer on the slope $[\partial\eta/\partial(1/\underline{c}_t)]_{i,T}$, in that for control at the metal-oxide interface the slope is increased only when the depletion layer resides at the metal-oxide interface.

3. Control at the Far Interface

In Section VI.D.1 it was explicitly assumed that the transport rate is not influenced by transport processes at the far interface (i.e., at $x = X$). This will always be true for cases in which diffusion can be neglected in comparison with migration. It is entirely possible, however, for defect injection to occur at the metal-oxide interface, with the highest transport barrier being associated with the oxide-solution double layer. To treat this problem generally would require

combining the general transport equation, Eq. (10) or (11), with
Poisson's equation and equations for transport processes at both
interfaces. As this problem cannot be solved analytically, a somewhat more restricted case, which can be solved analytically, will
be examined.

If the final transport barrier at $x = X$ is sufficiently high relative
to both the diffusion barrier and the entrance barrier at $x = 0$, the
defect distribution within the film will be very close to the equilibrium distribution. Thus in dimensionless variables

$$c = c_0 e^{-p} , \tag{299}$$

which with Poisson's equation becomes

$$\partial^2 p/\partial s^2 = -\alpha c_0 e^{-p} . \tag{300}$$

This equation integrates to give

$$\varepsilon = -\partial p/\partial s = 2(\alpha c_0/2)^{\frac{1}{2}}(e^{-p} + B^2)^{\frac{1}{2}} , \tag{301}$$

where B may be evaluated from the condition $\varepsilon = \varepsilon_S$ at $s = S$. Thus

$$B = [\varepsilon_S^2 - 2\alpha c_0 \exp(-p_S)]^{\frac{1}{2}} . \tag{302}$$

A second integration from $s = 0$ to $s = S$ then yields

$$\exp\left(\frac{p_S}{2}\right) = \exp\left[-B\left(\frac{\alpha c_0}{2}\right)^{\frac{1}{2}} S\right] - \frac{\sinh\left[B(\alpha c_0/2)^{\frac{1}{2}} S\right]}{B^2 + B(B^2 + 1)^{\frac{1}{2}}} . \tag{303}$$

If the interfacial rate process obeys an equation of the form of
Eq. (256), one may write

$$j = \exp(\varepsilon_S) \tag{304}$$

provided that the dimensionless flux density j and the dimensionless
position variable s are redefined according to

$$j = \frac{J}{M_p \nu_p} \exp\left(-\frac{Q_e}{kT}\right) , \tag{305}$$

$$s = \frac{x}{u_e^*/q} . \tag{306}$$

If the interfacial process involves the transport of mobile defects across the double layer, a transport equation of the form of Eq. (230) will apply at $s = S$, so that for high fields

$$j = c_S \exp(\varepsilon_S) = c_0 [\exp(-p_S)] \exp(\varepsilon_S) . \tag{307}$$

Equations (302), (303), and either (304) or (307) can be solved by numerical methods to yield j as a function of c_n and S. The boundary concentration $c_0 = n_0/N$ will be given by an equilibrium expression similar to Eq. (239).

The form of the resulting kinetics has not been investigated. Behavior in accord with this limiting treatment could only occur for thin films of very high ionic conductivity.

4. Low Fields

For sufficiently low field strengths the defect concentrations at the boundaries will be under equilibrium control, as noted in Sections VI. B and VI. C. If the moving defects alone give rise to the space charge in the film, the overall transport rate can be determined from the equations of Section IV. C. 1, with the boundary concentrations being evaluated from equilibrium considerations. The kinetics predicted on the basis of such a model have not been worked out in detail.

If a background space charge of opposite sign to that of the mobile defects exists within the film, and the film is thick compared with the characteristic space-charge length in the oxide (see Section

V. C), ionic conduction will be in accord with Ohm's law and independent of the boundary concentrations. If the film thickness is on
the order of, or less than, the characteristic space-charge length,
no simple form for the integrated transport equation results, and once
again the kinetics predicted on the basis of such a model have not
been worked out.

Finally, at intermediate field strengths the greatest degree of
complexity appears, since the linear transport equation cannot be
used and neither can diffusion be neglected in comparison with conduction. Furthermore, interfacial kinetics must in general be included
in the treatment. For such cases Poisson's equation can be solved
only by numerical methods.

VII. NATURE OF MOBILE DEFECTS IN
VITREOUS METAL OXIDES

A. Introduction

The nature of point defects responsible for ionic conduction in
crystalline solids is fairly well understood and has been fully treated
by a number of authors [1]. The situation with regard to vitreous
solids is, however, not nearly so well understood. Since the results
of structural studies on a number of anodically formed oxide films are
consistent with the interpretation that the films have a vitreous structure, a discussion of this subject seems in order.

In the book Modern Aspects of the Vitreous State edited by
MacKenzie [44] there appears the following statement in the editor's
preface:

"The structure of glassy materials has long been a topic of much
controversy. With the exception of perhaps organic polymers, com-
paratively little progress was made in the last three decades. This

is perhaps in part due to the unique character of materials in the vitreous state, viz., it possesses many of the mechanical properties of
crystalline solids and yet structurally resembles a liquid in its lack
of long range periodicity. "

In the light of this statement by an expert on the vitreous state,
it is indeed presumptuous to attempt to provide a detailed mechanism
of ionic transport in the glassy state. Nevertheless, such an attempt
will be made with the full realization that in detail any simple model
is bound to have serious shortcomings but hopefully will be correct
in broad outline.

Summarizing briefly the salient features of ionic conduction in
bulk vitreous oxides, most of the published data are concerned with
the effect of composition and past history on the conductivity of
mixed oxide glasses at fields low enough for Ohm's law to apply.
The temperature dependence is generally found to follow the normal
exponential form characteristic of a thermally activated process. For
$SiO_2 - B_2O_3$ glasses containing lithium or sodium the activation energies fall in the range 0.5 to 0.7 eV and vary in a complex manner with
composition [45]. The current carriers are believed to be the alkali-
metal ions.

For a single-component glass, such as B_2O_3, much higher activation energies are found, in this case about 2.2 eV [46]. Furthermore, the preexponential factor for ionic conduction in B_2O_3 glass is
about 10^8 ohm^{-1} cm^{-1}, which is of the same order as the corresponding quantity for the alkali halides and very much larger than that for
glasses containing sodium and lithium (about 10^2 ohm^{-1} cm^{-1}). These
numbers illustrate the substantial differences that exist between the
ionic conduction properties of a glass containing only network-forming
metal atoms and one containing in addition network-modifying metal
atoms.

It is generally accepted that the network modifiers exist as ions in the interstices of the three-dimensional network provided by the network-forming metal atoms and migrate through the network under the influence of the applied field. Such a mechanism, of course, cannot be applied to a glass like vitreous B_2O_3 or for that matter to any single-component glass. Otto [46] considered the question of conduction in B_2O_3 glass and concluded that the activation energy is too low to be accounted for by conduction via oxygen ions. Furthermore, the large preexponential factor observed for this system precludes an impurity mechanism and suggests rather a mechanism in which thermal dissociation of the vitreous network takes place to form defect pairs. If this were the case, the empirical activation energy would be composed of a term arising from the enthalpy of formation of the defects and a term equal to the activation energy for migration of the defects, in a manner analogous to that for ionic conduction in the alkali halides via a Schottky-defect mechanism (see Eq. (191)). The problem in accounting for ionic conduction in vitreous B_2O_3 in this manner rests in discovering defects whose enthalpy of formation will be sufficiently low to be in accord with the empirical activation-energy data. In the following section a number of possible defects are considered.

B. Nature of Network Defects

Since a large enthalpy of formation would be expected to accompany the creation of multicharged defects, as a result of the enthalpy contributions of the ionization potentials and electron affinities, the possibility of formation of singly charged network defects will be investigated first. A description of such defects was first proposed by the author in 1966 [47-49].

Within the framework of the random-network model of the vitreous state as proposed by Zachariasen [50] a defect can be imagined to form by the rupture of one of the metal-oxygen bonds, followed by separation of these "dangling bonds" through a succession of processes each of which may be described as an exchange of partners. In order to contribute to ionic conduction these dangling bonds must acquire or lose an electron. The ionized form of the defects should be strongly favored since vitreous oxides have fairly high dielectric constants. Such defects may be described alternatively as regions of local nonstoichiometry bearing a single charge, positive in the case of cation-network defects and negative in the case of anion-network defects, in order to satisfy valency requirements.

Thus one may imagine the formation of such defects in the following manner: Taking the example of an odd-valent metal oxide, B_2O_3, a pair of defects may be formed by transferring the molecular ionic group BO^+ from one region of the glass to another and then allowing the vitreous structure to relax in these two regions. The net result would be a stoichiometric excess of one-half an oxygen ion in the former region and one-third of a metal ion in the latter.

In this description there is a temptation to regard the defects as consisting of BO^+ and BO_2^-, which would be misleading in relation to many of the properties that these defects would be expected to display, and particularly so with respect to the mechanism of their migration. The value in this description lies in showing that regions that depart from stoichiometry by one ion equivalent can be separated by an indefinite distance from one another and still have the intervening region free of additional strain.

In the case of an even-valent metal oxide, such as SiO_2, this same description would lead to the conclusion that local departures from stoichiometry cannot be less than two ion equivalents, since a

singly charged molecular ionic group cannot be formed from silicon
and oxygen. Thus for even-valent metal oxides one can do no better
in minimizing the charge on the defects than to suppose their formation to be equivalent to the displacement of O^{2-} from one region of
the vitreous oxide to another. (Singly charged radical ions, such as
SiO^+ and O^- could conceivably be formed but would likely be less
favorable energetically than SiO^{2+} and O^{2-}.) The defect pair so generated might therefore be described simply as an oxygen-ion vacancy
and interstitial oxygen ion were it not for the fact that such terms
cannot be defined unambiguously for a vitreous solid.

In an attempt to gain further insight into the expected behavior
of the proposed singly charged network defects, the two-dimensional
model of vitreous B_2O_3, due to Zachariasen [50], is presented in
Fig. 3. The solid circles represent the locations of the metal atoms
and the open circles those of the oxygen atoms. Breaking a bond
between metal atom 1 and oxygen atom 1 in such a manner that the
metal atom loses an electron to the oxygen and then exchanging partners between metal atoms 1 and 2 transforms the structure from that
depicted in Fig. 3a to that shown in Fig. 3b. In this manner a pair
of singly charged defects are generated without disturbing the vitreous structure unduly except in the immediate vicinity of the defects.
(In producing these drawings the atoms positioned at the periphery
of the drawings have been maintained fixed, the remaining atoms being moved so as to maintain the bond lengths and bond angles more
or less constant.) Further partner exchanges can lead to the migration of the anionic and/or cationic defects as indicated in Fig. 3c,
where the anionic defect is shown displaced by one additional partner
exchange.

In terms of the more general description of the defects — that is,
the one that is not based on the random-network model for the vitreous state — it is difficult to ascertain what the transport behavior of

the defects would be. It is expedient, therefore, to depend heavily on the description based on the random-network model in attempting to obtain such insight. Before proceeding along these lines, however, the list of possible ionic point defects in vitreous oxides will be completed with the inclusion of defects that might loosely be described as Frenkel and Schottky defects, respectively.

The pseudo-Frenkel defects could presumably be generated by the displacement of a metal ion from one position in the vitreous network to another, so that the anionic and cationic defects will bear a charge of magnitude equal to the oxidation state of the metal. The Schottky-type defects, on the other hand, would be made up of regions of local nonstoichiometry amounting to a deficiency of one oxygen ion for the cationic defects and one metal ion for the anionic defects, the relative proportions being such as to maintain overall electrical neutrality and stoichiometry. Note that the singly charged defects of the type described for B_2O_3 can exist only for vitreous oxides, as their presence, though preserving short-range order, precludes long-range order.

It is clear both from an examination of Fig. 3 and from a more general consideration of the problem that the motion of either cationic or anionic defects involves a cooperative motion of both metal and oxygen species in the vicinity of the defect. In a sense the same can be said of point-defect migration in crystalline solids occurring via accepted mechanisms since the motion of the point defect causes a disturbance in the lattice in the vicinity of the defect. Once the defect has passed through a given region of the crystal, however, that region returns precisely to its original structure. This will not be the case for the network defects, a fact clearly illustrated in Fig. 3. The motion of these defects may thus be described as a propagating cooperative disturbance that on moving through a vitreous slab has the effect of transporting one ion equivalent of material from

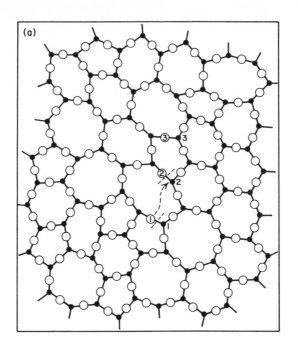

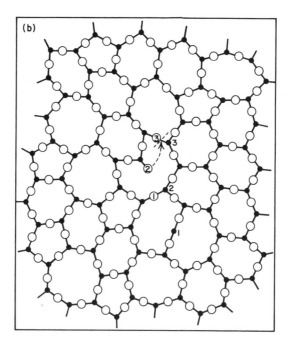

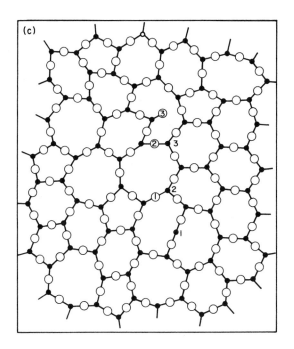

Fig. 3. Random-network model for vitreous B_2O_3 showing the steps in the formation (a, b) and the migration of network defects (b, c). Metal atoms are represented by solid circles and oxygen atoms by open circles. In producing these figures the atoms at the periphery have been maintained fixed, the remaining atoms being moved so as to maintain the bond lengths and bond angles approximately constant.

one interface to another. Furthermore, as the disturbance moves
through a given region of the vitreous network, the atoms are displaced irreversibly; no single atom, however, is displaced as far as
one interatomic distance.

An additional property that these defects should exhibit is also
illustrated in Fig. 3, but it can also be inferred from the more general
description of the defects. Defect motion preserves the order amongst
both the metal and oxygen atoms along the path of motion. Thus ion
transport by this mechanism maintains the order in both the anion and
cation subnetworks, except for a small amount of reordering rising
from the random nature of the transport process. Repeated motion of
network defects through a vitreous oxide has the effect of allowing
the anion and cation subnetworks to move through one another, so to
speak.

The question now arises as to the value of the transport numbers
expected for such a mechanism. Since there is no stationary matrix
through which the defects move (i. e., defect motion amounts to a
propagating disturbance in the matrix itself), one cannot define a
unique stationary point or plane within the vitreous solid against
which to measure the relative motion of the metal and oxygen atoms,
and hence the transport numbers. If the conduction mechanism is
correct, therefore, there can be no such thing as a unique marker that
can be inserted into the vitreous film to define a position against
which to measure unique transport numbers. Thus different atomic
markers are expected in general to lead to different values for the
apparent transport numbers.

Restricting attention to anodically formed vitreous metal-oxide
films, it is clear that, if such films grow by the motion of network
defects having the properties described here, then all of the metal
newly incorporated into the film will appear in the oxide next to the

metal-oxide interface, and all the oxygen newly incorporated will appear next to the oxide-electrolyte interface, regardless of whether it is the anionic, cationic, or both network defects that are mobile.

If an uncharged atomic marker is introduced into the oxide film, and further oxidation is carried out, as the structure of the oxide changes in the vicinity of the atomic marker, due to the passage of defects, the marker atom will take part in this structural change in a manner that cannot be predicted without a much more detailed description of defect migration than that presented here.   Thus any apparent transport number between 0 and 1 could conceivably arise from the model. If, however, the marker used were to introduce into the vitreous oxide a rigid (e.g., crystalline) structure, large compared with the atomic spacing in the vitreous oxide, the network defects would have to bypass such regions. If short-range order among the metal and oxygen atoms were maintained across the boundaries between such regions and the vitreous oxide matrix, then during film growth the metal and oxygen subnetworks would move relative to these regions at the same rate (on a chemical equivalence basis), but, of course, in opposite directions. Equal amounts of oxide film would therefore be formed on the metal and electrolyte side of such a region, leading to an apparent transport number of 0.5. A transport number of 0.5 cannot, however, be regarded as in any way characteristic of the transport mechanism proposed. As already noted, there is no characteristic transport number for the model.

The data that bear most directly on the matter of transport numbers are those obtained in the course of anodic oxidation studies on the valve metals since in most cases the films formed appear to have vitreous structures.

C. Comparison of Predicted and Observed
Transport Behavior

In most of the work aimed at determining the relative ionic mobilities during the formation of anodic oxide films, a technique involving
the formation of "duplex films" has been used. Thus Randall, Bernard,
and Wilkinson [51] observed that when tantalum was anodically oxidized in various concentrations of phosphoric acid, phosphorus became incorporated into the film at a concentration determined by the
electrolyte concentration, and, furthermore, film dissolution in hydrofluoric acid took place at a rate characteristic of the concentration of
incorporated phosphorus. On examining duplex films, making use of
the above, they observed that there were three distinct oxide layers:
a layer of (presumably) stoichiometric oxide next to the metal and two
outer layers containing different concentrations of phosphorus or phosphate. The concentration of phosphate in the outermost layer was
characteristic of the most recent anodizing process, regardless of
the order in which the solutions were used. At the same time the
thickness of the phosphate-free innermost layer increased during the
second anodization.

The data of Randall et al. are consistent with the interpretation
that the newly incorporated anions reside at the oxide-electrolyte
interface, whereas the newly incorporated metal ions reside at the
metal-oxide interface, with film growth occurring by a mechanism
that preserves the order in both the metal and oxygen subnetworks.
Thus the data are entirely consistent with ion transport via singly
charged network defects. As such a mechanism for ionic conduction
had not been proposed at the time of their publication, Randall et al.
concluded that both oxygen and metal must migrate through the film
by an interstitial-type mechanism, the transport numbers being fortuitously equal (i. e. , $t_+ \simeq t_- \simeq 0.5$).

Very similar results have been obtained by Dell'Oca and L. Young [52], who used ellipsometry to determine the thickness of the phosphate free inner film and phosphate-containing outer film for films formed on tantalum in dilute phosphoric acid. Assuming that the thickness of oxide formed per coulomb is the same for the two types of oxide, they calculated transport numbers from the ratio of the thicknesses of the two portions of the film. Thus in 0.23 N phosphoric acid at 25°C they obtained for the apparent transport number of the metal 0.51 to 0.52 at 1 mA/cm^2 and 0.56 at 10 mA/cm^2.

Amsel and Samuel [53] formed duplex films on aluminum in an ammonium citrate solution containing first oxygen-18 and then oxygen-16. The distribution of the oxygen isotopes was conserved to within 60Å, indicating that the order in the oxygen subnetwork is largely preserved during film growth. Pringle [54] has shown in a similar fashion that the oxygen order is largely preserved during anodic film growth on tantalum, with a small amount of scrambling occurring at the boundary between the oxygen-16 and oxygen-18 films, such scrambling being bound to take place if conduction occurs via the mechanism proposed by the author.

The only reported results that bear directly on the question of whether or not the mechanism of film growth preserves the order in the metal subnetwork are those due to Verkerk, Winkel, and De Groot [55], who introduced radioactive tantalum into two tantalum specimens, one of which was subsequently anodically oxidized. The outer layers of both specimens were dissolved away in successive increments with hydrofluoric acid, and thus the activity profile for the two specimens was determined. The activity profiles, though different, were not markedly so, indicating that the order of the metal atoms was largely preserved. Film growth by interstitial metal-ion migration would have produced inversion of the activity profile.

The experimental evidence suggests, therefore, that ionic conduction during anodic film growth on tantalum occurs predominantly,
if not entirely, by a mechanism that largely preserves the order in
both the metal and oxygen subnetworks. Such a transport mechanism
will lead to an apparent transport number, defined in terms of the relative thickness of the components of duplex films, of close to 0.5, as
observed, provided that the markers introduce rigid (crystalline)
regions into the oxide, as already discussed. Deviations from this
value of 0.5 can occur if the thickness per coulomb differs for the two
or more layers of oxide formed or if the incorporated anions do not
behave as postulated but instead act simply as charged markers of
atomic dimensions.

Most of the credit for focusing attention on the unusual transport
properties of anodically formed films is due to Davies, Pringle, and
co-workers, who in a series of papers [54, 56-59] describe experiments in which radioactive inert gases were used as markers. Metal
foils were bombarded with xenon-125 atoms to a known depth, and
the metal was subsequently anodized. Essentially monoenergetic
conversion electrons are ejected from the K shell of the xenon-125
atoms, the energy being partially absorbed in traversing the oxide.
Thus the energy loss can be used to determine the depth of the markers in the oxide. The authors used a calibration method to relate
the shape of the curve for counting rate versus electron energy to the
depth of the xenon-125 atoms.

In a detailed study on several metals [59] xenon-125 was deposited in a thin oxide layer on the surface of a metal foil, which was
then further anodized. The fractional burying of the xenon marker was
measured and set equal to the (apparent) transport number for the
metal. Some of the results are summarized in Table 1.

The fact that different results were obtained for aluminum anodized
in the different electrolytes is probably due at least in part to pore

Table 1

Apparent Metal-Ion Transport Numbers in Anodic Oxides[a, b]

Metal	Electrolyte	Transport Number	Current density (mA/cm^2)
Al	Sodium tetraborate in ethylene glycol	0. 54–0. 63	0. 1–10. 0
Al	Aqueous ammonium dihydrogen citrate	0. 33–0. 41 0. 57	0. 1 10. 0
Ta	Aqueous Na_2SO_4, 0.1 M	0. 26–0. 31	0. 1–10. 0
Nb	Aqueous Na_2SO_4, 0.1 M	0. 22–0. 33	0. 1–10. 0
Zr	0. 4 M KNO_3 0.04M HNO_3	0. 00–0. 05	0. 1–10. 0
Hf	0. 4 M KNO_3 0.04M HNO_3	~ 0. 05	1. 0
W	0. 4 M KNO_3 0.04 M HNO_3	0. 30–0. 37	0. 1–1. 0

[a] Based on xenon-125 marker.

[b] Data from Davies and Domeij [58].

formation on anodizing in the aqueous ammonium citrate electrolyte. Dissolution of metal was observed by a radiotracer technique. A disadvantage of the β-spectroscopy method is that the electrons are scattered at wide angles and the energy attenuation of an individual electron is not a simple function of the xenon-125 depth. This limitation was overcome by using α-spectroscopy [58], with radon-222 serving as the marker and α-particle emitter. α-Particles do not undergo significant wide-angle scattering, and the energy loss per particle is determined only by the distance between the radon-222 atom and the surface. The results obtained by this method were consistent with the β-spectroscopy determinations.

Pringle, using a precision sectioning technique [54], has since found that an initially normal distribution of xenon-125 markers remains in a normal distribution on thickening the oxide, but the standard deviation of the distribution increases with increasing film thickness. Pringle's observations can be explained, he stated, "by assuming that the charge transfer event involves the simultaneous movement of a group of atoms, both tantalum and oxygen, in such a way that charge and material is transported from one side of the group to the other. The charge being transported is therefore localized not on a single atom, but on a group of atoms, so that the charge is carried through the oxide in the form of a lattice disturbance." Ionic transport via network defects is, of course, just such a mechanism, though set out in somewhat more specific terms.

The most recent results by Pringle [54] give much the same apparent transport number for Ta_2O_5 as the earlier work, the precision being, however, vastly improved, so that small differences in the apparent transport numbers resulting from film formation at different temperatures or current densities or from using different inert-gas markers were easily detected. Of interest is the fact that inert-gas-atom markers lead to substantially smaller values for the apparent metal transport number for Ta_2O_5 (~ 0.3) than do markers consisting of anions incorporated from the electrolyte (~ 0.5). Such a difference is not surpris ng and can be accounted for either along the lines given or by assuming that the phosphate ions occupy the same kind of sites as do the inert-gas markers but move relative to these sites due to the fact that they carry a negative charge. If the latter explanation is correct, the fact that the apparent transport number for phosphate markers is close to 0.5 must be regarded as fortuitous.

Note that two of the metals that are expected to form even-valent metal oxides, zirconium (ZrO_2) and hafnium (HfO_2) display metal transport numbers (based on a [125]Xe marker) that are zero within experimental uncertainty. This is consistent with the conclusion that singly charged network defects cannot exist in even-valent vitreous metal oxides. Doubly charged defects would normally be expected to take the form of "interstitial" oxygen ions and/or oxygen ion "vacancies," which would give the above result. On the other hand, tungsten would also be expected to form an even-valent oxide, WO_3, yet its transport behavior is almost identical with that displayed by tantalum and niobium, which form odd-valent metal oxides (Ta_2O_5 and Nb_2O_5). A possible explanation for this is that the oxide formed on tungsten is not WO_3 but rather consists totally or in part of W_2O_5, since this oxide is known to exist. Alternatively, the doubly charged defects in WO_3 might consist of a more complex form of network defect, rather than simply of oxygen-ion "vacancies" and/or "interstitials."

In summary, the observed transport behavior of the odd-valent metal oxides is in good agreement with that deduced on the basis of the network-defect model:

1. Ion transport appears to preserve the order in both metal and oxygen subnetworks.

2. On the assumption that incorporated anions like SO_4^{-2} or PO_4^{-3} form local crystalline structures, an apparent transport number based on such markers close to 0.5 is predicted and observed.

3. Atomic markers will give rise to apparent transport numbers differing in general from 0 or 1, again in accord with observation.

4. Finally, the fact that the even-valent metal oxides do not appear to display unusual transport behavior (WO_3 being a possible exception) is expected from, though it is not a mandatory requirement of, the network-defect model.

Although the proposed network defects appear to this point to represent transport properties satisfactorily, it remains to be seen whether or not their formation is energetically favorable.

D. Enthalpy of Formation of Network-Defect Pairs

In this section an attempt is made to calculate the enthalpy of formation of network-defect pairs in vitreous oxides in order that it may be compared with activation energies for ionic conduction. For comparison purposes the enthalpies of formation of Frenkel and Schottky defects are also estimated.

The sequence of processes from which the enthalpy term can be estimated is described as follows: The initial state is taken to be a vitreous solid of an odd-valent metal oxide that is free of network defects of the type we have outlined. The oxide is then considered to be expanded into its component atoms, widely separated, the energy required being that to break all the metal-oxygen bonds. The defect is then created in the expanded network, the only energy term being that required to transfer an electron from a metal atom to an oxygen atom. The expanded network containing the defect is now condensed to the glassy state, with the consequent reforming of all the metal-oxygen bonds except one. Finally, the oxide in the vicinity of the charged defects is allowed to polarize under the influence of the field generated by the pair of charges. The enthalpy of formation of the defect pair, ΔH_d^0, is therefore given by

$$\Delta H_d^0 = \Delta H_b^0 + I_1 - E_1 + \Delta H_p^0 , \qquad (308)$$

where ΔH_b^0 is the mean bond energy, I_1 is the first ionization energy, for the metal atom, E_1 is the first electron affinity for an oxygen atom, and ΔH_p^0 is the polarization energy for the defect pair. The bond energy is simply the negative of the enthalpy of formation of one equivalent of oxide from its component atoms and can be calculated readily from standard thermochemical data, provided standard enthalpies of formation of the vitreous oxide are known or can be estimated. This

proves no major obstacle as the difference in the enthalpies of formation of vitreous and crystalline oxides is small, so that the use of
data for the crystalline oxide introduces only a small error in the
value calculated for ΔH_b^0.

The first ionization potentials for metals are quite accurately
known and readily available from many sources. The situation with
regard to the electron affinity of oxygen, however, is not so satisfactory. The most recent values appear to lie in the range 1 to 2 eV,
with perhaps the best value being 1.47 eV [60].

The problem of calculating the polarization energy for the defect
pairs is essentially equivalent to that of calculating solvation energies for ions in solution. The polarization free energy for a species
of charge q_e, on neglecting both the discrete nature of the solvent
and quadripole contributions [61], takes the form $(q_e^2/2r)(1 - 1/\epsilon)$.
Because of these approximations, r must be regarded as an effective
radius for the ion. The difficulty in applying this expression is of
course in knowing what value to use for the effective radius. To side-
step this problem the following procedure is used: Noting that the
negatively charged defect can be regarded as an O^- ion singly bonded
to a metal, the anion network defect is in a sense isoelectronic with
F^- and bears the same charge. Thus its polarization energy can perhaps be estimated from that of F^- in some medium. In an analogous
manner the polarization energy for the cationic network defect in
B_2O_3, say, can perhaps be estimated from that for Li^+ in some medium,
and in Al_2O_3 from that for K^+ in some medium, and so on. Writing the
total standard polarization free energy in the form

$$\Delta G_p^0 = \Delta G_p^\infty (1 - 1/\epsilon) , \qquad (309)$$

where ΔG_p^∞ is the standard polarization free energy in the limit as
$\epsilon \rightarrow \infty$, and applying the thermodynamic relation $\partial(\Delta G^0)/\partial(1/T) = \Delta H^0$,

Eq. (309) becomes

$$\Delta H_p^0 = \Delta H_p^\infty (1 - 1/\epsilon) - \Delta G_p^\infty (1/\epsilon)(\partial \ln \epsilon / \partial \ln T) \,. \tag{310}$$

There is no evidence that the dielectric constant for vitreous B_2O_3 (or for anodically formed Ta_2O_5, Nb_2O_5, or Al_2O_3) varies significantly with temperature near room temperature. Furthermore, as an estimate of ΔH_p^∞ for the defect pair, the standard enthalpy of hydration, ΔH_h^0, of the appropriate metal fluoride will be used, since there appears to be no better method for estimating it at present. In addition, however, one can hope that water has similar properties to a vitreous oxide insofar as the polarization energy is concerned, since both consist mainly of oxygen in terms of volume occupancy and have amorphous structures. To obtain a better estimate for ΔH_p^∞ would be extremely difficult and hardly justified in the present application. Making the appropriate substitutions and approximations in Eq. (310) leads to the result

$$-\Delta H_p^0 \approx -\Delta H_h^0 (1 - 1/\epsilon) \,. \tag{311}$$

In Tables 2 and 3 the relevant quantities are given along with the calculated values for the enthalpies of formation of the defect pairs. For B_2O_3 thermochemical data are available for the calculation of the mean bond energy in both the crystalline and vitreous solids. For the other oxides, however, only data for the crystalline solids are available. It was therefore assumed that the bond energy for the vitreous solids would be lower than that for the crystalline solids by the same percentage as was calculated for B_2O_3 (i. e., $0.6_6\%$). As can be seen, any error introduced by this approximation is indeed small.

As noted in the introduction to Section VII, the ionic conduction data obtained for vitreous B_2O_3 suggests that the equilibrium concentration of carriers is maintained via thermal dissociation of the vitreous network, the principal evidence supporting this being the large

Table 2

Enthalpy of Hydration of Selected Singly Charged Ions[a]

Ion	Ionic radius r (Å)	$-\Delta H_h^0$ (eV)	Isoelectronic network defect
Li^+	0.68	5.392	$>B^+$
Na^+	0.97	4.206	$>Al^+$
K^+	1.33	3.337	
Tl^+	1.45[b]	3.14_6[c]	$>Bi^+$
Tm^+	1.45_4[d]	3.14_0[c]	$>Ta^+$
Rb^+	1.47	3.118	$>Nb^+$
Cs^+	1.67	2.866	
F^-		5.238	$-O^-$

[a] Data from Halliwell and Nyburg [62] except as noted.

[b] Data from Harvey and Porter [63].

[c] Obtained by interpolation; $-\Delta H_h^0$ is a smooth function of ionic radius.

[d] Obtained by interpolation; thus

$$r(Tm^+) = r(Xe4f^{14})^+$$

$$\simeq r(Xe^+) + [r(Xe4f^{14}5d^{10})^+ - r(Xe^+)](14/24)$$

$$= r(Cs^+) + [r(Au^+) - r(Cs^+)](7/12) = 1.45_4$$

for $r(Au^+) = 1.30$ [63].

Table 3

Estimation of Enthalpy of Formation of Network-Defect Pairs

Vitreous oxide	ΔH^0_b (eV)	I_1 (eV)	$-\Delta H^0_h(MF)$ (eV)	$\epsilon \underline{\underline{a}}$	ΔH^0_d (eV)
B_2O_3	4.839	8.361	10.630	~9$\underline{\underline{b}}$	2.28_1
Al_2O_3	5.114	6.049	9.444	9.8	1.21_2
Bi_2O_3	2.978	7.355	8.384	62	0.61_4
Ta_2O_5	5.022	7.958	8.378	27.6	3.43_6
Nb_2O_5	4.868	6.947	8.356	41.4	2.19_0

$\underline{\underline{a}}$The ac dielectric constants have been used here since very slow polarization processes would presumably not contribute to ΔH^0_p under the dynamic conditions of defect generation.

$\underline{\underline{b}}$Estimated from the value for Al_2O_3, noting the trend of increasing ϵ for increasing cationic radius found for the Group IIA metal oxides.

value for the preexponential factor, which rules out the possibility
that the current is being carried by a small impurity concentration.

Thus assuming the mobile defects to be network defects, their
equilibrium concentration will be proportional to $\exp(-\Delta H_d^0/2kT)$ in
a manner precisely analogous to that for Schottky-defect formation in
the alkali halides (see Eqs. (188) through (191) and related discussion). The low-field activation energy for ionic conduction will then
be given by $(\Delta H_d^0/2 + Q)$, where Q is the activation energy for the
migration of the mobile network defect. The activation energy for the
migration of an anionic or cationic defect would be expected to be
approximately one-half the enthalpy of formation of a network-defect
pair, since a migration step involves the partial breaking of a metal-
oxygen bond in an ionic fashion, followed by a partner exchange.

Using this estimate for Q, the activation energy for ionic conduction in B_2O_3 will be on the order of ΔH_d^0, which from Table 2 is
2.28 eV. This is to be compared with the experimental value of
2.16 eV [64]. The agreement between these two results is clearly
well within the uncertainty in the estimate of Q. The postulate of
these network defects is seen, therefore, to account for the ionic
conduction properties of pure vitreous B_2O_3, both in terms of the
large preexponential factor, which for the model described should
have a value close to that for the alkali halides, and with regard to
the magnitude of the empirical activation energy.

It is worthwhile to attempt to estimate the enthalpy of formation
of a Frenkel defect in a vitreous oxide in a manner analogous to the
above so that the two may be compared. As not all the required information is available for B_2O_3, the calculation will be carried out instead for Al_2O_3. Proceeding as before, the enthalpy of formation of
Frenkel-defect pairs may be calculated through Eq. (312), with an
estimation of the polarization energy being given in Eq. (313):

$$\Delta H_f^0 = 3\Delta H_d^0 + I_1 + I_2 + I_3 - 3E_1 + \Delta H_p^0 \; , \tag{312}$$

$$-\Delta H_p^0 \approx [-\Delta H_h^0(AlF_3)](1 - 1/\epsilon) \; , \tag{313}$$

where I_1, I_2, and I_3 are the first, second, and third ionization energies of aluminum and $\Delta H_h^0(AlF_3)$ is the enthalpy of hydration of AlF_3.
This estimate of the polarization energy $-\Delta H_p^0$ should be too large in
that it takes no account of the fact that the three O^- ions are adjacent to one another and to a cation vacancy. Insofar as this part of
the calculation is concerned, therefore, the calculated enthalpy of
formation for the Frenkel defect will be a lower limit.

The various values are given in Table 4 along with the calculated value for the enthalpy of formation of the Frenkel defect, 8.70 eV.
By comparison, the enthalpy of formation of a network-defect pair is
1.22 eV. Admittedly there is in all likelihood a considerable error in
both of these estimates. It is extremely improbable, however, that
the errors are sufficient to overcome the large difference calculated
for the enthalpies of formation.

An estimate of the enthalpy of formation of Schottky defects in
Al_2O_3 leads to a result slightly above that for Frenkel-defect pairs.
On the basis of these calculations, therefore, it appears that network
defects should predominate over Frenkel and Schottky defects in odd-
valent vitreous oxides. Thus on this basis alone it seems probable
that ionic conduction in vitreous oxides containing only network-
forming metal atoms will occur via a network-defect mechanism rather
than an interstitial or vacancy mechanism. It is not certain, however,
that the valve-metal oxides produced by anodic oxidation have vitreous rather than microcrystalline structures, although diffraction data
are certainly in accord with the former supposition, at least for the
films formed on aluminum and tantalum.

Again it is pointed out that it is probably not possible for singly
charged network defects to exist in even-valent vitreous oxides. In

Table 4

Estimation of Enthalpy of Formation of Frenkel-Defect

Pairs in Vitreous Al_2O_3

ΔH_b^0 (eV)	$I_1 + I_2 + I_3$) (eV)	$-\Delta H_h^0(AlF_3)$ (eV)	$\epsilon \underline{\underline{a}}$	ΔH_d^0 (eV)
15.34_2	53.54_7	62.12	9.8	8.70

$\underline{\underline{a}}$The ac dielectric constants have been used here since very slow polarization processes would presumably not contribute to ΔH_p^0 under the dynamic conditions of defect generation.

such oxides it appears that the defects must be doubly charged in order for the cationic and anionic defects to be independent of one another (i.e., not connected by a region of excess strain). For even-valent metal oxides, therefore, it is not unreasonable to expect that the defects responsible for conduction would resemble rather closely anion vacancies or interstitial anions, as already noted. Such defects, of course, may also exist in crystalline oxides.

An interesting prospect therefore emerges. Since during anodic oxidation the oxide film is formed under conditions that are kinetically controlled (i.e., far from equilibrium conditions), the structure of the oxide might well be related to the mechanism of ionic transport. Thus in odd-valent metal oxides energy considerations lead to the conclusion that ionic transport can take place much more readily via a network-defect mechanism than it can via an interstitial or vacancy mechanism. In order that a network-defect mechanism be operative, however, the oxide must form in the vitreous state, a state slightly higher in free energy than the crystalline one. Hence by forming in the vitreous state at the cost of a small increase in free energy (a matter of no particular concern since the oxide is formed under conditions far from equilibrium) the system is able to maintain a relatively easy path for further growth of the oxide.

According to this reasoning, the amorphous or glassy nature of these oxides can be regarded as a direct consequence of the favorable mechanism available for ionic transport in odd-valent vitreous oxides. As this argument does not apply to even-valent metal oxides, there is no obvious reason for them to form in a vitreous rather than a crystalline state. A greater degree of crystallinity for even-valent metal-oxide systems is therefore perhaps to be expected. This does appear to be the case, although exhaustive data relating to this matter are, to the author's knowledge, not available.

In Section V. B it was shown that ionic conduction through thin films at high field strengths is unlikely to take place by a mechanism in which the defect pairs are generated homogeneously throughout the film. The procedure used to estimate the activation energy for ion transport through bulk vitreous B_2O_3 cannot be used, therefore, to estimate the activation energy for the anodic oxidation of valve metals. However, the defects must be generated somewhere (most probably at either the oxide-electrolyte or metal-oxide interface in the light of the material of Section VI), and this generation must involve at least the incipient formation of a network-defect pair. Thus some fraction of the enthalpy of formation of a network-defect pair might be expected to make up at least a portion of the activation energy for anodic film growth. A correlation between the empirical zero-field activation energy Q_e and the enthalpy of formation of network-defect pairs, ΔH_d^0, might therefore be expected. A test of this hypothesis is left until the steady-state kinetics of film growth on valve metals are discussed in Section VIII.

VIII. MECHANISM OF ANODIC OXIDATION
OF THE VALVE METALS

A. Introduction

Tantalum, niobium, and aluminum, when anodically oxidized under conditions in which the current efficiency for film formation is close to 100% and electrolyte incorporation into the film is minimal, appear to exhibit the same kinetic behavior. It seems likely, therefore, that the same mechanism is operative for these three systems, and no doubt for others as well. In this section the properties that are common to all three systems are examined with the purpose of determining the mechanism.

Perhaps the most fundamental property common to these systems is the very large field strengths required for film formation near room temperature, being typically on the order of 10^6 to 10^7 V/cm. According to the general arguments presented in Sections V and VI, it is, a priori, unlikely that ion transport through thin films at high field strengths would be controlled within the film. In the absence of any direct, unequivocal experimental evidence in support of space-charge effects, an explanation for the kinetic behavior of these systems must therefore be sought, initially at least, from a mechanism involving interfacial control.

An extensive study of the data in the literature, combined with the results of recent unpublished work, have led the author to the conclusion that the rate is most likely controlled at the oxide-electrolyte interface, with the rate-controlling step possibly being anion-network-defect injection into the oxide.

All of the main features displayed by these systems, including the magnitudes of the activation energies and distances, can be accounted for at least qualitatively, and frequently quantitatively, in terms of such a model. This, of course, does not preclude the possibility of another model doing as well or better in accounting for the data.

In the following sections the above hypothesis is developed and expanded, first by considering steady-state behavior and then transient behavior.

B. Steady-State Film Growth

1. Rate-Controlling Step

The possibility that the rate is controlled either by defect injection at the metal-oxide or the double layer-oxide interface, with equilibrium conditions prevailing across the oxide-electrolyte double layer, is examined first.

In Section VI. D. 2 the relationships between overpotential and electrode capacitance were derived for just these cases, and the influence of a depletion layer in the oxide next to one or other of the interfaces was also determined. The overpotential was predicted to be a linear function of reciprocal capacitance, and such has, indeed, been found to be the case, generally.

Thus L. Young [65] found that chemically polished specimens of tantalum, if leached in boiling water before being anodically oxidized in dilute sulfuric acid, gave zero overpotential intercepts to within experimental uncertainty, ± 0.2 V. If the film present after chemical polishing was not removed, however, the intercept was up to 0.5 V more negative, the slope being essentially unaltered. Furthermore, by examining the effect of the film left by chemical polishing on the

adhesion of the anodically formed oxide to the metal, Young concluded
that this film must remain at the metal-oxide interface. Assuming that
the film forms a depletion layer next to the oxide-metal interface,
these results may be compared with the predictions for such a case
given in Section VI. D. 2. The presence of such a depletion layer is
predicted to produce a decrease in the intercept, as observed, no
matter at which interface the rate-controlling defect-injection occurs.
If the rate is controlled at the metal-oxide interface, a depletion
layer producing a decrease in the intercept of 0.2 V is predicted to
increase the slope, $[\partial \eta / \partial (1/\underline{c})]_{i,T}$, by about 10%, contrary to observation. No such change of slope is predicted, however, if the rate
is controlled at the double layer-oxide interface.

Further evidence in support of rate control at the double layer-
oxide interface, rather than the metal-oxide interface, is provided by
the results of Vermilyea [66], who showed that the relationship between ion current and field strength for tantalum oxide formation is
not influenced by the crystallographic orientation of the metal surface.
A similar result was obtained for aluminum by Basinska et al. [67],
who showed that all crystals on polycrystalline aluminum are anodically oxidized to equal oxide thicknesses in dilute aqueous ammonium borate. On the other hand, if the rate-controlling step involves
the formation of a positively charged network defect at the metal-
oxide interface via the rupture of a metal-oxygen bond (cf. Section
VII), one would not expect the kinetics to be significantly dependent
on crystallographic orientation. The evidence in favor of control at
the oxide-electrolyte, rather than the metal-oxide, interface is clearly
not very compelling.

If the rate were controlled by ion transport within the film, space-
charge effects should be apparent at least in the case of thin films.
Space-charge effects have been investigated for the tantalum system

by L. Young [68, 69] and Vermilyea [70-72]. Young [69] found the
differential field strength $\tilde{E}$ to be constant to within 0.5% for the
growth of films from a few tens of angstroms to greater than 3000Å
(i. e., approaching the thickness where a prebreakdown decrease in
current efficiency occurs). These results limit the characteristic
space-charge length for the system either to a value on the order of
10Å or less, or to a value very much larger than 3000Å (cf. Section
VI. D). If the former is the case, the concentration of mobile ionic
defects must be on the order of the lattice-ion concentration, which
is unlikely, whereas if the latter is the case, the rate must be controlled at one or both of the interfaces.

Additional evidence against control within the film is provided
by the excellent linearity of graphs of η versus $1/\underline{c}$ [65] over the
entire range of measurements, limiting the dimensions of any possible
interfacial space-charge zone once again to the order of 10Å or less.

The case for interfacial control is therefore strong, at least for
tantalum. Dignam and Ryan [34] have presented data that show a
decrease of about 6% in $\tilde{E}$ during the formation of films up to about
200Å in thickness on aluminum in a glycol borate electrolyte. For
films thicker than 200Å, $\tilde{E}$ was constant. These results were analyzed
on the basis of the high-field Frenkel-defect theory, assuming the effect to be due to space charge. As will emerge shortly, the results
can also be interpreted as arising from a change in the mean dielectric
constant of the film with thickness, which will occur in general if the
film has a duplex structure. Dorsey [73] has shown that films formed
on aluminum in a 2.0 M aqueous boric acid electrolyte at 60°C consist
of an inner layer with an approximately constant thickness of about
200Å and an outer layer whose thickness increases with increasing
formation voltage. The thickness of the inner layer was found to
increase with increasing formation temperature. If such a film structure were to result from formation in a glycol borate electrolyte, the

observed decrease in $\tilde{E}$ with thickness could be accounted for in this way. In any event the original interpretation of the data as a space-charge effect arising out of the mechanism for ionic transport is probably incorrect.

2. Higher Order Field Terms

In Section III. C it was shown that for very high field strengths the net activation energy for ion transport across an activation barrier is expected to contain higher order terms in the field strength. It was L. Young [74] who first showed that many of the previously reported anomalies relating to the steady-state anodic oxidation of tantalum in dilute sulfuric acid were removed simply by including a quadratic term in the field strength in the expression for the net activation energy. Since other forms for the net activation energy have been suggested subsequently [34, 75], Dignam and Gibbs [20] undertook to examine the available data for tantalum and niobium to try and determine the form of the net activation energy $W_e(E)$ that best represents the data. In order to have an effective test for the best form for $W_e(E)$, precise data covering a wide range of field strengths are required. By far the best set of data to date in this regard are those obtained by L. Young for tantalum [74]. The films were formed in 0.2 N sulfuric acid, their thickness being measured by the minimum-reflectance technique [2]. The analysis of these data (tantalum data, set A) is summarized in Table 5.

The method of analysis was as follows: The data in the form log i as a function of T and E were processed by a 7094 IBM computer using a nonlinear least-squares program that chose the parameters to minimize the variance in log i. It also calculated both the root-mean-square (rms) deviation of the experimental log i values from those calculated using the least-squares parameters as well as the standard deviations of the parameters themselves.

Table 5

Values[a] for the Parameters and rms Deviation of log i Obtained on Fitting

Young's Tantalum Data (Set A [74]) to the Equation $i = i_0 \exp[-W_e(E)/kT]$

for Various Assumed Forms for $W_e(E)$

Functional form of $W_e(E)$	Rms deviation of [log (i/A cm⁻²)]	Log (i₀/A cm⁻²)	Q_e (eV)	b_e (A/V)	$u_e^* = b_e Q_e$ (eÅ)
1. Linear field expansion[b]	0.136 (0.012)	8.19 (0.22)	1.62 (0.02)	9.50 (0.16)	15.4 (0.3)
2. Quadratic field expansion[c]	0.035 (0.003)	8.33 (0.06)	2.21 (0.02)	16.14 (0.36)	35.7 (0.8)
3. Cubic field expansion[d]	0.034 (0.003)	8.33 (0.06)	2.57 (0.17)	21.20 (3.70)	54 (10)
4. Parabolic potential energy[e]	0.078 (0.007)	8.25 (0.12)	1.90 0.01	13.18 (0.07)	25.0 (0.2)
5. Cosine potential energy[f]	0.093 (0.009)	8.24 (0.15)	1.81 0.02	12.09 (0.07)	21.9 (0.3)
6. Morse potential energy[g]	0.074 (0.007)	8.27 (0.12)	2.13 (0.02)	3.63 0.02	
7. Exact Schottky law[h]	0.099 (0.009)	8.26 (0.16)	2.31 0.02	1.53 0.01	
8. Approximate Schottky law[i]	0.088 (0.008)	8.27 (0.14)	2.51 0.02	2.18 (0.06)	

Note: for functions 1 through 5, $b_e = u_e^*/Q_e$.

a. Values in parentheses represent standard deviations.

b. $W_e(E) = Q_e[1 - (b_e E)]$.

c. $W_e(E) = Q_e[1 - (b_e E) + (b_e E)^2/C_2]$.

d. $W_e(E) = Q_e[1 - (b_e E) + (b_e E)^2/C_2 + (b_e E)^3/C_3]$.

e. $W_e(E) = Q_e[1 - (b_e E) + (b_e E)^2/4]$; cf. Eq. (70).

f.
$$W_e(E) = Q_e\left\{1 - \left[1 - \left(\frac{2b_e E}{\pi}\right)^2\right]^{\frac{1}{2}}\right\} - b_e E\left[1 - \frac{2}{\pi}\arcsin\left(\frac{2b_e E}{\pi}\right)\right] \quad \text{from } U_1(u) = (Q_e/2)[1 - \cos(u/u^*)];$$

cf. Eqs. (42) through (48).

g.
$$W_e(E) = Q_e(1 - 2b_e E)^{\frac{1}{2}} - b_e E \ln\left[\frac{1 + (1 - 2b_e E)^{\frac{1}{2}}}{1 - (1 - 2b_e E)^{\frac{1}{2}}}\right] \quad \text{from } U_1(u) = Q_e[1 - \exp(-u/w^*)]^2, \quad b_e = w^*/Q_e.$$

h. $W_e(E) = Q_e[1 - 2(b_e E)^{\frac{1}{2}} - bE]$, from the potential-energy function for a mobile ion of charge q bound to a trapping site of charge q'; $b_e = (X_0 q)/Q_e$, $Q_e = qq'/\epsilon X_0$, where X_0 is the distance of closest approach of the two charges (see Ref. [20]).

i. $W_e(E) = Q_e[1 - 2(b_e E)^{\frac{1}{2}}]$, approximation of case 7 valid for $(bE)^{\frac{1}{2}} \ll 1$.

It is apparent from the first two columns of Table 5 that the rms deviation for set A of the tantalum data is markedly dependent on the form assumed for $W_e(E)$, varying from a maximum of 0.136 for case 1 to a minimum of 0.034 for case 3. All of the two parameter forms of $W_e(E)$ (cases 4 to 8) give substantially better fits to the data than does the simple linear field expansion (case 1), which also contains two adjustable parameters. A very much better fit is achieved, however, on representing $W_e(E)$ as a quadratic function of the field strength (case 2), which involves three adjustable parameters. No significant improvement is achieved on representing $W_e(E)$ as a cubic function of the field strength (case 3).

In the original paper [20] data for niobium obtained by L. Young [76] (niobium-A data), again by using the minimum-reflectance technique, were analyzed in a similar fashion, along with data for both tantalum and niobium obtained by L. Young and Zobel [75] through ellipsometry (tantalum-B and niobium-B data). These data did not display quite as sensitive a dependence of the rms deviation on the choice of $W_e(E)$, due to the fact that they covered a narrower range of field strengths. The general trend of rms values versus the form chosen for $W_e(E)$ was, however, very much the same for all sets; for all cases the quadratic and cubic forms for $W_e(E)$ gave the best fits. Furthermore, it was concluded that the quadratic form for $W_e(E)$ fitted the data so well that there was no detectable systematic deviation between the data and the computed functions. The improvement on going to the cubic form for $W_e(E)$ is clearly not statistically significant.

It is worth noting that there is no evidence for $W_e(E)$ being temperature dependent. If ion transport were controlled within the film itself, which is amorphous, $W_e(E)$ would be expected to be temperature dependent as a consequence of the distribution of values for the activation energy and activation distance (see Sections II. C and III. D).

Values for the activation-energy parameter Q_e and the activation-dipole parameter u_e^* are included in Table 5 primarily to show the dependence of the numerical value of these on the choice of the form for $W_e(E)$. Thus the apparent or effective activation energy Q_e is seen to vary from a low of 1. 62 to a high of 2. 57 eV. A similar variation exists for u_e^*. It is clear from this that great care must be taken in comparing parameters for different systems.

In Table 6 the kinetic parameters for all four sets of data are presented for the cases in which $W_e(E)$ is represented as linear, quadratic, and cubic functions of the field strength, respectively. From the values for the standard deviations of the coefficients it can be seen that only in the case of the tantalum-A data are the experimental results sufficiently precise and extensive to permit an evaluation of the parameters in the cubic expansion of $W_e(E)$, and even here the standard deviation in the coefficient C_3 is almost as large as the coefficient itself. For the quadratic expansion, on the other hand, the parameters are determined with reasonable precision for all four sets of data. A final column for $Q^* \equiv C_2Q_e$ has been added for future use and for comparison with the same quantity determined for aluminum anodically oxidized in a glycol borate · electrolyte [77, 78]. Using a steady-state method, Dignam, Goad, and Sole [77] obtained a value for Q^* of 7. 65 ± 0. 23 eV, whereas Dignam and Goad [78] obtained 7. 08 ± 0. 35 eV by a potentiostatic method. In both cases the measurements were performed only at room temperature, so that the values of C_2 and Q_e could not be resolved.

These results are now examined in the light of the material presented in Section III and on the assumption that the rate is controlled by network-defect injection at the oxide-double layer interface in

Table 6

Values $\underline{a}$ for the Parameters Calculated from Young's Data for Ta and Nb

Assuming $W_e(E)$ To Be Represented by Linear, Quadratic, and Cubic

Functions of E

Type of function	Log i_0 (A/cm^2)	Q_e (eV)	b_e (Å/V)	C_2	C_3	$Q^* = C_2 Q_e$ (eV)
Data for Ta, set A:						
Linear	8.189 (0.217)	1.616 (0.023)	9.505 (0.163)	– –	– –	– –
Quadratic	8.339 (0.057)	2.210 (0.022)	16.134 (0.357)	3.313 (0.179)	–	7.32 (0.40)
Cubic	8.334 (0.055)	2.570 (0.167)	21.203 (3.705)	2.314 (1.002)	-13.180 (8.888)	– –
Data for Ta, Set B:						
Linear	4.155 (0.540)	1.482 (0.050)	11.708 (0.370)	– –	– –	– –
Quadratic	4.473 (0.228)	2.365 (0.095)	20.695 (1.636)	3.535 (0.622)	–	8.36 (1.51)
Cubic	4.459 (0.237)	1.949 (1.221)	13.379 (34.996)	-2.694 (28.867)	1.860 (15.476)	– –

Data for Nb, set A:						
Linear	7.728 (0.235)	1.433 (0.025)	11.607 (0.226)	— —	— —	— —
Quadratic	7.844 (0.161)	1.802 (0.053)	18.411 (1.324)	3.262 (0.636)	— —	5.88 (1.56)
Cubic	7.844 (0.163)	1.900 (0.431)	21.058 (16.214)	2.444 (6.112)	-14.778 (72.722)	— —
Data for Nb, set B:						
Linear	7.025 (0.442)	1.500 (0.045)	13.547 (0.460)	— —	— —	— —
Quadratic	7.177 (0.337)	1.913 (0.145)	21.393 (3.918)	3.358 (1.646)	— —	6.42 (3.18)
Cubic	7.295 (0.352)	2.720 (1.451)	38.044 (47.667)	2.107 (5.709)	-10.940 (42.213)	— —

[a] Values in parentheses represent the standard deviations.

accord with the material of Section VI. C. It is emphasized once
again, however, that the details of this mechanism are uncertain.
On writing Eqs. (253) or (256) in the form

$$i = i_0 \exp[-W_e(E)/kT] \tag{314}$$

the following expressions for i_0 and $W_e(E)$ are obtained:

$$i_0 = \frac{q_e \nu_0 [\exp(\Delta S^0_{pa}/k)]}{X(H^+)} \tag{315}$$

$$W_e(E) = (\Delta H^0_{pa} - q_e dE_d) + (Q_0 - q_e b_+ E) , \tag{316}$$

where

$$\epsilon E = \epsilon_d E_d . \tag{317}$$

In deriving Eq. (253) higher order terms in the field strength were
deliberately neglected in the interest of avoiding unnecessary complexity. The manner in which these higher order terms might arise
must now be considered. In Eq. (316) only $(Q_0 - q_e b_+ E)$ is an activation energy, $(\Delta H^0_{pa} - q_e dE_d)$ being the standard enthalpy change
(including the electrical term) for proton abstraction from the surface
hydroxyl groups. Thus the material of Section III applies directly
only to the term $(Q_0 - q_e b_+ E)$. It was shown, however, that for
$E < 10^7$ V/cm and $b_+ \sim 2\text{Å}$ no significant contribution from higher
order field terms is expected. Even if a contribution to the quadratic
field term were to arise from the term $(Q_0 - q_e b_+ E)$, its contribution to
$W_e(E)$ would not likely be the major one since $b_+ E$ is presumably
appreciably smaller than dE_d. The explanation for the higher order
field term will therefore be sought in the behavior of the enthalpy
term.

The arguments presented in Section III apply strictly to a consideration of net activation energies and cannot be applied to enthalpy

changes without at least some modifications. The one effect that would appear to survive, in modified form, is that arising from a consideration of the energy of polarization. Thus if each surface proton is assumed to contribute a polarizability α_p, then, on transporting a proton from the surface into the bulk of the solution, there will be an increase in enthalpy arising from the polarization energy equal to $\frac{1}{2}\alpha_p E_d^2$. On replacing α_p by $\chi_p/4\pi(M_p/\delta)$, where M_p/δ is the concentration of sites for bound protons in the double layer, δ is the thickness of the region containing the surface protons, and χ_p is the contribution of the protons to the local electric susceptibility (see Section II. C. 2), and including this enthalpy term in the expression for $W_e(E)$, Eq. (316) becomes

$$W_e(E) = Q_e - u_e^* E(1 - u_e^* E/Q^*) , \tag{318}$$

where

$$Q_e = Q_0 + \Delta H_{pa}^0 ,$$

$$u_e^* = q_e[d(\epsilon/\epsilon_d) + b_+] ,$$

and

$$Q^* \approx 8\pi M_p q_e^2 \delta/\chi_p .$$

In obtaining the expression for Q^*, the term b_+ has been neglected in comparison with $d(\epsilon/\epsilon_d)$. Setting $M_p \approx 10^{15}$ cm^{-2}, $\delta \approx 1.5$Å, and $\chi_p \approx 10$ gives a value for Q^* that is in agreement with the experimental results. If this explanation accounts correctly for the major portion of the quadratic field term, Q^* should be roughly the same for all oxide systems involving the same or similar electrolytes. From the very limited data presented here, this appears to be the case. It is not immediately obvious, however, why Q^* should be the same for aluminum anodically oxidized in a glycol borate

electrolyte as it is, say, for tantalum anodically oxidized in 0.2 N
sulfuric acid. On a mole-fraction basis, however, the glycol borate
electrolyte is one-half water [79], and hence the composition of the
region immediately next to the oxide surface may not differ appreciably from that for an aqueous electrolyte.

Two other phenomena that might contribute to higher order field
terms suggest themselves, one being dielectric saturation and the
other electrostriction. If ϵ is assumed to be a linearly decreasing
function of field strength, due to dielectric saturation, then it is
clear from Eqs. (316) and (317) that this will contribute to a quadratic
term in $W_e(E)$ (i. e., make a positive contribution to $1/C_2$). On the
other hand, it would appear equally likely that ϵ_d would decrease
with increasing field strength. In any event no estimate of the magnitude of this effect can be made. If d is assumed to decrease
linearly with increasing electrostriction pressure (see Section III.C.2),
a cubic term in the field strength will arise. Unfortunately no attempt
has been made to fit the data to a form of $W_e(E)$ that includes a linear
and cubic term, but no quadratic term. The analysis of Table 6 indicates, however, that if a cubic term is included in the expansion of
$W_e(E)$, then it has the opposite sign to that predicted on the basis
of electrostriction.

Dell'Oca and L. Young [80] have offered yet another possible
explanation for the quadratic field term, an explanation that is based
on the influence of incorporated electrolyte on the kinetics of film
growth. As this effect appears to be operative only for those electrolytes that give rise to substantial incorporation within the film (i. e.,
dilute or concentrated phosphoric acid and concentrated sulfuric acid),
a discussion of the effect properly belongs in the next section.

3. Influence of Electrolyte Incorporation

The first systematic investigation of the effect of electrolyte incorporation on the kinetics of anodic film growth is due to Vermilyea [81], who studied film formation on tantalum in 1% Na_2SO_4 and 80% H_2SO_4. Similar studies were subsequently undertaken by L. Young [74] and Draper [82], who used a range of concentrations of H_2SO_4. In all three studies it was found that as the H_2SO_4 concentration is increased, particularly above 60%, the formation field strength increases.

Thus Draper found that at 1.0 mA/cm^2 the cell voltage rose at the rate of about 0.5 V/sec in dilute H_2SO_4, increasing to about 2 V/sec in 100% H_2SO_4. That sulfate should be incorporated into the film when present in the electrolyte in high concentration is not too surprising, since under such conditions a significant concentration of sulfate ions chemisorbed on the metal atoms of the surface is to be expected (see Section VI.C). The interesting result is that in all three studies the capacitance of films formed at the same current to the same final overpotential were nearly the same irrespective of the H_2SO_4 concentration. Thus from Draper's work, the capacitance for films formed at 1 mA/cm^2 to 100 V in dilute H_2SO_4 and 100% H_2SO_4, respectively, differed by only about 10%. The data for all concentrations plotted in the form of overpotential η versus reciprocal capacitance (1 kHz) $1/\underline{c}_t$, fell nearly on a common straight line.

In Section VI.D.2 it was shown that η versus $1/\underline{c}$ for fixed formation-current density will be a straight line provided that steady-state film growth is controlled entirely by defect injection at one of the interfaces. Furthermore, on neglecting $q_e b_+$ in comparison with $q_e d(\epsilon/\epsilon_d)$ (or alternatively assuming $b_+ = b'_+(\epsilon/\epsilon_d)$, see Eq. (228)),

the slope $[\partial\eta/\partial(1/\underline{c})]_{i,T}$ is predicted to be independent of the film
properties (see Eq. (292)) within the limitations imposed by the re-
quirement of rate control by defect injection at the oxide-electrolyte
interface, depending only on the properties of the interface. To the
extent that the properties of the interface are nearly independent of
electrolyte concentration, the graph of η versus $1/\underline{c}$ will also be
nearly independent of it. On the other hand, the field coefficient
u_e^*/kT is predicted to vary almost directly proportionally with the film
dielectric constant ϵ (see Eq. (318)), so that for the same current
density the field strength will vary inversely proportionally with the
film dielectric constant (assuming that the interfacial properties
remain essentially constant). The proposed model appears, therefore, to account for these results. The reader should be cautioned,
however, since in addition to the transport mechanism itself two
assumptions are required in relation to the above explanation.

The first assumption is that the interfacial properties change by
only a small (though not negligible) amount on going from dilute to
100% H_2SO_4. At first glance this appears to be improbable. However, 100% H_2SO_4 is not a stoichiometric compound but simply an
intermediate member of a continuous series of solution of H_2O and
SO_3. It would be interesting to pursue this sort of experimental investigation with even higher concentrations of SO_3 than is represented
by 100% H_2SO_4, in an attempt to effect a drastic change in the double-
layer structure.

The second assumption is that the dielectric constant measured
at 1 kHz, ϵ^t say, and the static dielectric constant ϵ are either
essentially one and the same quantity or their ratio ϵ^t/ϵ remains
nearly constant with increasing electrolyte incorporation.

This and other questions are pursued further in Section VIII. C. 4.
Note, however, that the first assumption is eliminated if the rate is

assumed to be controlled by defect injection at the metal-oxide interface.

Essentially the same behavior is found for films formed on tantalum in electrolytes containing a large proportion of H_3PO_4 or ethylene glycol (see Ref. [2], Section 8.11). However, significant phosphate incorporation occurs for much lower electrolyte concentrations than is required for sulfate incorporation, suggesting that phosphate is more readily chemisorbed on the surface metal atoms than is sulfate.

A phenomenon that is closely related to the above described behavior of the η versus $1/\underline{c}_t$ plots is that connected with the distribution of field strength through duplex films during their formation. Dell'Oca and L. Young [80] found that ellipsometric data for a film formed first in 14.7 M phosphoric acid and then in 0.23 N phosphoric acid could be interpreted by assuming the film to consist of an inner layer of essentially pure oxide (optical constants the same as for formation in dilute H_2SO_4), a second layer containing a high concentration of phosphate, and a third outer layer containing a lower concentration of phosphate. Furthermore, they showed that the mean formation-field strength calculated with this model agrees with the experimental value if it is assumed that the formation-field strength in each individual layer is characteristic of that layer and independent of whatever other layers are present. A similar result had been obtained earlier by Vermilyea [81].

This result is essentially identical with that involving capacitance measurement and can be accounted for in the same way. Thus if the rate is assumed to be controlled entirely at the oxide-electrolyte interface, the electric displacement must be assumed to be constant throughout the film (i. e., negligible space charge). The field strength in the Kth oxide layer, of dielectric constant ϵ_K, is then given by $E_K = (\epsilon_d/\epsilon_K)E_d$, where E_d is determined by the properties of the double

layer, the temperature, and the current density. The field strength
in the Kth oxide layer is thus characteristic of that layer and independent of whatever other layers are present. An identical result is
obtained if the rate is assumed to be controlled at the other interface.

Dell'Oca and Young, and Vermilyea, placed a different interpretation on the data. They assumed that film growth is controlled by
ionic conduction through the film, the formation-field strength in any
one layer being therefore characteristic of the intrinsic conduction
properties of that layer and hence independent of whatever other layers are present. In this model the continuity condition on the flux
density at the boundary between two oxide layers requires the establishment of a space-charge zone near the boundary, in precisely the
same way that it is required within the oxide at either the metal-
oxide or the oxide-electrolyte interface (cf. Section V. B. 5 and the
following) when ion transport is assumed to be controlled within the
film. Once again, since any such space-charge zone must, from the
data, be very narrow, the mobile-defect concentration must be
assumed to be on the order of the lattice-ion concentration.[*]

The two different interpretations of the above result arise, therefore, from postulating opposite limiting conditions. The author's
interpretation, which is in accord with Cabrera and Mott's original
concept of the mechanism of anodic oxidation [19], is based on
assuming that the mobile-defect concentration in the oxide is so low
that the characteristic space-charge length is very much larger than
the film thickness. The interpretation of Dell'Oca, Young, and

---

[*]The alternative of making the ad hoc assumption that a surface
charge is established at the interface between the oxide layers and is
somehow adjusted so as to produce a discontinuity in the electric displacement of just the correct amount so that no volume space-charge
zone is required to equalize the fluxes in the two layers [80] must be
rejected as being totally unsupported (and in the author's view
unsupportable) by current theory.

Vermilyea, on the other hand, is based implicitly on the assumption
that the mobile-defect concentration is so high that the characteristic
space-charge length is only a few angstroms. According to the
author's interpretation, the field strength in the oxide layers should
be inversely proportional to their static dielectric constants.
Dell'Oca and Young found that the field strengths were roughly
inversely proportional to their dielectric constants, measured presumably at 1 kHz.

Since the matter has been raised in the literature [80], it is
worth noting that the requirement that J be continuous across the
boundary between two oxide layers introduces no additional con-
straints if film growth is controlled at the oxide-electrolyte (or metal-
oxide) interface. The continuity of J coupled with the continuity of
the normal component of the electric displacement D requires in general that the defect concentration change discontinuously at the
boundary. Provided that the defect concentration in both layers is so
low that the space charge can be neglected (which of course must be
the case if the rate is controlled entirely at the interface), this discontinuity will give rise to no observable kinetic effects. If ion
transport within the film occurs via a momentum-retention mechanism
(see Section VI. D. 1), then the defect velocity would be expected to
be approximately proportional to the field strength and hence the
defect concentration approximately proportional to the dielectric
constant.

It is clear from the results of Dell'Oca and Young [80] that
electrolyte incorporation into a film during formation can contribute
to the curvature of ln J versus $\bar{E}$ plots. They showed this beyond
any doubt in the case of tantalum anodically oxidized in a 0.23 N
phosphoric acid electrolyte. Two contributing effects are identified.
With increasing J, there is an increase in both the fraction of the
film containing phosphate and the phosphate concentration in that

fraction. Both these effects produce a curvature in the plots of $\ln J$ versus $\bar{E}$ of the same sign and order of magnitude as the curvature found for formation in dilute sulfuric acid. In addition, they lead to a dependence of the $\ln J$ versus $\bar{E}$ curves on the formation conditions. Since such a dependence has not been detected for films formed in dilute sulfuric acid [80], it is unlikely that the curvature in $\ln J$ versus $\bar{E}$ plots observed for tantalum oxide formation in 0.2 N H$_2$SO$_4$ is due to electrolyte incorporation, though the possibility cannot be ruled out.

4. Correlation of the Empirical Activation Energy with ΔH_d^0

Before any attempt can be made to compare empirical activation energies with calculated values, some decision must be reached as to the precise manner in which the activation energy is to be extracted from the data, since the value calculated for Q_e depends on the form assumed for $W_e(E)$ (see Tables 5 and 6, Section VIII. B. 2). It will be assumed, partly on the basis of the material presented in Section VIII. B. 2, but largely arbitrarily, that $W_e(E)$ is given by Eq. (318). Furthermore, it will be assumed that for the anodic oxidation of tantalum, niobium, aluminum, and bismuth in dilute aqueous electrolytes, Q^* is nearly the same for all four systems and accordingly equal to approximately 7.3 eV. The values for Q_e for tantalum and niobium are thus taken from Table 6 for the case of a quadratic field expansion of $W_e(E)$. The only data for the anodic oxidation of aluminum in an aqueous electrolyte in which measurements were made over a range of temperature are those due to Harkness and L. Young [83]. Measurements in aqueous borate solution (20 g/l orthoboric acid adjusted to pH 7 with concentrated aqueous NH$_4$OH) were fitted within experimental error by the equation

$$i(A/cm^2) = 10^{7.3} \exp\{-[1.3 - 8.8_5 E(V/\text{Å})]/kT(eV)\} \ .$$

If it is assumed that the data actually fit an equation of the form of Eq. (318) with $Q^* \approx 7.3$ eV, the results $Q_e \simeq 1.3_9$ eV and $u_e^* \simeq$ 10.6 eÅ can be calculated without difficulty, given that the mean field strength for the set of measurement is about 0.08 V/Å. On this basis it is seen that the quadratic term for aluminum is not as important as it is in the case of tantalum and niobium.

Data for bismuth anodically oxidized in aqueous 0.1 N NaOH have been obtained by Masing and L. Young [84] over a range of temperatures. They were found to fit the same form of equation as do the data for aluminum. In this case fitting the data to Eq. (318) with $Q^* \approx 7.3$ eV produces no significant change in either the activation energy or activation dipole, their values being $Q_e = 1.14$ eV, $u_e^* = 37$ eÅ.

In Fig. 4 these values for Q_e have been plotted against the estimated enthalpies of formation of the network-defect pairs, ΔH_d^0

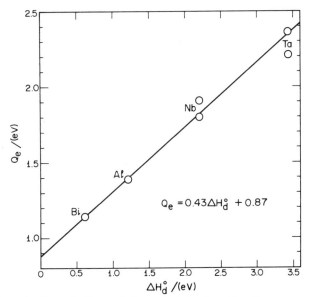

Fig. 4. Correlation of the empirical activation energy Q_e for the anodic oxidation of bismuth, aluminum, niobium, and tantalum, with the calculated enthalpy of formation of network-defect pairs, ΔH_d^0.

(Table 3). On the basis of this rather limited sample, Q_e appears to be a linear function of ΔH_d^0, a result that may be rationalized in terms of the model leading to Eqs. (314) through (318) as follows:

The injection of an anion defect at the oxide-double layer interface involves the partial breaking of a metal-oxygen bond in an ionic fashion, followed by a partner exchange with a surface O^- species. The activation energy for this process would therefore be expected to be on the order of half the standard enthalpy change for breaking the bond, that is, half the standard enthalpy change for the formation of a network-defect pair. Assuming that the standard enthalpy change for proton abstraction, ΔH_{pa}^0, is not too different from one oxide film to the next, Q_e should be a linear function of ΔH_d^0 with a slope of about 0.5 and an intercept on the order of the mean value of ΔH_{pa}^0 for the oxide films. The empirical result, $Q_e = 0.43\ \Delta H_d^0 + 0.87$, implies on this basis a value for ΔH_{pa}^0 of about 0.8_7 eV, which may be compared with the value 0.58 eV for the standard enthalpy change for the reaction $H_2O_\ell \rightarrow H_{aq}^+ + OH_{aq}^-$. Dignam and Gibbs [9] obtained a value for the activation energy for proton abstraction at the equilibrium potential of 0.84 eV for the formation of Cu_2O on copper. If the activation energy for the reverse reaction at zero overpotential is assumed to be nearly zero (i.e., $Q_- \approx 0$, see Eq. (245)), an assumption consistent with the data [10], this gives $\Delta H_{pa}^0 \approx 0.84$ eV, in close agreement with the above value. Once again the results can also be interpreted on the basis of defect injection at the other interface.

C. Transient Phenomena

1. Nature of Transients

When subjected to either a changing current density or field strength, the relationship between the ion-current density and field strength is not in general that found for steady-state conditions. There are, of course, an unlimited number of types of transient measurements that can in principle be carried out. The majority of measurements that have been made can be described as either constant-current transients (galvanostatic transients) or constant-voltage transients (potentiostatic transients). Galvanostatic transients have been observed for tantalum by Dewald [33], Vermilyea [85], and L. Young [86]; for indium antimonide by Dewald [33]; and for aluminum by Goad and Dignam [87]. The technique used by Dewald, Young, and Goad and Dignam was to change abruptly the total current density I for the system, originally under steady-state conditions, and follow the overpotential as a function of time. A remarkable feature of these transients is that the field strength during the transient decays toward its new steady-state value, not in a characteristic time, but rather for a characteristic charge passed per unit area; that is to say, the duration of the transient is inversely proportional to the current density.

The results for aluminum (glycol borate electrolyte) presented in Fig. 5 illustrate this point, since the data for all three runs performed for different current densities and current ratios are seen to be superimposible when plotted against the total charge passed per unit area, $\Sigma = \int_0^t I \, dt$. The data for tantalum and indium antimonide show precisely the same behavior. Thus for these systems the decay of $(E - E_2)/(E_2 - E_1)$ depends only on the charge passed, being independent of I_1, I_2, and, to a first approximation at least, the temperature [33], E_1 and E_2 being the initial and final steady-state field strengths and I_1 and I_2 being the corresponding current densities. The initial very rapid rise in $(E - E_2)/(E_2 - E_1)$ is a consequence of having to charge up the metal-oxide-electrolyte capacitor, the subsequent

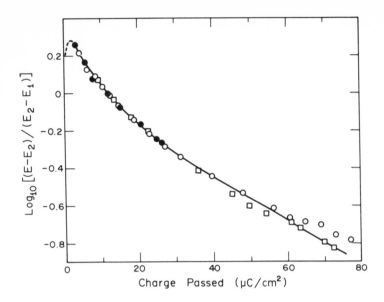

Fig. 5. Decay of the excess field, $E - E_2$, with charge passed during galvanostatic transients performed on the system Al/Al_2O_3 in glycol borate electrolyte. o, $I_1 = 208.6$ $\mu A/cm^2$, $I_2/I_1 = 2$; ●, $I_1 = 207.5$ $\mu A/cm^2$, $I_2/I_1 = 0.5$; ◻, $I_1 = 40.9$ $\mu A/cm^2$, $I_2/I_1 = 11$. Data from Goad and Dignam [87].

relatively slow decay arising presumably out of the kinetics of ion transport.

Potentiostatic transients are usually performed on electrodes possessing sufficiently thick films, so that the film thickness and hence the field strength can be regarded as constant during the transient. Transient measurements of this kind have been carried out for tantalum by Vermilyea [85], L. Young [86], and Taylor and Dignam [88]; for niobium by Vermilyea [85]; for aluminum by Vermilyea [85] and Dignam and Ryan [89]; and for bismuth by Masing and L. Young [84]. The experiments have been performed either by changing the cell potential abruptly from its initial steady-state value or reapplying a potential to a preformed electrode initially close to zero overpotential and then following the change in current density with time.

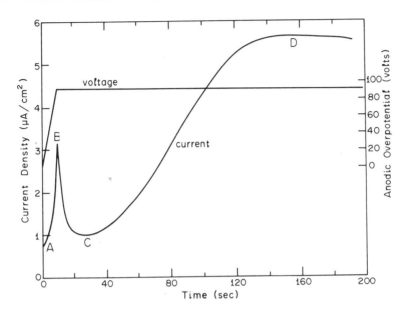

Fig. 6. Time dependence of the current density on reapplication of a potential, as illustrated, to an aluminum electrode formed initially at 209.5 $\mu A/cm^2$ in a glycol borate electrolyte and then annealed at 60°C. Data from Dignam and Ryan [90].

An example of the latter type of measurement is presented in Fig. 6, the data being for aluminum anodically oxidized in a glycol borate electrolyte. The current in the region A to C is presumably predominantly charging current, whereas predominantly ion current exists beyond C. From a series of such measurements it was found that the region beyond C was not characterized by the elapsed time but rather by the charge passed, much as for the constant–current transients. Thus for films prepared initially in the same way the quantity $\ln(I/I_{ss})/\ln(I_0/I_{ss})$ was found to depend only on the charge passed per unit area, Σ, and not on the initial current density I_0 or the final steady–state current density I_{ss}. The formation of the films at different current densities did, however, influence to a small but significant extent the functional dependence of

$\ln(I/I_{ss})/\ln(I_0/I_{ss})$ on Σ [89]. Tantalum behaves in a similar fashion [88].

One parameter that is fairly readily extracted from either galvan-ostatic or potentiostatic transient measurements, and is not depend-ent on the choice of model, is the so-called transient differential field coefficient β_t defined as follows:

$$\beta_t = \lim_{(dE/dt) \to \infty} \left(\frac{\partial \ln i}{\partial E}\right)_T , \tag{319}$$

where i is the ion-current density. Usually β_t is evaluated close to $E = E_{ss}$, where E_{ss} is the steady-state field strength correspond-ing to i. Ideally $\beta_t(E = E_{ss}) \equiv \beta_t(E_{ss})$ would be measured by first establishing steady-state conditions, then making an infinitesimal and infinitely rapid change in E (or ln i), and measuring the result-ant instantaneous change in ln i (or E), the ratio of the changes giving $\beta_t(E_{ss})$. In practice the field (or total current density I) is changed rapidly, and an extrapolation procedure is employed in an attempt to eliminate the effect of the charging current i_c on the meas-urement, to obtain $(\Delta \ln i)/(\Delta E) \simeq \beta_t(E_{ss})$. For evaluating β_t Dignam and Ryan [90] have used a potential-sweep method that in principle at least avoids the problem of correcting for the charging current, pro-vided that $i_c = (\epsilon^t/4\pi)(dE/dT)$, with ϵ^t independent of i and E. Unfortunately these assumptions do not seem to be borne out [90].

A steady-state differential field coefficient β_{ss} may be defined in a manner analogous to β_t:

$$\beta_{ss} = \lim_{(dE/dt) \to 0} \left(\frac{\partial \ln i}{\partial E}\right)_T . \tag{320}$$

It can be seen from the material presented in Section VIII. B. 2 that β_{ss} is a function of the steady-state field strength (or ion-current density), being in fact given by

$$kT\beta_{ss} = -[\partial W_e(E)/\partial E]_T \qquad (321)$$

or

$$kT\beta_{ss} = u_e^*(1 - 2u_e^*E/Q^*) \qquad (322)$$

(cf. Eqs. (314) and (318)). L. Young [86] has shown that $kT\beta_t(E_{ss})$ for tantalum (0.2 N H_2SO_4) is also a function of the steady-state field strength. An interesting feature of these results is that although $kT\beta_{ss}$ and $kT\beta_t(E_{ss})$ are both field dependent, their ratio, calculated at the same field strength, is independent of field strength within experimental error [20, 91]. This same behavior has been observed for aluminum (glycol borate electrolyte) [90]. The limited data available suggest, therefore, that

$$\beta_{ss}(E_{ss})/\beta_t(E_{ss}) = \gamma_\beta , \qquad (323)$$

where γ_β is a constant independent of E_{ss} and T for a given metal-oxide system. Combining Eqs. (323) and (321) gives

$$\beta_t(E_{ss}) = -(1/\gamma_\beta)(1/kT)[\partial W_e(E)/\partial E]_{T, E=E_{ss}} . \qquad (324)$$

Dignam and Gibbs [20] checked the validity of Eq. (323) against L. Young's data for tantalum (0.2 N H_2SO_4) [74, 86]. The procedure used was to calculate $[\partial W_e(E)/\partial E]_{T, E=E_{ss}}$ from the constants provided in Tables 5 and 6 (tantalum data, set A) and then to carry out a regression analysis of $\beta_t(E_{ss})$ against $(1/kT)[\partial W_e(E)/\partial E]_{T, E=E_{ss}}$ to obtain $(1/\gamma_\beta)$ and the rms deviation of $\beta_t(E_{ss})$. The analysis was performed for each assumed functional form of $W_e(E)$. For comparison purposes $\beta_t(E_{ss})$ was also fitted to a quadratic function in the field strength, namely, $kT\beta(E_{ss}) = A_0 + A_1 E_{ss} + A_2 E_{ss}^2$. The results are presented in Table 7. For either the quadratic or the cubic form of $W_e(E_{ss})$ the one-parameter fit to the data, using Eq. (323), is essentially as good as that obtained by using an independent

Table 7

Comparison of the One-Parameter Fit of L. Young's $\beta_t(E_{ss})$ Values for
Ta [86], Using Eq. (323), with an Independent Three-Parameter Fit[a]

Functional form of $W_e(E_{ss})$	Rms deviation for $\beta_t(E_{ss})$ (Å/V)	γ_β
Linear field expansion	9.61	4.48
Quadratic field expansion	3.91	3.21
Cubic field expansion	3.64	2.33
Parabolic potential energy	5.31	3.77
Cosine potential energy	6.41	3.98
Morse potential energy	7.57	4.62
Exact Schottky law	6.91	3.12
Approximate Schottky law	6.13	2.87
Independent three-parameter fit	3.51	

[a] See Table 5 for the forms of $W_e(E_{ss})$.

three-parameter fit. The dependence of the value for γ_β on the
choice of the form for $W_e(E_{ss})$ should be noted. Values for γ_β for
a number of metal-oxide systems are given in Table 8. It can be
seen that γ_β is on the order of 3 for all four of the metal-oxide
systems.

In summary, it may be said that the general features of the
transient response are very similar from system to system, indeed
remarkably so. The characteristic charge per unit area for the transient decays, as well as γ_β, is nearly the same for tantalum, indium
antimonide, aluminum, and bismuth [33, 84, 87, 91]. These, then, are
the essential features that any kinetic theory must account for.

Table 8

Values for the Ratio of the Steady-State to Transient Differential Field Coefficient, $\gamma_\beta = \beta_{ss}(E_{ss})/\beta_t(E_{ss})$, for a Number of Metal-Oxide Systems

Oxide	Electrolyte	γ_β	Refs.
Ta_2O_5	2% Aqueous Na_2SO_4	2.6_3	[85]
	0.2 N Aqueous H_2SO_4	3.1_3	[74, 86]
	0.2 N Aqueous H_2SO_4	3.21	Table 7 [a]
	Dilute Aqueous H_2SO_4	2.5_1	[33]
	Dilute Aqueous H_2SO_4	2.9_0	[87]
	Mean	2.8_8	
Nb_2O_5	5% Aqueous ammonium borate, pH 7	2.1_0	[85]
Al_2O_3	5% Aqueous ammonium borate, pH 7	3.2 [b]	[85]
		2.6_9 [c]	[85]
	Mean	2.9_5	
Al_2O_3	Saturated glycol borate electrolyte	3.0_9	[90]
		3.3_5	[87]
	Mean	3.2_2	
Bi_2O_3	0.1 M Aqueous NaOH	3.0	[84]

[a] Value obtained for the quadratic field expansion of $W_e(E_{ss})$.

[b] $i \approx 2 \times 10^{-3} A/cm^3$.

[c] $i \approx 2 \times 10^{-5} A/cm^2$.

2. Outline of Existing Models

Of the four basic models that have been put forward to explain
the valve-metal transient kinetics, only two have been developed to
the point where quantitative comparison with data is possible.

The first to be proposed was the high-field Frenkel-defect model,
due to Bean, Fisher, and Vermilyea [32]. This model has been treated
extensively in Section V. B. Its principal features are that ion transport is controlled within the film, the defect concentration being,
however, field dependent under steady-state conditions and adjusting
only slowly to sudden changes in the field or current.

This mechanism was shown to be improbable (Section V. B) on
general grounds. More specific evidence against it, however, can
be cited. Thus from Eqs. (216) and (221), and noting that for this
theory $\beta = \beta_t$, it follows that the characteristic space-charge length
X_c' is proportional to $(1/I_{ss})^{1-1/\gamma_\beta}$ for fixed temperature. L. Young's
data for tantalum [74] show adherence to the same conduction equation
for a film 600Å thick over a three-decade range in current density.
Setting $\gamma_\beta \approx 3$, this corresponds to a two-decade range for X_c',
which is only just possible, since X_c' must be very much less than
the film thickness yet greater than its value obtained on setting
$n_\infty \approx 10^{21}$ cm^{-3} (see Eq. (216)), which is approximately 1Å. Again
Tvarusko [92] observed transients for 10- to 100-Å films on tantalum
that were not distinguishable from those for thick films. A mechanism
involving bulk control for such thin films is unlikely. Finally,
although the high-field Frenkel-defect theory accounts moderately ·
well for galvanostatic transients [33], it has been shown, first by
L. Young for tantalum [86] and then by Dignam and Ryan for aluminum
[89], that it predicts the wrong kind of behavior for potentiostatic
transients.

L. Young has proposed an ionic avalanche model, specifically
to account for potentiostatic transients [35, 86], which is in some
respects similar to the above model. The central features of his
model are that ion transport is controlled within the film and that
under constant-field conditions the number of mobile defects increases
toward its steady-state value at a rate proportional to i^2. The model
appears to involve the interaction of one ionic defect in motion with
another to form additional mobile ionic defects — a kind of second
order avalanche process. It accounts, qualitatively, for potentiostatic transients [35]; however, the fitting parameter varies with
field strength. Since the mechanism has not been worked out
in detail, it is not possible to make an independent estimate of the
fitting parameter nor to predict the form of its dependence on field
strength. In fact the model in its present form cannot be tested
against data for any transients except potentiostatic ones.

For this model (or any other model involving rate control within
the film) to be valid, the characteristic space-charge length must be
$\tilde{<} 10 \text{Å}$. It seems to the author that this is likely to be the case only
if virtually every ion in the vitreous oxide becomes a conduction ion
under steady-state conditions, in which case the system would behave
rather like a molten salt as far as ionic conduction is concerned.
Whether or not this could be made compatible with the ionic avalanche
model is not clear.

Adams, Van Rysselberghe, and Willis (Ref. [2], Section 20.05),
on observing transients for very thin films on zirconium, suggested
a mechanism based on control by cation injection at the metal-oxide
interface, with the number of metal ions at suitable locations for
entering the oxide controlled in turn by the characteristics of the
screw dislocations at the metal surface. These were assumed to
depend on the current density and to adjust slowly with changing

conditions. This model has not been quantitatively formulated, but it appears to be inconsistent with the high degree of reproducibility achieved for both steady-state and transient measurements. The same experiment gives the same results with extremely good reproducibility, particularly for tantalum and niobium.

The most recent models are due to the author [20, 47-49, 93]. In the original version [93] the amorphous oxide was considered to be composed of small crystallites or polymeric units (dielectric mosaic model), with the principal rate-controlling step being ion transport between such units. The field strength in such a region will be a function of the polarization of the oxide. Dielectric relaxation processes will accordingly give rise to ion-current transients. To account quantitatively for the transients it was assumed that the dielectric polarization relaxes to its equilibrium, or steady-state, value at a rate that is proportional to the ion-current density (provided that $i \neq 0$). This latter assumption was rationalized in terms of the large amount of energy dissipated per ion jump, this energy effectively bringing about local polarization equilibrium.

The model on which the equations were developed must be rejected, since it cannot possibly be valid for films only a few tens of angstroms in thickness. However, the same equations can be very easily recast in terms of a model involving interfacial control [20, 47-49], retaining the essential feature of the earlier model, namely, that ion-current transients arise as a result of dielectric relaxation processes in the film. Thus if the ion-current density is controlled in total or in part by the field strength in the double layer, as per Section VI.C, the continuity condition on the electric displacement makes E_d a function of the polarization of the film for fixed E. Hence if Eq. (317) is assumed to be valid, $\epsilon_d E_d = E + 4\pi P$, where P is the polarization of the oxide, and since $i = f^n(E_d)$, then a relaxation in P produces a similar relaxation in i (or rather $\ln i$). (The same result is obtained if the rate is assumed to be controlled at the metal-oxide interface.) In the following section this model is developed and shown to account quantitatively for a variety of transient kinetics. Furthermore, a study of the kinetics of the dielectric polarization process reveals a behavior that is precisely that required to account quantitatively for the ion-current transients.

3. Dielectric Relaxation Model for Transients

a. Basic Equations. In this development we employ the model of Section VI. C. The form of the kinetics predicted, however, is not changed if the rate is assumed to be controlled at the metal-oxide interface. Equations (314) through (318) may be written as follows:

$$i = i_0 \exp\left[-W(E_d)/kT\right] , \tag{325}$$

$$W(E_d) = Q_e - u^* E_d (1 - u^* E_d / Q^*) , \tag{326}$$

$$u^* = q_e (d + b'_+) , \qquad b'_+ = (\epsilon_d/\epsilon) b_+ , \tag{327}$$

$$\epsilon_d E_d = E + 4\pi P . \tag{328}$$

The activation process giving rise to the effective activation distance b_+ occurs at the oxide-double layer interface. The "proper" field strength to use in relation to this process could therefore be E, E_d, or some linear combination of the two. As the choice has no effect on the form of the kinetics deduced on the basis of this model, it will be assumed (in order to simplify the algebra somewhat) that the correct field to use is E_d, in which case the "true" activation distance becomes $b'_+ = (\epsilon_d/\epsilon) b_+$, as given in Eq. (327). Again, to simplify the algebra, Eq. (325) may be written

$$i = A_i \exp(\beta_d E_d) , \tag{329}$$

where

$$\beta_d = -(1/kT)[\partial W(E_d)/\partial E_d]_T \tag{330}$$

and

$$A_i = i_0 \exp\left\{-Q_e \frac{[1 - (u^*E_d)^2/Q_e Q^*]}{kT}\right\}. \tag{331}$$

As noted in Section IV. A, β_d and A_i vary only slowly with E_d and hence for the purpose of most of the following development are considered to be constant at a given temperature.

All that is required to complete the model is to prescribe the equations that define the relaxation behavior of the dielectric polarization of the film, P, in the presence of an ion-current density i. The total polarization will consist of the normal "fast" contribution P_1, arising from electronic displacements and small nuclear displacements, and one or more "slow" contributions, P_2, P_3, ... , with large relaxation times. Thus

$$4\pi P = \chi_1 E + 4\pi \sum_{K=2,3,...} P_K , \tag{332}$$

where χ_1 is the electric susceptibility corresponding to P_1, that is, for the fast processes.

It is well known that under conditions of zero ion-current there are many thermally activated slow polarization processes [2, 89, 90]. The rate of change of a thermally activated polarization process is usually expressed as

$$4\pi dP'_\ell/dt = (1/\tau_\ell)(\chi'_\ell E - 4\pi P'_\ell) , \tag{333}$$

where τ_ℓ is the appropriate relaxation time and $4\pi P'_\ell = \chi'_\ell E$ for equilibrium (or steady-state) polarization conditions. Relaxation in accord with Eq. (333) will not produce the correct form for the transients. To generate transients of the correct form one must assume such a polarization mechanism that the rate of polarization change is controlled by the passage of ion current:

$$4\pi dP_K/dt = B_K i \left(\chi_K E - 4\pi P_K\right) , \qquad (334)$$

where B_K is a constant that determines the extent of coupling between the ion current and the "polarization current," dP_K/dt. The coupling process may be rationalized in one of two ways. If the defects responsible for ionic conduction have the properties outlined in Section VII. B (i. e., are network defects), then their passage through the film will be accompanied by local irreversible structure changes. Such structure changes constitute ion-current-driven polarization changes. Alternatively, or additionally, the passage of a defect through the film could produce localized vibrational excitation, which would in turn lead to rapid achievement of polarization equilibration within a small volume centered on the moving defect. In either case the effect may be approximated by assuming that defect motion brings about equilibrium polarization within a distance r of the moving defect. The rate of change of the polarization via the ion-current-driven relaxation process is then given by Eq. (334), where $B_K = \pi r_K^2/q$. Thus B_K, expressed in units of cm^2/e, is a measure of the cross-sectional area that is influenced by the passage of a defect through the oxide. For both of these models, but particularly for the former, B_K would be expected to be essentially temperature independent (i. e., B_K would not possess a Boltzmann factor), its magnitude being determined by structural factors.

b. Steady-State and Transient Field Coefficients. Certain features of this model can be tested without reference to the detailed kinetics of the polarization process(es). Thus an equation for the steady-state ion conduction, as well as one for the ratio of the steady-state to the transient differential field coefficient, $\beta_{ss}(E_{ss})/\beta_t(E_{ss}) = \gamma_\beta$, can be derived, and furthermore a test of the magnitude of the field coefficients can be carried out. From Eq. (332) it follows that

$$\lim_{(dE/dt) \to \infty} 4\pi \left(\frac{\partial P}{\partial E} \right)_T = \chi_1 E \tag{335}$$

and

$$\lim_{(dE/dt) \to 0} 4\pi \left(\frac{\partial P}{\partial E} \right)_T = \chi_{ss} E , \tag{336}$$

where $\chi_{ss} = \sum_{K=1,2,\dots} \chi_K$. Combining these results with Eqs. (328) and (314), and the definitions of $\beta_{ss}(E_{ss})$ and $\beta_t(E_{ss})$ (Eqs. (319) and (320)), then gives

$$kT\beta_{ss} = - \left[\frac{\partial W(\epsilon E/\epsilon_d)}{\partial E} \right]_{T, E=E_{ss}} = - \left[\frac{\partial W_e(E)}{\partial E} \right]_{T, E=E_{ss}} = kT \left(\frac{\epsilon}{\epsilon_d} \right) \beta_d , \tag{337}$$

$$kT\beta_t(E_{ss}) = - \left[\frac{\partial W(\epsilon^t E/\epsilon_d)}{\partial E} \right]_{T, E=E_{ss}}$$

$$= - \left(\frac{\epsilon^t}{\epsilon} \right) \left[\frac{\partial W_e(E)}{\partial E} \right]_{T, E=E_{ss}}$$

$$= kT \left(\frac{\epsilon^t}{\epsilon_d} \right) \beta_d , \tag{338}$$

where $\epsilon = 1 + \chi_{ss}$ and is the static dielectric constant for the oxide and $\epsilon^t = 1 + \chi_1$ and is the dynamic dielectric constant. From Eqs. (337) and (338) it is apparent that γ_β is a constant, independent of field strength, in agreement with the experimental findings (see Section VIII. C. 1); that is,

$$\gamma_\beta \equiv \beta_{ss}(E_{ss})/\beta_t(E_{ss}) = \epsilon/\epsilon^t . \tag{339}$$

Furthermore, from Eq. (337), $i_{ss} = i_0 \exp[-W_e(E)/kT]$, which on making the approximation contained in Eq. (329) becomes $i_{ss} = A_i \exp(\beta_{ss} E)$.

Table 9

Estimates from Available Data of $q_e(d+b'_+)/\epsilon_d = u_e^*/\gamma_\beta \epsilon^t$

for a Dilute Aqueous Electrolyte

Oxide	u_e^* (eÅ)	γ_β	ϵ^t	$u_e^*/\gamma_\beta \epsilon^t$ (eÅ)
Al_2O_3	10.6	2.9$_5$	9.8	0.37
Bi_2O_3	37	3.0	62	0.20
Ta_2O_5	35.7	2.8$_8$	27.6	0.45
Nb_2O_5	33.2	2.1$_0$	41.4	0.38

Mean 0.35 ± 0.11

By making use of Eq. (339), $u^*/\epsilon_d = q_e(d+b'_+)/\epsilon_d$ can be calculated for those systems for which data are available, which in turn can lead to a rough estimate of the double-layer capacitance. Thus from Eqs. (318) and (327), $u_e^* = (\epsilon/\epsilon_d)u^*$, which with Eq. (339) gives

$$u_e^*/\gamma_\beta \epsilon^t = u^*/\epsilon_d = q_e(d+b'_+)/\epsilon_d \ . \tag{340}$$

For formation in dilute aqueous electrolytes, $q_e(d+b'_+)/\epsilon_d$ and hence $(u_e^*/\gamma_\beta \epsilon^t)$ should be approximately the same for all the metal-oxide systems whose behavior is in accord with the proposed model.

In Table 9, the relevant data of Tables 3, 6, and 8, along with the values for u_e^* for aluminum and bismuth given in Section VIII. B. 4, have been collected and $(u_e^*/\gamma_\beta \epsilon^t)$ has been evaluated for aluminum, bismuth, tantalum, and niobium. (The values of u_e^* for tantalum and niobium are those calculated from the A data sets (see Table 6), as they show a considerably smaller standard deviation than do those calculated from the B data sets.) With the possible exception of

Bi_2O_3, the quantity $(u_e^*/\gamma_\beta \epsilon^t)$ is the same within experimental uncertainty for the four metal-oxide systems, the mean value being 0.35 ± 0.11 eÅ. The double-layer capacitance, excluding the contribution of the diffuse double layer (and adsorption pseudocapacitance) is given by $\underline{c}_{dl} = \epsilon_d/4\pi d$, which on eliminating ϵ_d by using Eq. (340) becomes

$$\underline{c}_{dl} = \frac{q_e(1 + b'_+/d)}{4\pi(u_e^*/\gamma_\beta \epsilon^t)} \approx 25\left(1 + \frac{b'_+}{d}\right) \frac{\mu F}{cm^2} , \qquad (341)$$

where the mean value for $(u_e^*/\gamma_\beta \epsilon^t)$ has been used. Setting $b'_+ \approx \tfrac{1}{2}d$ then gives $\underline{c}_{dl} \approx 38\,\mu F\ cm^2$.

Vermilyea [42], from capacitance measurements at 1 kHz over a range of pH, calculated a value of 50 $\mu F/cm^2$ for the double-layer capacitance associated with anodically formed Ta_2O_5 in aqueous solution. The agreement appears to be satisfactory considering the approximations and assumptions involved in obtaining both estimates. A change in the functional form assumed for $W_e(E)$ can by itself bring the present estimate of the double-layer capacitance into agreement with Vermilyea's value. Thus if it is assumed that $W_e(E)$ for the tantalum system takes the form arising from a parabolic potential-energy barrier (see Table 5), then from Table 5, $u_e^* = 25.0$ eÅ, and from Table 7, $\gamma_\beta = 3.77$, which with Eq. (341) and the assumption $b'_+ \approx \tfrac{1}{2}d$ gives $\underline{c}_{dl} \approx 55\ \mu F/cm^2$.

The significance of the agreement achieved here is, of course, that it shows the values for the field coefficients β_{ss} and β_t to be consistent with those estimated from the model.

c. Galvanostatic Transients. In order to account for potentiostatic and galvanostatic transients, the form of the kinetics of the polarization process(es) must be included. The case of galvanostatic transients, performed as per Fig. 5, is treated first.

In this derivation it is assumed that i is large enough to make the rate of change of the "slow" components of polarization via mechanisms involving thermal activation negligible compared with that induced by ion-current flow. From Eqs. (333) and (334), this assumption takes the form

$$\sum_{K=2,3} B_K i (\chi_K E - 4\pi P_K) \gg \sum_{\ell=2,3} \left(\frac{1}{\tau_\ell}\right) (\chi_\ell' E - 4\pi P_\ell') . \tag{342}$$

Note that Eq. (342) cannot be written simply as $B_K i \gg 1/\tau_K$, $K = 2, 3, \ldots$, since a one-to-one relationship does not exist between the components P_K and P_ℓ'. The division of P into components P_ℓ' is governed by the activation energies for the polarization processes, whereas the division into components P_K is governed by structural factors and is not influenced by the activation-energy distribution.

To show that this model predicts that galvanostatic transients are a function of the total charge passed, Σ, and do not depend on time explicitly, an expression for the charging-current density i_c is first obtained. The charging current has the following form:

$$4\pi i_c = dE/dt + 4\pi dP/dt , \tag{343}$$

which with Eq. (328) reduces to

$$4\pi i_c = \epsilon_d dE_d/dt . \tag{344}$$

The total current density, $I = i + i_c$, can then be written in the form

$$I = i + I(\epsilon_d/4\pi) dE_d/d\Sigma \tag{345}$$

Substituting for E_d from Eq. (329) and rearranging, Eq. (345) becomes

$$\frac{di}{i(1 - i/I)} = \frac{4\pi\beta_d}{\epsilon_d} d\Sigma . \tag{346}$$

Integrating for the conditions $i = I_1$ for $\Sigma = 0$, and $I = I_2$ = a constant leads to

$$1 - I_2/i = (1 - I_2/I_1) \exp(-4\pi\beta_{ss}\Sigma/\epsilon) \tag{347}$$

where the substitution $\beta_d = \epsilon_d\beta_{ss}/\epsilon$ has been made. Dividing through by I_2, Eq. (334) can be written

$$4\pi dP_K/d\Sigma = B_K(i/I_2)(\chi_K E - 4\pi P_K) . \tag{348}$$

Using the boundary conditions

$$4\pi P_K(\Sigma = 0) = \chi_K E_1 , \quad K = 2, 3, \ldots , \tag{349}$$

Eq. (348) may be solved for P_K to obtain

$$4\pi P_K = \left\{ E_1 + B_K \int_0^\Sigma (i/I_2)E \exp\left[B_K \int_0^\Sigma (i/I_2)d\Sigma\right] \right\} \chi_k$$

$$\times \exp\left[-B_K \int_0^\Sigma (i/I_2)d\Sigma\right] . \tag{350}$$

Noting that

$$I_2 = A_i \exp\beta_{ss}E_2 , \tag{351}$$

Eq. (329) can be written

$$\beta_d E_d - \beta_{ss}E_2 = \ln(i/I_2) . \tag{352}$$

Eliminating E_d and P from Eqs. (352), (328) and (332) and simplifying leads to the equation

$$4\pi \sum_{K=2,3,\ldots} P_K = \epsilon E_2 - {}^t\epsilon E + (\epsilon/\beta_{ss}) \ln(i/I_2) . \tag{353}$$

Eliminating P_K, $K = 2, 3, \ldots$, from Eq. (353) using Eq. (350), and (i/I_2) from the resulting equation using Eq. (347), an integral equation in E and Σ results. This may be transformed to a differential equation by the usual method of integral isolation and differentiation, the order of the equation being equal to the number of distinguishable, non-zero, B_K values. In principle, this may be solved to yield $(E - E_2)/(E_2 - E_1)$ as a function of (I_2/I_1) and Σ. As an exact solution to this equation has not been found, the following approximation is investigated.

From Eq. (346), it follows that for $\Sigma \gg \epsilon/4\pi\beta_{ss}$, $|i - I_2| = |i_c| \ll I_2$. Thus for $\Sigma \gg \epsilon/4\pi\beta_{ss}$, the charging current may be neglected, and i set equal to I_2. For the tantalum system $\epsilon/4\pi\beta_{ss} = \epsilon'\gamma_\beta/4\pi\beta_{ss} \simeq 1.5\ \mu C/cm^2$, a value close to this applying to the aluminum, niobium and the bismuth systems. In Fig. 5, it is seen that $(E - E_2)/(E_2 - E_1)$ reaches its peak value at about $2\ \mu C/cm^2$. The precise position of the peak depends on the current ratio, I_2/I_1.

Making the approximation $i = I_2$, Eqs. (353) and (350) become

$$4\pi \sum_{K=2,3,\ldots} P_K = \epsilon E_2 - \epsilon' E \tag{354}$$

$$4\pi P_K = \chi_K E_1 e^{-B_K \Sigma} + B_K \chi_K e^{-B_K \Sigma} \int_0^\Sigma E e^{B_K \Sigma} d\Sigma . \tag{355}$$

The combination of Eqs. (354) and (355) represents an integral equation for E in Σ. This can be reduced to a homogeneous differential equation in $(E - E_2)$ with constant coefficients through the successive application of the following operations: multiply through by $e^{B_K \Sigma}$, differentiate with respect to Σ, then multiply through by $e^{-B_K \Sigma}$ (K = 2 the first time, K = 3 the second time, etc.) Such a differential equation can be readily integrated. The result assuming only two current-driven relaxation processes (i. e., setting $B_K = 0$ for $K > 3$) is given by [88]

$$\frac{E - E_2}{E_1 - E_2} = A_1 e^{-a_1 \Sigma} + A_2 e^{-a_2 \Sigma} , \tag{356}$$

where A_1, A_2, a_1, a_2 are related to χ_2, χ_3, B_2, B_3 as follows: B_2 and B_3 are the roots of the quadratic equation

$$B^2 - T^* B + S^* = 0 \tag{357}$$

and χ_K, $K = 2, 3$, is given by

$$\chi_K = \frac{Q^* - P^* B_j}{B_K - B_j} , \tag{358}$$

where

$$P^* = \epsilon^t (A_1 + A_2) ,$$

$$Q^* = \frac{\epsilon^t (A_1 a_1 + A_2 a_2)}{1 + A_1 + A_2} ,$$

$$S^* = \frac{a_1 a_2}{1 + A_1 + A_2} ,$$

$$T^* = \frac{a_1 (1 + A_2) + a_2 (1 + A_1)}{1 + A_1 + A_2} .$$

Including additional polarization terms leads to additional exponential terms in Eq. (356), the total number of terms being equal to the number of distinguishable, nonzero B_K values.

From Fig. 5 it follows that for the aluminum system at least two polarization mechanisms involving ion-current induction are required to account for the data, since $\ln [(E - E_2)/(E_2 - E_1)]$ is not a linear function of Σ beyond the maximum. The solid line in Fig. 5 is calculated according to Eq. (356) by using the following values (chosen to minimize the variance in E): $B_2 = 1.22 \times 10^4$ cm^2/C, $\chi_2 = 1.91 \epsilon^t$, $B_3 = 9.32 \times 10^4$ cm^2/C, $\chi_3 = 0.44 \epsilon^t$. The values for B_2 and B_3 correspond to cross-sectional radii for the ion-current-driven polarization mechanisms of 2.4_8 and 6.8_7 Å, respectively, and the values for χ_2 and χ_3 give $\gamma_\beta = 3.3_5$, in reasonable agreement with independent estimates of this parameter (see Table 8).

Dewald's data for Ta_2O_5 [33] have also been fitted very accurately [87], assuming two ion-current-driven relaxation processes, the parameters leading to the best fit being $B_2 = 6.57 \times 10^3$ cm^2/C, $\chi_2 = 36.0_2$, $B_3 = 4.34 \times 10^4$ cm^2/C, $\chi_3 = 16.4_0$, where ϵ^t has been set equal to 27.6. These correspond to cross-sectional radii of 1.8_2 and 4.6_9 Å, and to a value for γ_β of 2.90.

d. Potentiostatic Transients. The case of potentiostatic transients is now examined. The general features of the results presented in Fig. 6 can be interpreted in terms of the present model as follows: The current in the region A to C is presumed to be almost entirely charging current, the form of the curve in this region being therefore determined by the kinetics of polarization change occurring via mechanisms involving thermal activation. These mechanisms lead to an increase in P and hence in E_d in the constant-field region, B to C, until in the vicinity of C, E_d is large enough for the ion current to be comparable to the charging current. Beyond C, the rate of change of P induced by mechanisms involving thermal activation can presumably be neglected compared with that induced by the ion current, and furthermore in this region $i \approx I$. In the vicinity of D, P has essentially achieved its steady-state value, the drop in current beyond D arising from the slow decrease in the field strength as the film thickens.

The following approximations should be reasonably good in the region beyond C, up to D: $i \approx I$, Eq. (342) valid, and

$$E = \text{constant} = (1/\beta_{ss}) \ln (I_{ss}/A_i) , \tag{359}$$

where I_{ss} is the steady-state current density corresponding to E. By making these approximations, an analytical solution can be obtained. Thus Eq. (334) can be integrated for E constant to give

$$4\pi P_K = \chi_K E + (4\pi P_K^0 - \chi_K E)e^{-B_K \Sigma} , \tag{360}$$

where P_K^0 is the value of P_K for $\Sigma = 0$. Furthermore, on eliminating E_d and P from Eqs. (328), (329), and (332), and setting $i = I$ and $\beta_d = (\epsilon_d/\epsilon)\beta_{ss}$, the following result is obtained:

$$4\pi \sum_{K=2,3} P_K = \frac{\epsilon}{\beta_{ss}} \ln\left(\frac{I}{A_i}\right) - \epsilon^t E . \tag{361}$$

Eliminating $\sum_{K=2,3} P_K$ and E from Eqs. (360) and (361) and making use of Eq. (359) then gives

$$\ln\left(\frac{I_{ss}}{I}\right) = \sum_{K=2,3} C_K e^{-B_K \Sigma} , \tag{362}$$

where the C_K are constants determined by the initial conditions and are given by

$$C_K = \beta_{ss}(\chi_K E - 4\pi P_K^0)/\epsilon . \tag{363}$$

Since for both the aluminum and tantalum systems, $B_3 \approx 7 \times B_2$ and χ_3 is from one-half to one-fourth of χ_2, the term containing B_2 will dominate except during the early stages of the transient, when $\Sigma \widetilde{<} 1/B_3$. The condition $\Sigma \approx 1/B_3$ is satisfied in the vicinity of C, Fig. 6, where the charging current is not yet negligible in comparison with the ion current. Thus for the region of the curve that can be analyzed according to Eq. (362), $C_3 e^{-B_3 \Sigma} \ll C_2 e^{-B_2 \Sigma}$, so that Eq. (362) reduces to

$$\ln(I_{ss}/I) = \ln[I_{ss}/I(\Sigma' = 0)]e^{-B_2 \Sigma'} . \tag{364}$$

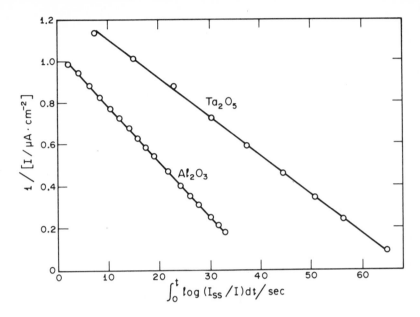

Fig. 7. Analysis of potentiostatic ion-current transients for Al_2O_3 (Fig. 6) and Ta_2O_5 (L. Young's data [86]) according to Eq. (365). The results are according to Dignam and Ryan [90].

On differentiating this with respect to t, the resulting equation may be integrated in the following alternative form:

$$\frac{1}{I} = \frac{1}{I(t=0)} - B_2 \int_0^t \ln(I_{ss}/I)\, dt \tag{365}$$

where $t = 0 = \Sigma'$ corresponds to a time beyond C of Fig. 6.

In Fig. 7 the data of Fig. 6 lying beyond C are plotted according to Eq. (365), along with similar data obtained by L. Young for Ta_2O_5 [86]. The results confirm Eq. (365). From a series of such measurements [90] on Al_2O_3 formed initially at about 2×10^{-4} A/cm^2 in a glycol borate electrolyte, a mean value for B_2 of $(1.22 \pm 0.07) \times 10^4$ cm^2/C was obtained, in exact accord with the value of B_2 obtained from the galvanostatic transients (Fig. 5), which were also

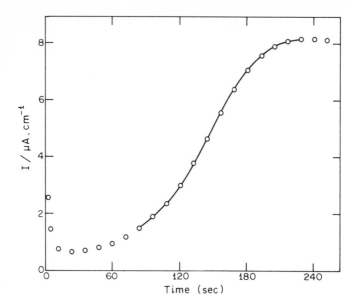

Fig. 8. Analysis of potentiostatic ion-current transients for Ta_2O_5 (0.2 N H_2SO_4) according to Eq. (364). The data are represented by open circles, the least-squares computed curve by the solid line. The results and analysis are taken from Taylor and Dignam [88].

measured in a glycol borate electrolyte at current densities of about 2×10^{-4} A/cm^2. In another series of experiments on Al_2O_3 formed initially at about 2×10^{-6} A/cm^2 in a glycol borate electrolyte [90], a mean value for B_2 of $(1.46 \pm 0.04) \times 10^4$ cm^2/C was obtained, this being significantly larger than that obtained for films formed at 2×10^{-4} A/cm^2. Thus B_2 appears to depend somewhat on formation conditions, a result that is not unexpected since the link between the extraordinary polarization mechanisms and the passage of ion current must be structure dependent.

A similar result has been found for Ta_2O_5. Taylor and Dignam [88] carried out potentiostatic transient measurements on Ta_2O_5 formed at 2.4×10^{-4} A/cm^2 in 0.2 N H_2SO_4 and then stored for about 1 month at room temperature in air. The data were analyzed in accord with

Eq. (364), the value of B_2 being chosen to minimize the variance in I. An example of the quality of fit achieved is shown in Fig. 8. The calculated curve is seen to reproduce the data almost exactly over the experimental region in which the approximations leading to Eq. (364) are expected to be valid. A series of such measurements yielded a mean value for B_2 of $(6.6_3 \pm 0.3_6) \times 10^3$ cm^2/C, once again in excellent accord with the value obtained from Dewald's galvanostatic transient measurements $(6.57 \times 10^3$ cm^3/C), which were performed in 0.1 N H_2SO_4 at current densities of about 10^{-4} A/cm^2. Analysis of L. Young's data [86] by Dignam and Ryan [90] (see Fig. 7), on the other hand, led to a value for B_2 of 8.0×10^3 cm^2/C. The specimen from which this result was obtained, however, was formed at 100 V (constant potential) for 3.5 h and then annealed by immersion in boiling water for 5 min. Again, the value of B_2 appears to depend on formation conditions.

Finally, it is worth noting that by including the polarization mechanisms involving thermal activation and those involving ion-current induction, as well as the charging current, in the analysis of potentiostatic transients, Dignam and Ryan [90] were able to reproduce all of the features of Fig. 6. A quantitative fit of these data over the entire time period can only be attempted, however, with a knowledge of the kinetics of the polarization processes involving thermal activation and the way in which these processes interact with those involving ion-current induction. Similarly the potentiostatic transient results analyzed by L. Young [94] in terms of the high-field Frenkel-defect model cannot be tested against the dielectric relaxation model without first determining the kinetics of the polarization changes occurring via mechanisms involving thermal activation [35].

e. Alternating-Current-Impedance Measurements under Galvanostatic Conditions. The measurement of ac impedance during film formation has been carried out for Al_2O_3 and Ta_2O_5 by Winkel, Pistorius, and Van Geel [95]; for Al_2O_3 by Goad and Dignam [87]; and for Ta_2O_5 by Taylor and Dignam [88]. The results of Winkel et al. included only the in-phase component of the ac impedance, both the in-phase and out-of-phase components being reported by Goad, Taylor, and Dignam.

Taylor and Dignam [88] used an automatic system that involved the application to the cell of a constant current on which was superimposed a small ac current of fixed amplitude and angular frequency ω. A phase-sensitive detector was used to measure alternately the ac

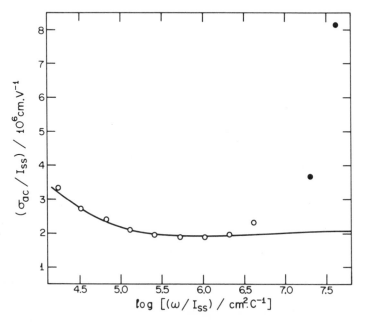

Fig. 9. Variation in ac conductance σ_{ac} for the system Ta/Ta_2O_5 in 0.2 N H_2SO_4 as a function of dc current density I_{SS} and angular frequency ω. o, $I_{SS} = 223\,\mu A/cm$; •, $I_{SS} = 1.16\ mA/cm^2$; the solid line is the calculated curve. The data and calculations are taken from Taylor and Dignam [88].

component of the cell potential, which was in and out of phase, respectively, with the ac current. These ac components, as well as the dc component of the cell potential, were recorded automatically in digital form for computer processing. From a single run the ac conductivity σ_{ac} and the "effective" dielectric constant ϵ_{eff} were obtained for a single dc current density I_{ss} and frequency, the equivalent circuit used for the calculations being a resistor and capacitor in parallel. The derivation of expressions for σ_{ac} and ϵ_{eff} in terms of the present model is involved, as are the expressions themselves [88]; the equations are therefore not given here. The form of the equations is such that ϵ_{eff} and σ_{ac}/I_{ss} are predicted to be functions of ω/I_{ss} and of the system constants (B_K, χ_K, ϵ^t, β_{ss}), but independent of ω and I_{ss} individually.

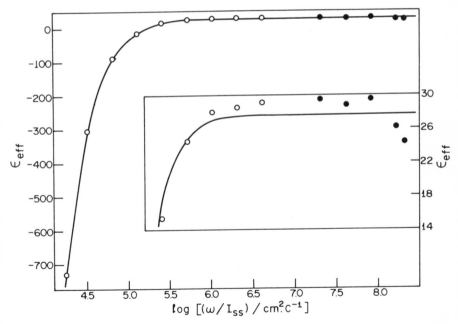

Fig. 10. Variation in the effective ac dielectric constant ϵ_{eff} as a function of dc current density I_{ss} and angular frequency ω for the same system and conditions as for Fig. 9. The data and calculations are taken from Taylor and Dignam [88].

Taylor and Dignam's results, along with curves computed by
using values for β_{ss} calculated from the constants of Table 6 (quadratic function), the values for B_2, χ_2, B_3, χ_3 obtained from Dewald's
data, and L. Young's value for ϵ^t (27.6) are presented in Figs. 9 and
10. For frequencies below about 1 kHz the agreement between the
calculated and measured results is good; above 1 kHz, however, the
ac conductivity increases rapidly to values very much larger than
those deduced from the model. A further process is apparently contributing to the ac conductivity at high frequencies. The discrepancy cannot be removed by including a third independent ion-current-
driven dielectric relaxation process [88].

The data of Dignam and Goad for Al_2O_3 show good agreement
with the model over the range of frequencies investigated, 1.5 to
100 Hz.

f. Direct Investigation of the Kinetics of Polarization Change Induced
by Ion Current. Using a voltage-pulse technique, Taylor and Dignam
[88, 96] carried out a direct investigation of the kinetics of polarization change induced by an ion current in order to ascertain first of
all whether such mechanisms exist at all and, in the event of their
existence, to compare the values of the constants B_K, χ_K obtained
from such a study with those obtained from ion-current-transient
measurements.

The problem in detecting directly the extraordinary polarization
current (i. e., that arising from polarization change induced by an ion
current) is that it is predicted to be at most only a few percent of the
ion current producing it. It is therefore necessary to observe the
effect of the extraordinary polarization current in the absence of an
ion current.

In general both the mechanisms involving thermal activation and
those involving ion-current induction operate to change the polarization

of the film when an appreciable ion current is flowing. In the absence of an ion current, however, all polarization changes must be thermally activated. Thus when a square-wave voltage pulse of sufficient amplitude to induce ion-current flow is applied across a preformed film, the polarization of the oxide will be driven upward during the pulse by both the extraordinary and the thermal mechanisms. After the pulse, the ion current is again zero, so that the polarization must decay toward its original value by thermal mechanisms only. The integrated discharge current (after the pulse) measures the total polarization induced during the pulse. Increasing the total integrated ion current that flows during the pulse, while maintaining the pulse height and duration constant, will lead to an increase in the integrated discharge current if mechanisms involving ion-current induction are indeed operative, but not in general otherwise.[*]

Figure 11 shows the data obtained for a series of such voltage-pulse experiments performed on Ta_2O_5 films formed at 2.4×10^{-4} A/cm^2 in 0.2 N H_2SO_4 [96]. For all three runs the field pulse ΔE and pulse duration Δt are the same, and the initial field E_1 is sufficiently low for the ion current to flow only during the charging pulse. The data show the result expected if polarization mechanisms involving ion-current induction are operative.

In a more extensive investigation of this phenomenon [88] the results of similar experiments were compared quantitatively with the predictions of the model under study, the one essential difference in the experimental technique being that the total charge passed per unit area during and after the pulse was recorded directly on an

---

[*]It is possible that some other unspecified extraordinary polarization process could produce a similar result, since although the pulse height ΔE and pulse duration are kept constant, the actual field values must of course be increased in order to produce an increase in i.

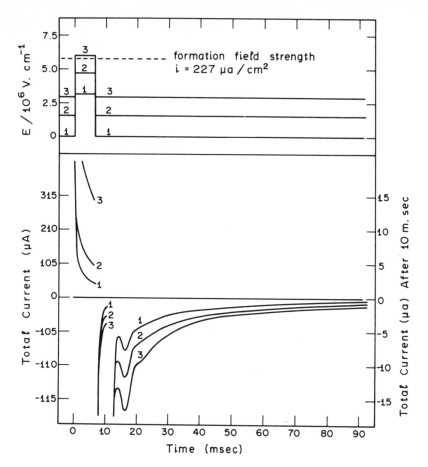

Fig. 11. Current-time curves illustrating increasing rate of dielectric polarization for the system Ta/Ta$_2$O$_5$ in 0.2 N H$_2$SO$_4$ with increasing ion current for a constant polarizing field increment and time interval. The effect is made apparent through the behavior of the discharge current. (The oscillations are of instrumental origin.) Data from Dignam and Taylor [96].

oscilloscope by using electronic integrators, from which the total charge passed per unit area during the charging pulse, Σ_+, and that during the discharge period, Σ_-, could be obtained. Assuming that the polarization of the oxide returns to its initial value within the period of time in which the measurements are made, the total ion charge passed during the pulse, Σ_i, is given by

$$\Sigma_i = \Sigma_+ + \Sigma_- . \qquad (366)$$

If the pulse duration is sufficiently short, the amount of polarization induced will be small, so that the various polarization mechanisms may be treated as being independent of one another. Integration of Eqs. (333) and (334) over the charging period then gives [88, 96]

$$4\pi\Sigma_+ = \left\{ \epsilon^t + \sum_\ell \chi_\ell' \left[1 - \exp\left(-\frac{\Delta t}{\tau_\ell} \right) \right] \right.$$

$$\left. + \sum_{K=2,3} \chi_K \left[1 - \exp\left(-B_K \Sigma_i \right) \right] \right\} \Delta E + 4\pi\Sigma_i , \qquad (367)$$

which on making the approximation $B_K \Sigma_i \ll 1$, $K = 2, 3$, and utilizing Eq. (366), can be written

$$-\Sigma_- = \left\{ \epsilon^t + \sum_\ell \chi_\ell' \left[1 - \exp\left(-\frac{\Delta t}{\tau_\ell} \right) \right] + \left(B_2\chi_2 + B_3\chi_3 \right) \Sigma_i \right\} \left(\frac{\Delta E}{4\pi} \right) . \qquad (368)$$

For Δt and ΔE held constant, Eq. (368) predicts that $-\Sigma_-$ should be a linear function of Σ_i with slope $(B_2\chi_2 + B_3\chi_3)(\Delta E/4\pi)$.

Two sets of experiments were performed. In the first set the test electrode was held at E_1 for 2 h before applying the voltage pulse. The results of these experiments are summarized in Fig. 12. In the second set the electrodes were subjected to 10 to 15 pulses, delivered at 1-min intervals, with only the last charge-time being recorded (Fig. 13). For values of Σ_i larger than those appearing in Figs. 12 and 13, the discharge current decayed over a long period of time, making the measurement of Σ_- effectively impossible. Using the values for B_2 and B_3 obtained from Dewald's [33] galvanostatic transient measurements, the approximation made in deriving Eq. (368)

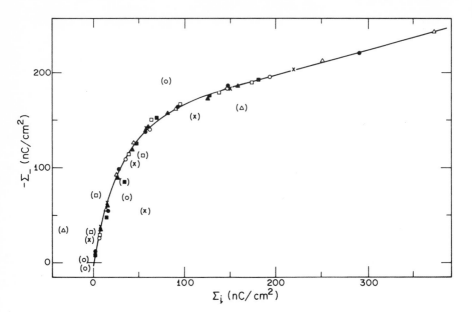

Fig. 12. Dependence of the charge passed during discharge, Σ_-, on the ion charge passed during the charging pulse, Σ_i. The contribution from the "fast" polarization processes ($\epsilon^t \Delta E/4\pi$) has not been included. $\Delta E = 3.20 \times 10^6$ V/cm for all runs; for runs marked $\times$, $\triangle$, $\blacksquare$, and $\bullet$, $\Delta t = 25.9$ msec; for runs marked $\blacktriangle$, o, $\square$, $\Delta t = 10.6$ msec. For further details refer to the original paper by Taylor and Dignam [88].

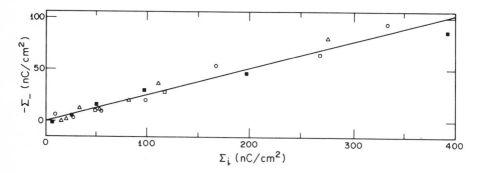

Fig. 13. Dependence of the charge passed during discharge, Σ^-, on the ion charge passed during the charging pulse, Σ_i, as per Fig. 12, except that a repetitive cycling procedure was used. $\Delta E = 3.2 \times 10^6$ V/cm, $\Delta t = 25.9$ msec. The data are according to Taylor and Dignam [88].

(i. e. , $B_K \Sigma_i \ll 1$, $K = 2, 3$) is justified for both sets of data, so they should obey Eq. (368) if the model is correct.

The data of Fig. 13 do indeed obey this equation very precisely, the experimental slope divided by $\Delta E/4\pi$ being (9.30 ± 0.42) × 10^5 cm^2/C , which is to be compared with the theoretical value, $(\chi_2 B_2 + \chi_3 B_3) = (2.37 + 7.12) \times 10^5 = 9.49 \times 10^5$ cm^2/C , where again the values for B_K, χ_K obtained from Dewald's data have been used. The data lying on the linear portion of Fig. 12 yield for the slope divided by $\Delta E/4\pi$ a value of 9.65×10^5 cm^2/C , which is also in agreement with the theoretical value.

The curved portion of Fig. 12, however, is not predicted by Eq. (368). The solid line in Fig. 12 was calculated [88] from an equation derived by assuming a third independent ion–current–driven relaxation process, with constants $B_4 = 3.3 \times 10^7$ cm^2/C and $\chi_4 = 0.53$. In one sense this explanation for the curved region of Fig. 12 cannot be ruled out, since including the additional polarization term will have virtually no effect on galvanostatic and potentiostatic ion–current transients. This follows from the fact that B_4 is so large that $e^{-B_4 \Sigma_i}$ can be neglected compared with unity over essentially the entire period of the transient. On the other hand, the interaction radius corresponding to this value for B_4 is 130Å, which appears to be too large to be physically meaningful. It is possible that the curved portion of Fig. 12 is due to the presence of certain nonlinear thermally activated polarization mechanisms (e. g., $4\pi dP_j/dt \propto (E\chi_j)^2 - (4\pi P_j)^2$). The problem of unraveling all of the details of these charging–current measurements is clearly a challenging one. For further discussion the original paper [88] may be consulted.

The essential result of these experiments is that polarization processes involving ion–current induction appear to take place in

anodically formed Ta_2O_5 films, and furthermore, the kinetic constants
for these processes are in accord with those calculated from ion-
current transients assuming the dielectric relaxation model to be
valid. Despite the fact that certain features of the polarization
mechanisms remain unexplained, the author believes that these
results represent strong support for the basic tenets of the dielectric
relaxation model, with the details of the model, however, being in
considerable doubt.

g. Transients Following Thermal Treatment. A number of authors [85,
97, 98] have carried out experiments in which anodically formed films
have been subjected to a thermal annealing treatment, after which
some sort of ion-current-transient measurement is performed.
Vermilyea [85] has shown that thermal annealing of anodic Ta_2O_5
films brings about a change in the film structure. Thus he observed
a reduction in the diffuseness of x-ray diffraction patterns and a
reduction in the rate of film dissolution in hydrofluoric acid on
annealing at temperatures above 200°C. The dissolution rate was
found to decrease linearly with the logarithm of the annealing time
at a given annealing temperature. Furthermore, the magnitudes of
the activation energies for the processes bringing about the structure
changes (determined from the dissolution rates) were found to be the
same as those extracted from the changes in galvanostatic transients
brought about by annealing.

As a number of the parameters for the model being investigated
are certain to be at least somewhat structure sensitive (in particular
B_2, χ_2, B_3, χ_3, but also ϵ^t), the results obtained by Vermilyea [85]
can perhaps be accounted for in this way. Since a substantial amount
of reordering of the amorphous film toward a crystalline structure,
brought about by annealing, would be expected to block more or less
effectively ion transport via the proposed mechanism involving

network defects (see Section VII), the major effect of annealing might instead be due to this. In any event it is clear that our present knowledge of the structure of anodically formed films is insufficient to attempt a quantitative analysis of transients after thermal treatment.

4. Shortcomings of the Dielectric Relaxation Model for Transients and Possible Alternatives

Although some features of the polarization processes (Fig. 12) and of ac impedance measurements during film formation (Fig. 9) are not explained in terms of the dielectric relaxation model, in total the model gives an excellent quantitative and self-consistent account of the data. In particular, the author knows of no data that appear to be in contradiction with the model, though there are certainly many phenomena that are not in any obvious way encompassed by it. This said, the model in its present formulation nevertheless seems to have several important shortcomings.

In a recent publication Siejka, Nadai, and Amsel [99], using nuclear methods of analysis for oxygen-16 and oxygen-18, appear to have shown that the contribution of the oxide-solution double layer to the overpotential during film formation is independent of current density, and in fact is zero within experimental error for both tantalum and aluminum anodized in an aqueous solution of 5 wt % ammonium citrate of pH 6. The experimental uncertainty was on the order of 0.1 V. Such a result is not expected for the interfacial model outlined in Section VI. C (cf. Section VI. D. 2)1 Neither is it expected for a model involving rate control at the metal-oxide interface, nor for any model involving control within the bulk of the film. The implication of the result is that the oxide surface and the outer Helmholtz plane are essentially coincident. If this is correct, then the detailed model for the interfacial reactions at the oxide-electrolyte interface presented in Section VI. C cannot be correct. Rate control by defect injection at either the metal-oxide or the oxide-electrolyte interface, however, would not be precluded, leaving most of the equations of Section VIII unchanged in form.

The data provide strong support for the existence of at least one polarization mechanism involving ion-current induction. Furthermore, the proposed mechanism for the migration of network defects leads naturally to one such mechanism. The apparent necessity for postulating a second such mechanism (for all three systems: tantalum, aluminum, and indium antimonide) appears to the author a significant weakness. A number of attempts to replace one of the extraordinary polarization mechanisms with an equation or equations relating to interfacial processes (see Section VI.C) have, however, been totally unsuccessful. Two difficulties arise in any such attempt. The first stems from the condition that in order to account for the basic form of the transient kinetics (see Section VIII.C.1) the kinetic equation replacing that for one of the extraordinary polarization mechanisms must take a form that does not contain i or t explicitly but only the time integral of i, Σ_i. No conventional treatment known to the author leads to such a kinetic equation. The second difficulty is a consequence of the apparent verification of the two extraordinary polarization mechanisms afforded by the charging-current experiments (i.e., $\sum_K B_K \chi_K$ from the charging-current measurements equals that calculated from the ion-current transients). Thus any new kinetic process must manifest itself in some manner as a polarization change. It is apparent that a time variation in the boundary charge $4\pi(m - m^*)q_e$ (see Eq. (248)) would behave in all important respects as a polarization process.

An approach to the problem along these lines, however, was unsuccessful. An alternative way of eliminating one of the polarization mechanisms is to suppose that there is only a single mechanism involving ion-current induction, the kinetics of which is given by

$$4\pi \frac{dP_2''}{d\Sigma_i} = B(\chi_2'' E - 4\pi P_2'') + \gamma_1 \frac{d}{d\Sigma_i}(\chi_2'' E) - \gamma_2 4\pi \frac{d^2 P_2''}{d\Sigma_i^2} , \qquad (369)$$

where γ_1 and γ_2 are positive constants. This equation is equivalent to Eqs. (332) and (334) and may be obtained from them on setting

$$P_2'' = P_2 + P_3 ; \quad B = B_2 B_3/(B_2 + B_3) ; \quad \chi_2'' = \chi_2 + \chi_3 ;$$

$$\gamma_1 = (\chi_2 B_2 + \chi_3 B_3)/(\chi_2 + \chi_3)(B_2 + B_3) ; \quad \gamma_2 = 1/(B_2 + B_3) .$$

A further difficulty with the model concerns the material discussed in Section VIII. B. If the dielectric relaxation model is accepted, then it must be concluded that electrolyte incorporation into anodic films changes ϵ and ϵ^t by approximately the same factor. This implies that $(\chi_2 + \chi_3)$ is approximately proportional to ϵ^t, so that as ϵ^t changes by electrolyte incorporation $(\chi_2 + \chi_3)$ changes proportionately, ϵ/ϵ^t therefore remaining approximately constant. The author can think of no convincing reason why this should be the case.

The major problem in attempting to refine the model or develop a new one lies in our inadequate knowledge of the structure of both the oxide films themselves, and of the film-electrolyte interface. An intensive study of the properties of this interface along the lines pioneered by Vermilyea [42] but including ion-conduction measurements as well would appear to be valuable in this regard.

IX. APPLICATIONS TO OTHER METAL-FILM SYSTEMS

The purpose of this section is to indicate some of the other areas in which the material presented in this chapter might be applied with success. No attempt is made to develop these areas, however.

If, as has been suggested, the detailed kinetic behavior of the valve-metal-oxide systems is due in no small measure to the fact that the films take on an amorphous structure and grow by transport of singly charged network defects, then very different kinetic behavior is expected for the metals that form crystalline oxides when oxidized anodically. Thus metals that form even-valent metal oxides (MO, MO_2, and MO_3) on anodic oxidation are not expected to behave like tantalum and aluminum. It is probable that for metals forming oxides of composition MO the oxides will be essentially crystalline and ion transport will involve predominantly the metal ion, either by an interstitial or a vacancy mechanism. In either case, from the material presented in Sections V and VI, rate control at the interface away from which the defects move is expected, at least for thin films.

Films consisting of silver oxide or silver salts probably belong in a class by themselves due to the unusually high metal-ion defect concentration and mobility found in such systems. There is evidence that ion conduction through the system $Ag/AgCl$/aqueous solution is space-charge limited [100] and hence is in accord with the equations developed in Section IV. C. 1. The kinetics of anodic oxide-film growth on silver are complex and not well understood [101]. There is some evidence that the rate processes are controlled primarily at the oxide-electrolyte interface for Ag_2O formation, with the rate of diffusion of neutral species through the film controlling the limiting thickness achieved before the onset of AgO formation [10]. It is possible, however, that the results can be accounted for in terms of the equations of Section IV. C. 1, or even possibly those of Section VI. D. 3, which apply to the case of rate control at the far interface. The conversion of Ag_2O to AgO, on the other hand, appears to be controlled by O^{2-} ion migration across the Ag_2O/AgO interface [102], with apparently no contribution arising from the transport of oxygen or metal through the films.

For crystalline oxide films of composition MO_2, ion transport will in all likelihood involve oxygen-ion migration, either along grain boundaries or by a vacancy mechanism, since the enthalpy of formation of such defects at the appropriate interface is almost certain to be substantially less than that for the formation of a quadruply charged interstitial metal ion. The same argument applies, of course, to oxides of composition MO_3.

The application of the material in Sections VI.C and VII to data for the formation of porous anodic oxide films on aluminum is straightforward and appears to be consistent with recent interpretations of such data [101]. Thus oxide-film dissolution has been shown to be field assisted [103, 104]. From the material of Section VI.C, the field coefficient for the dissolution process should be given approximately by $(q_e d_+ \epsilon/\epsilon_d)/kT$, which is substantially smaller than that for film growth, $[q_e(d_+ + d_- + b'_+)\epsilon/\epsilon_d]/kT$, so that steady-state anodic oxidation conditions should lead to a stable barrier-layer oxide-film thickness. Thus if the dissolution rate exceeds the formation rate, the barrier-layer-film thickness will decrease, causing the field strength to increase and hence the ratio of film-formation rate to film-dissolution rate to increase, until the two rates become equal.

Much of the material presented in this chapter can be applied, with little modification, to the oxidation of metals in a gaseous environment [16, 36], and to plasma and solid-state oxidation.

In selecting material for this chapter an attempt has been made to present self-consistent models, hopefully of fairly general applicability, rather than to provide explanations for specific phenomena, the main exception being the material of Section VIII. It is clear that much sifting of current and new data is required before the range of applicability of the material presented can be ascertained.

ACKNOWLEDGMENTS

The author wishes to express his appreciation for the invaluable contribution to the substance of this chapter made by his co-workers, Drs. D. B. Gibbs, D. G. W. Goad, P. J. Ryan, D. F. Taylor, and D. J. Young, and by those who have read the manuscript and offered suggestions, J. P. S. Pringle, D. F. Taylor, and D. J. Young. He is also indebted to the National Research Council of Canada for supporting his own researches in this field.

GLOSSARY OF COMMONLY USED SYMBOLS

a	Activation distance
A	General constant; Helmholtz free energy
b_+, b_-	Effective activation distances at the oxide-film boundary, $x = 0$
b'_+, b'_-	Activation distances corresponding to b_+ and b_-
b_e	$= u^*_e/Q_e$
B	General constant
B_K	Constant of proportionality between (dP_K/dt) and $i \Delta P_K$ $(K = 1, 2)$
c	Dimensionless concentration, $= n/N$
c_0, c_S	Value of c at $s = 0$, $s = S$
c_∞	Value of c in the space-charge-free region of the film
C_2, C_3	Coefficients in the field expansion of $W(E)$ or $W_e(E)$; general coefficients
$\underline{c}$	Low-frequency capacitance of the electrode system
$\underline{c}^t$	High-frequency (~ 1 kHz) capacitance of the electrode system
$\underline{c}_{dl}$	Contribution of the Helmholtz double layer to the electrode capacitance
d	Jump distance; thickness of the Helmholtz double layer
d_+, d_-	Activation distances for the Helmholtz double layer, $d_+ + d_- = d$

D	Diffusion coefficient; electric displacement
e	Charge on the proton; $\exp(1)$
E	Field strength
E_d, E_0, E_X	Value of E in the Helmholtz double layer, at $x = 0$ and $x = X$, respectively
E_∞	Value of E in the space-charge-free region of the film
$\bar{E}$	Mean field strength, $= -V_X/X$
$\tilde{E}$	Differential field strength, $= -(\partial V_X/\partial X)_{J,T}$
E_L	Local electrostatic field strength
$f(z)$	Function of z
$F(z)$	Function of z
$g(z)$	Inverse of $f(z)$, that is, $g[f(z)] = z$
G	Gibbs free energy
ΔH^0	Standard enthalpy change: ΔH_d^0, for defect-pair formation; ΔH_f^0, for Frenkel-defect formation; ΔH_h^0, for hydration; ΔH_{pa}^0, for proton abstraction; ΔH_p^0, for polarization; ΔH_s^0, for defect injection
i	Ion (or Faradaic) current density
i_c	Charging-current density, $= (dE/dt)/4\pi + dP/dt$
I	Total current density, $= i + i_c$
j	Dimensionless flux density, $= J/2a\,N\exp(-Q/kT)$ or $= J/A_J$
J	Flux density
J_{ss}	Steady-state flux density
k	Boltzmann's constant
K	Summing index
K_d	Equilibrium constant for defect-pair formation
K_p	Coefficient of compressibility
ℓ	Summing index; exponent
m	Concentration of surface species $(-O^-)_{surface}$
M	Surface concentration of metal ions in a position to enter the film

M_p Concentration of sites for protons at the oxide-electrolyte interface

n Mobile-defect concentration

n_0, n_X Value of n at $x = 0$, $x = X$

n_∞ Value of n in the space-charge-free region of the film

n^* Immobile-defect concentration

N Concentration of sites for mobile defects

p Dimensionless potential, $= qV/kT$; pressure

P Polarization

P_1, P_2 Polarization components, $P = P_1 + P_2 + \cdots$

P_f Probability of lattice dissociation

q Charge on the mobile ion

q_e Charge on the electron

Q Activation energy ($E = 0$) for ionic transport: Q, within the film; Q_0, across the film interface at which the defects enter; Q_+, Q_-, across the Helmholtz double layer in the forward and reverse directions, respectively

Q_0, Q_1, Q_2 Coefficients in the field expansion of Q

Q_e Effective activation energy; for rate control at the oxide-electrolyte interface, $= \Delta H_{pa}^0 + Q_0$

ΔQ Peak height of diffusion barriers within the film less that of the entrance barrier ($E = 0$)

Q^* $= C_2 Q_e$

r Dimensionless space-charge density, $= (4\pi q a^2/\epsilon kT)\rho$; ionic radius

r_K Interaction radius, $= (B_K q/\pi)^{\frac{1}{2}}$

s Dimensionless position coordinate, $= x/a$ or $= x/(kT\beta/q)$ or $= x/(u_e^*/q)$

s' Dimensionless position coordinate, $= (8\pi q \beta_{ss} n_\infty/E)x$

s_0' Integration constant

S Dimensionless film thickness, $= X/a$

S_c Dimensionless, characteristic space-charge length, $= X_c/a$

t	Time
T	Absolute temperature; with subscript, a term in a series
$\underset{\sim}{T}$	Symmetry transformation
u	Charge-displacement coordinate
u_0, u_m	Value of u corresponding to a minimum and a maximum in the potential energy, respectively
u^*	Activation dipole for ionic migration
u_0^*, u_1^*	Coefficients in the field expansion of u^*
u_e^*	Effective activation dipole; for rate control at the oxide-electrolyte interface, $= (\epsilon/\epsilon_d)q_e d + qb_+$
U	Internal energy; potential energy of mobile defect
v	Ionic mobility
$\bar{v}$	Specific volume
V	Electrostatic potential
V_0, V_X, V_{aq}	Value of V at $x = 0$, $x = X$ and in the electrolyte, respectively
ΔV^0	$V_X - V_{aq}$ at the equilibrium electrode potential
$W(E)$	Net activation energy for migration in the positive x direction and in the presence of a field E
$W_f(E)$	Net activation energy for the formation of defect pairs
$W_e(E)$	Effective net activation energy for anodic film growth
x	Position variable
X	Film thickness
X_c	Characteristic space–charge length
y	General variable
z	General variable; oxidation number
α	From $r = \alpha c$, $\alpha = 4\pi q^2 a^2 N/\epsilon kT$; polarizability
β	Differential field coefficient, $= (\partial \ln J/\partial E)_{T,n}$
β_{ss}	Steady-state differential field coefficient, $= (\partial \ln J_{ss}/\partial E)_T$
β_t	Transient differential field coefficient, $$= \lim_{(dE/dt)\to\infty} \left(\frac{\partial \ln J}{\partial E}\right)_T$$

γ — Ratio of field coefficients, $= (2\beta_{ss} - \beta)/\beta$ for $qE_0 < qE_\infty$, $= \beta/(2\beta_{ss} - \beta)$ for $qE_0 > qE_\infty$; geometrical factor

γ_β — $= \beta_{ss}/\beta_t$

γ_e — $= u^*/u_e^*$

ε — Dimensionless field strength, $= -\partial p/\partial s = qaE/kT$

$\varepsilon_0, \varepsilon_S$ — Value of ε at $s = 0$, $s = S$

ε_∞ — Value of ε in space-charge-free region of the film

$\bar{\varepsilon}$ — Mean dimensionless field strength, $-p_S/S$

$\tilde{\varepsilon}$ — Differential dimensionless field strength, $= -(\partial p_S/\partial S)_j$

ϵ — Dielectric constant (generally for zero frequency) of the film

ϵ_0 — Dielectric constant at $x = 0 = S$

$\epsilon_0^!$ — Dielectric constant at optical frequencies

ϵ_d — Effective dielectric constant of the Helmholtz double layer

ϵ^t — High-frequency (generally ~ 1 kHz) dielectric constant

ϵ_{eff} — Effective dielectric constant calculated from the ac impedance of film-electrode system (parallel equivalent circuit)

η — Overpotential

θ — Polar angle

μ — Chemical potential

μ^0 — Standard chemical potential

$\bar{\mu}$ — Electrochemical potential

ν — Kinetic frequency factor

ρ — Volume-charge density

σ — Surface-charge density

$\bar{\sigma}$ — Cross-sectional area for Frenkel-defect recombination

σ_{ac} — Alternating-current conductance of film-electrode system (parallel equivalent circuit)

Σ — $= \int_0^t I \, dt$

Σ_i — $= \int_0^t i \, dt$

τ	Relaxation time
ϕ	Longitudinal angle
χ	Electric susceptibility
χ_0	Electric susceptibility at optical frequencies
χ_1	High-frequency ($\sim 1\,kHz$) electric susceptibility
χ_s	Static electric susceptibility, $= \chi_1 + \chi_2 +$
ω	Angular frequency of ac signal; general coefficient
Ω'	Volume per electrochemical equivalent weight of the oxide
Ω	Volume per atom equivalent weight

REFERENCES

[1] N. F. Mott and R. W. Gurney, Electronic Processes in Ionic Crystals, 2nd. ed., Oxford University Press, London, 1957.

[2] L. Young, Anodic Oxide Films, Academic Press, London, 1961.

[3] W. S. Goruk, L. Young, and F. G. R. Zobel, in Modern Aspects of Electrochemistry, No. 4 (J. O'M. Bockris, ed.), Plenum Press, New York, 1966.

[4] D. A. Vermilyea, in Advances in Electrochemistry and Electro-chemical Engineering (P. Delahay and C. Tobias, eds.), Vol. 3, Interscience, New York, 1963.

[5] T. P. Hoar, in Modern Aspects of Electrochemistry, No. 2 (J. O'M. Bockris, ed.), Butterworths, London, 1959.

[6] N. G. Bardina, Russian Chem. Rev., 33, 286 (1964).

[7] J. W. Diggle, T. C. Downie, and C. W. Goulding, Chem. Rev., 69, 365 (1969).

[8] A. K. Reddy, M. Genshaw, and J. O'M. Bockris, J. Electroanal. Chem., 8, 406 (1964).

[9]   M. J. Dignam and D. B. Gibbs, Can. J. Chem., 48, 1242 (1970).

[10] D. B. Gibbs, Anodic Films on Copper and Silver in Alkaline
      Solution, Ph. D. thesis, University of Toronto, 1968.

[11] G. C. Wood and A. J. Brock, Nature, 209, 773 (1966).

[12] C. Parsons and G. C. Wood, Corros. Sci., 9, 367 (1969).

[13] S. W. Khoo, M. Sc. Dissertation (1966) and Ph. D. Thesis
      (1968), Univ. of Manchester.

[14] G. C. Wood and S. W. Khoo, J. Appl. Electrochem., 1, 189 (1971).

[15] A. Sanfeld, Introduction to the Thermodynamics of Charged and
      Polarized Layers, Wiley-Interscience, London, 1968, Chapter 8.

[16] M. J. Dignam, D. J. Young, and D. Goad, submitted to J. Phys.
      Chem. Solids.

[17] A. T. Fromhold, Jr., J. Phys. Chem. Solids, 24, 1081 (1963).

[18] M. J. Dignam, J. Phys. Chem. Solids, 29, 249 (1968).

[19] N. Cabrera and N. F. Mott, Rept. Prog. Phys., 12, 163 (1948-49).

[20] M. J. Dignam and D. B. Gibbs, J. Phys. Chem. Solids, 30, 375
      (1969).

[21] M. J. Dignam, Can. J. Chem., 42, 1155 (1964).

[22] L. B. W. Jolly, Summation of Series, 2nd ed., Dover, New York,
      1960, p. 30.

[23] S. G. Christov and S. Ikonopisov, J. Electrochem. Soc., 116,
      56 (1969).

[24] L. Young, J. Electrochem. Soc., 110, 589 (1963).

[25] N. Ibl, Electrochim. Acta, 14, 1043 (1967).

[26] A. T. Fromhold, Jr., and E. L. Cook, J. Appl. Phys., 38, 1546
      (1967).

[27] W. F. Ames, Nonlinear Ordinary Differential Equations in Trans-
      port Processes, Academic Press, New York, 1968.

[28] G. T. Wright, Solid State Electron., 2, 165 (1961).

[29] J. F. Dewald, J. Electrochem. Soc., 102, 1 (1955).

[30] L. Young, Can. J. Chem., 37, 276 (1959).

[31] M. J. Dignam and D. F. Taylor, Can. J. Chem., 49, 416 (1971).

[32] C. P. Bean, J. C. Fisher, and D. A. Vermilyea, Phys. Rev., 101, 551 (1956).

[33] J. F. Dewald, J. Phys. Chem. Solids, 2, 55 (1957).

[34] M. J. Dignam and P. J. Ryan, Can. J. Chem., 41, 3108 (1963).

[35] L. Young, Can. J. Chem., 50, 574 (1972); also in Extended Abstr., Corrosion-Dielectrics and Insulation Division, Electrochemical Society Meeting, Cleveland, October 1971, abstract No. 91, p. 234.

[36] D. J. Young and M. J. Dignam, submitted to J. Phys. Chem. Solids.

[37] M. J. Dignam, W. R. Fawcett, and H. Bohni, J. Electrochem. Soc., 113, 656 (1966).

[38] S. M. Ahmed, J. Phys. Chem., 73, 3546 (1969).

[39] Y. G. Bérubé and P. L. de Bruyn, J. Colloid Interface Sci., 28, 92 (1968).

[40] H. P. Boehm, Discussions Faraday Soc., 52 (1972), in press.

[41] S. Levine and A. L. Smith, Discussions Faraday Soc., 52 (1972), in press.

[42] D. A. Vermilyea, Surface Sci., 2, 444 (1964).

[43] R. S. Nelson and M. W. Thompson, Proc. Roy. Soc. (London), A259, 458 (1961).

[44] Modern Aspects of the Vitreous State, Vol. 1 (J. D. Mackenzie, ed.), Butterworths, London, 1960.

[45] G. W. Morey, The Properties of Glass, Reinhold, New York, 1954, p. 465.

[46] K. Otto, Phys. Chem. Glasses, 7, 29 (1966).

[47] M.J. Dignam, in Extended Abstr., Chemical Institute of Canada
      Symposium on Physical Chemistry of Solids, University of
      Montreal, Quebec, August 1966, p. 35.

[48] M.J. Dignam, in Extended Abstr., Dielectric and Insulation
      Division, Electrochemical Society Meeting, Dallas, Texas,
      May 1967, Vol. 4, No. 1, p. 33.

[49] M.J. Dignam, in Extended Abstr., Dielectric and Insulation
      Division, Electrochemical Society Meeting, New York, May
      1969, Vol. 6, No. 1, p. 11.

[50] W.H. Zachariasen, J. Am. Chem. Soc., 54, 3841 (1931).

[51] J.J. Randall, W.J. Bernard, and R.R. Wilkinson, Electrochim.
      Acta, 10, 183 (1965).

[52] C.J. Dell'Oca and L. Young, J. Electrochem. Soc., 117, 1545
      (1970).

[53] G. Amsel and D. Samuel, J. Phys. Chem. Solids, 23, 1707
      (1962).

[54] J.P.S. Pringle, in Extended Abstr., Dielectric and Insulation
      Division, Electrochemical Society Meeting, New York, May
      1969, Vol. 6, No. 1, p. 17.

[55] B. Verkerk, P. Winkel, and D.G. De Groot, Philips Res.
      Rept., 13, 506 (1958).

[56] J.A. Davies, J.P.S. Pringle, R.L. Graham, and F. Brown, J.
      Electrochem. Soc., 109, 999 (1962).

[57] R.L. Graham, F. Brown, J.A. Davies, and J.P.S. Pringle, Can.
      J. Chem., 41, 1686 (1963).

[58] J.A. Davies and B. Domeij, J. Electrochem. Soc., 110, 84
      (1963).

[59] J.A. Davies, B. Domeij, J.P.S. Pringle, and F.J. Brown, J.
      Electrochem. Soc., 112, 675 (1965).

[60] V. I. Vedeneyev, L. V. Gurvich, V. N. Kondrat'yev, V. A. Medvedev, and Y. L. Frankevich, Bond Energies, Ionization Potentials, and Electron Affinities, Arnold, London, 1966.

[61] A. D. Buckingham, Quart. Rev., 13, 183 (1959).

[62] H. F. Halliwell and S. C. Nyburg, Trans. Faraday Soc., 59, 1126 (1963).

[63] K. B. Harvey and G. B. Porter, Introduction to Physical Organic Chemistry, Addison-Wesley, Reading, Mass., 1963.

[64] S.A. Schtschukarew and R. L. Müller, Z. Phys. Chem., A150, 439 (1930).

[65] L. Young, Trans. Faraday Soc., 53, 841 (1957).

[66] D. A. Vermilyea, J. Electrochem. Soc., 103, 690 (1956).

[67] S. J. Basinska, J. J. Polling, and A. Charlesby, Acta Met., 2, 313 (1954).

[68] L. Young, Trans. Faraday Soc., 50, 153 (1954).

[69] L. Young, Proc. Roy. Soc. (London), A244, 41 (1958).

[70] D. A. Vermilyea, Acta Met., 3, 106 (1955).

[71] D. A. Vermilyea, J. Electrochem. Soc., 102, 655 (1955).

[72] D. A. Vermilyea, J. Electrochem. Soc., 104, 140 (1957).

[73] G. A. Dorsey, Jr., J. Electrochem. Soc., 116, 466 (1969).

[74] L. Young, Proc. Roy. Soc. (London), A258, 496 (1960).

[75] L. Young and F. G. R. Zobel, J. Electrochem. Soc., 113, 277 (1966).

[76] L. Young, Trans. Faraday Soc., 52, 502 (1956).

[77] M. J. Dignam, D. Goad, and M. Sole, Can. J. Chem., 43, 800 (1965).

[78] M. J. Dignam and D. Goad, J. Electrochem. Soc., 113, 381 (1966).

[79] M. J. Dignam and P. J. Ryan, Can. J. Chem., 46, 535 (1967).

[80] C. J. Dell'Oca and L. Young, J. Electrochem. Soc., 117, 1545 and 1548 (1970).

[81] D.A. Vermilyea, Acta Met., 2, 482 (1954).

[82] P.H.G. Draper, Electrochim. Acta, 8, 847 (1963).

[83] A.C. Harkness and L. Young, Can. J. Chem., 44, 2409 (1966).

[84] L. Masing and L. Young, Can. J. Chem., 40, 903 (1962).

[85] D.A. Vermilyea, J. Electrochem. Soc., 104, 427 (1957).

[86] L. Young, Proc. Roy. Soc. (London), A263, 395 (1961).

[87] D. Goad and M.J. Dignam, submitted to Can. J. Chem., 1972.

[88] D.F. Taylor and M.J. Dignam, submitted to J. Electrochem. Soc., 1972.

[89] M.J. Dignam and P.J. Ryan, Can. J. Chem., 46, 549 (1968).

[90] M.J. Dignam and P.J. Ryan, Can. J. Chem., 46, 535 (1968).

[91] M.J. Dignam, J. Electrochem. Soc., 112, 729 (1965).

[92] A. Tvarusko, U.S. Atomic Energy Comm. NP-16828 (1966).

[93] M.J. Dignam, J. Electrochem. Soc., 112, 722 (1965).

[94] L. Young, J. Electrochem. Soc., 111, 1289 (1964).

[95] P. Winkel, C.A. Pistorius, and W.Ch. Van Geel, Philips Res. Rept., 13, 277 (1958).

[96] M.J. Dignam and D.F. Taylor, Can. J. Chem., 48, 1971 (1970).

[97] M.J. Dignam, J. Electrochem. Soc., 109, 184 (1962).

[98] L. Young, J. Electrochem. Soc., 111, 1289 (1964).

[99] J. Siejka, J.P. Nadai, and G. Amsel, J. Electrochem. Soc., 118, 727 (1971).

[100] H.A. Hoyen, Jr., J.A. Strozier, Jr., and Che-Yu Li, Appl. Phys. Letters, 14, 104 (1969).

[101] M.J. Dignam, H.M. Barrett, and G.D. Nagy, Can. J. Chem., 47, 4253 (1969).

[102] R.G. Barradas and G.H. Fraser, Can. J. Chem., 42, 2488 (1964).

[103] J.P. O'Sullivan and G.C. Wood, Proc. Roy. Soc. (London), A317, 511 (1970).

[104] J.W. Diggle, T.W. Downie, and C.W. Goulding, J. Electrochem. Soc., 116, 737 (1969).

Chapter 3

ELECTRONIC CURRENT FLOW THROUGH

IDEAL DIELECTRIC FILMS

C.A. Mead

California Institute of Technology

Pasadena, California

I. INTRODUCTION

During the past few decades a large literature has accumulated on the subject of current flow through dielectric films. Much of this material contains detailed analyses of many physical effects and a great deal of multiparameter curve fitting. Until recently all this activity had given the field a rather bad name, since it appeared that all effects were very complicated and nothing could be understood in a first-principles way. It is true, in fact, that in many thin-film systems the current flow is dominated by impurities, trapping

287

processes, and so on, so that no simple, clear picture emerges for the mechanism of current flow. However, in the past few years it has become clear that certain insulating materials behave in a nearly ideal fashion and can be understood in a very simple and fundamental way.

In this chapter I shall not attempt to discuss the mass of literature dealing with data on dielectrics that were not well characterized and well understood. Instead, I shall concentrate on examples in which nearly ideal behavior was observed and in which the simple physics of the current-flow processes is clear. In retrospect it seems obvious that much of the previous data is also understandable on rather simple grounds and that there were a number of conceptual errors that led to the belief that vastly complicated processes were involved. This is by no means true for all the data in the literature, but certainly with good hindsight resulting from a clear understanding of ideal materials, a much better understanding of the nonideal cases is also possible. Since the details of all the results I shall cite are available in the published literature, I shall discuss only the ideas and basic principles involved and give references where a more complete discussion may be found.

II. THE METAL-VACUUM INTERFACE

Electronic states in a metal are filled up to some maximum energy, the Fermi level. Electrons in vacuum have a certain minimum energy: the rest energy in vacuum, or "vacuum level." Both of these energies are usually discussed in terms of some arbitrary reference energy. As long as one is working in a homogeneous medium (either metal everywhere or vacuum everywhere), the reference energy does not matter. However, when we discuss the interface between the metal and the vacuum, we must use the same reference energy for electrons in the metal and for electrons in the vacuum. In other words, we must know how the energies in the metal, and in particular the Fermi energy, are related to the energy of an electron in vacuum. In terms of an energy diagram we need to know at what position to draw the vacuum level relative to the Fermi level in the metal. The various metals have very different electron energies, and one might expect that for certain metals the Fermi level would lie above the vacuum level, and for others the Fermi level would lie below the vacuum level. However, what is actually found is that for all known metals the Fermi level lies below the vacuum level; that is, it requires energy to remove an electron from any metal and take it into vacuum.

There is a very simple experiment we can do to convince ourselves that this is in fact the case. Suppose we place two metal plates facing each other in a vacuum, as shown in Fig. 1, and apply a voltage between the two metal plates with an ammeter connected so that we can measure how much current flows between the metal plates

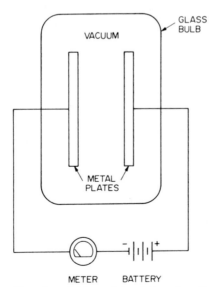

Fig. 1. Schematic of experiment to determine the energy of electrons in metal relative to vacuum.

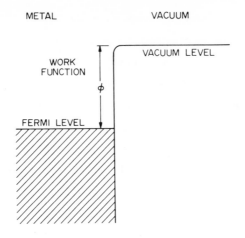

Fig. 2. Relative energies of electrons in metal and in vacuum.

through the vacuum. If the Fermi level in the left-hand (negative)
plate were higher than the vacuum level, electrons would spontane-
ously spill out into the vacuum and be carried across by the electric
field to the positive plate, and we would find that we could draw a
large current through the vacuum. In fact, when we do the experiment
we find that, no matter which metal we use, at room temperature
essentially no current flows through the vacuum. It is often said that
vacuum is an excellent insulator. That really is not a statement about
vacuum at all, but rather about the energy barrier between the Fermi
level of the metal and the vacuum level.

With this information we can draw an energy diagram of the
metal-vacuum interface as shown in Fig. 2. The difference in energy
between the Fermi level in the metal and the vacuum level is just the
well-known work function ϕ of the metal. It is the amount of work
necessary to get an electron from the Fermi level of the metal out into
the vacuum. At very low temperatures electrons in the metal have
insufficient energy to surmount the barrier and make their way out of
the metal into the vacuum. However, as the temperature of the metal
is raised, more and more electrons are evaporated into the vacuum.

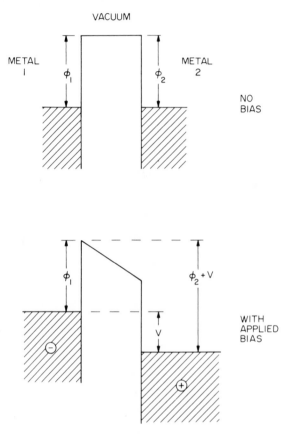

Fig. 3. Energy diagram of the experiment of Fig. 1, showing the conditions with and without applied bias.

Thus, if we take the experiment of Fig. 1 and heat the plate on the left, we find that, as the temperature is raised, the amount of current we can draw through the vacuum becomes larger and larger. We can illustrate this situation on an energy diagram, as shown in Fig. 3. Two identical metal plates are shown facing each other, each with a work function and the vacuum level shown between. If there is no voltage applied between the metal plates, the vacuum level is horizontal; that is, the energy of an electron at rest in the vacuum is independent of its position between the two metals. However, if we apply a voltage by connecting a battery between the two metal plates, the energy of electrons in the negative (left) plate is raised relative to those of the positive metal. Any electrons that are now liberated from the negative plate will be accelerated toward the positive plate until they smash into it and lose their energy by collisions with the other electrons and atoms in the metal. The current that can flow in this situation is governed by how many electrons in metal 1 have enough energy to surmount the work-function barrier ϕ_1. Once they have overcome this obstacle, they are drawn across through the vacuum by the electric field into metal 2 . Electrons in metal 2 are also evaporated out into the vacuum. However, they must not only surmount the work-function barrier, but in addition climb uphill against the electric field until they make their way over the top and arrive at metal 1. This, of course, requires a great deal more energy. In fact, the amount of energy required over and above the work function of metal 2 is just the applied voltage V. If there are electrons in metal 2 with enough energy to not only surmount the work function ϕ_2 but also the potential barrier created by the battery, they must have energy $\phi_2 + V$.

For energies well above the Fermi level, the energy distribution of electrons in a metal follows the Boltzmann law

$$N \propto \exp\left(-\frac{E}{kT}\right) , \tag{1}$$

where N is the density of electrons at any given energy, T is the absolute temperature, and k is the Boltzmann constant. The energy E of the electrons is measured from the Fermi level. Thus if the applied voltage is more than a few kT, the current from right to left will be small compared with the current from left to right, since very few electrons will have the energy to overcome both the work function and the applied voltage. In this case the current will be given very nearly by

$$J = J_0 \exp\left(-\frac{\phi_1}{kT}\right). \tag{2}$$

III. CURRENT-VOLTAGE CHARACTERISTIC
OF THE THERMIONIC DIODE

So far we have been discussing the symmetrical situation in which the work functions of both metals were equal. In such a situation the current for either direction of applied bias will saturate at the value given by Eq. (2). In more practical situations the work functions of the two metals may be quite different. Let us consider the practical device[*] shown in Fig. 4. The structure contains a barium-coated oxide cathode with work function ϕ_1 and a titanium anode with work function ϕ_2. We shall operate the device at

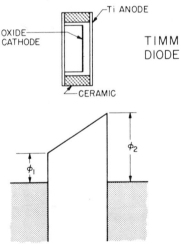

Fig. 4. Schematic and energy diagram of practical vacuum diode.

---

[*]These devices were originally designed to work as part of an integrated vacuum-tube electronic system and were kindly supplied by the General Electric Company, Owensboro, Kentucky.

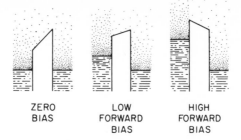

ZERO
BIAS

LOW
FORWARD
BIAS

HIGH
FORWARD
BIAS

Fig. 5. Schematic representation of the Boltzmann distribution of electrons in two metal electrodes of the device of Fig. 4. Note how the application of applied bias shifts one distribution relative to the other and thus causes current flow.

elevated temperatures, so that electrons may be emitted from the barium cathode, which we shall make negative with respect to the high-work-function titanium anode. Under these conditions we would expect a preponderance of current flow from the low-work-function metal to the high-work-function metal. In fact, by just knowing the Boltzmann law we can predict how the current that flows through the vacuum should depend on the voltage we apply between the two metals.

The energy diagram for the device under various bias conditions is shown in Fig. 5 together with a schematic representation of the Boltzmann distribution of electrons in the electrodes. With zero applied bias the device is in equilibrium, and the same number of electrons cross the peak of the barrier from left to right and right to left. As the oxide cathode is made more negative, its electron population is displaced toward higher energies, and a net current flows from left to right. In the low-forward-bias range the maximum height of the energy barrier that electrons must cross to get from the left to the right is $\phi_2 - V$. Thus, as the applied voltage is raised, the number of electrons with enough energy to surmount this potential barrier should increase exponentially

$$J \propto \exp\left(-\frac{\phi_2 - V}{kT}\right).$$ (3)

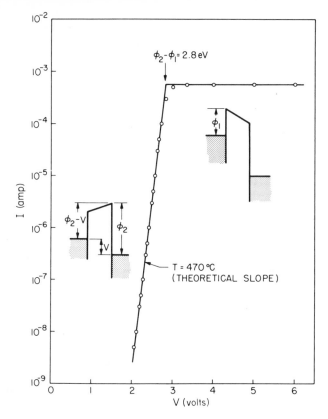

Fig. 6. Current–voltage characteristic of the thermionic diode of Fig. 4. The device was run in an oven at 470°C. Note the knee in the slope when the applied bias reaches $\phi_2 - \phi_1$.

Thus for applied voltages less than the difference between the two work functions the current through the device will increase exponentially with the applied voltage. When the voltage becomes equal to the difference between the work functions, the height of the potential barrier that the electrons must surmount becomes equal to ϕ_1 and does not decrease with further increase in the applied voltage. Therefore in high forward bias (voltages greater than $\phi_2 - \phi_1$) the current saturates and does not continue to increase.

Data taken on the actual device running at 470°C are shown in Fig. 6. For potentials less than 2.8 V the current does in fact increase

exponentially and saturates for larger voltages. We have thus determined the difference between the two metal work functions, $\phi_2 - \phi_1 =$
2.8 eV. In addition, from the slope of the exponential part of the
curve, we can determine the temperature of the device (or if we know
the temperature of the device, we can check to see whether the
Boltzmann law is being obeyed). The line drawn through the experimental points is such that the current increases by a factor of e for
every kT increase in the applied voltage, where the temperature T
was taken from a thermocouple attached to the device in the oven.
It is clear that the current increases in exactly the way predicted by
the Boltzmann factor.

We encounter many devices in which the current increases exponentially with the voltage. Most of these operate on the same fundamental principle as the vacuum diode we have just discussed; that is,
the height of an energy barrier which thermally excited electrons
must surmount depends directly on the voltage applied. In any device
in which this situation is found, the current will depend exponentially
on the voltage, increasing by a factor of e for every kT increase in
the applied bias.

So far we have shown that the current flowing through our vacuum
dielectric is thermionic in origin, and we have measured the difference
between the two work functions. In order to complete our understanding of the device, we must know the absolute value of either ϕ_1 or ϕ_2.

If we bias the unit just above the knee in the I-V curve (e.g.,
to 5 V), the current is given by Eq. (2) and we can determine ϕ_1 by
varying the temperature of the diode and observing the current. From
Eq. (2) we would expect a plot of the logarithm of the current as a
function of reciprocal absolute temperatures to be a straight line
whose slope is the barrier energy divided by Boltzmann's constant.
This thermal-activation-energy plot is a direct way of measuring the

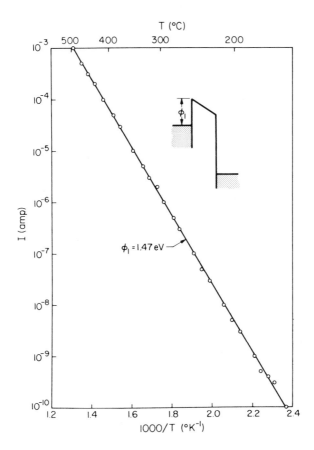

Fig. 7. Dependence on temperature of the thermionic current through the device of Fig. 4. The data were taken at an applied voltage of 5 V.

energy barrier ϕ_1. Data taken on the thermionic diode are shown in Fig. 7. We obtain a straight line as expected. The slope of this line gives directly $\phi_1 = 1.47$ eV.

We have now determined all parameters that are important for the operation of the device. The value we infer for the titanium work function is 4.27 eV and can be checked against published values, which range from 3.95 to 4.75 eV. Thus we have confidence that our energy diagram and concept of current flow are not only internally consistent but also correct.

IV. THERMIONIC CURRENTS IN THIN INSULATORS

The behavior of a thin insulating film in which the mobility is reasonably high and the density of donors (or acceptors) and traps is low enough for the band bending over the thickness of the film to be negligible represents a situation conceptually identical with that encountered in the vacuum tube of the preceding section [1].

In principle, the only change is that the work function is replaced by the energy barrier from the Fermi level of the metal to either the conduction or the valence band of the dielectric, whichever is lower. The study of metal-semiconductor and metal-insulator barriers is a highly refined topic in its own right and has been reviewed elsewhere [2]. We need here only to recognize that such barriers exist and that they can be measured by a number of well-known techniques.

Although a vast literature exists concerning the electrical behavior of thin insulating films, there has been a persistent problem in the characterization of the film itself. For this reason quantitative agreement between models similar to that used for the vacuum diode and real measurements on insulating films have only recently been achieved. In this section we discuss a straightforward approach to obtaining an insulating film that is well characterized and in which the requirements for behavior similar to that of a vacuum have been met. One starts with a single crystal of highly pure and well-characterized material for which the carrier concentration, trapping concentration, and barrier energies for various metals are known. This single crystal is then cleaved thin enough to permit making thin-film-type measurements.

This approach has been successfully employed recently by using gallium selenide single crystals with carrier concentrations of less than 10^{15} cm^{-3}. A schematic representation of the crystal structure of gallium selenide is shown in Fig. 8. It will be noted that each

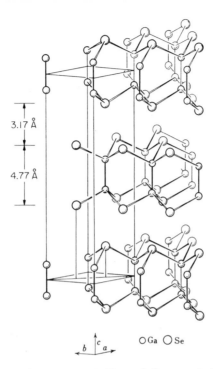

Fig. 8. Schematic representation of the crystal structure of gallium selenide. A small section of three layers is illustrated.

layer consists of a highly regular array of gallium and selenium atoms in a simple packing arrangement. Adjacent layers are held together by van der Waals forces, which permits easy cleavage of the crystal into very thin layers. Because of its low carrier concentration, the dielectric constant of the material could be determined from samples of macroscopic thickness [3]. The carrier concentration was then determined from measurements of $1/C^2$ versus V made on Schottky diodes formed on cleaved surfaces of the bulk crystal. By observing the behavior of the capacitance under incident illumination, an upper limit on the deep-trap concentration could also be set. The total of net carrier concentration and deep-trap concentration for this material was less than 10^{15} cm^{-3}. Therefore for samples less than 1000 Å thick the band bending under any conceivable bias conditions will be negligible.

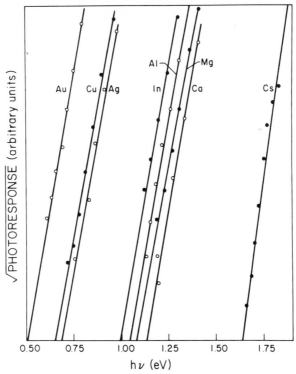

Fig. 9. Photoresponse of metal contacts to p-type gallium selenide. The intercepts on the hν axis are the barrier energies, that is,
the location of the Fermi level for the particular metal relative to the
valence band.

Photoemission measurements on the same Schottky barriers give
values for the barrier energies relative to the valence band [4].   A
plot of the square root of the photoyield from such an experiment as
a function of photon energy is shown in Fig. 9.   The intercepts on the
hν axis are the barrier energies relative to the valence band of the
gallium selenide.   Thin-film samples can be prepared by evaporating
aluminum onto one side of a freshly cleaved gallium selenide flake a
few microns in thickness.   The metallized side of the sample is then
mounted with a conducting epoxy to a metal block, and the opposite
side of the sample is removed by successive application and removal

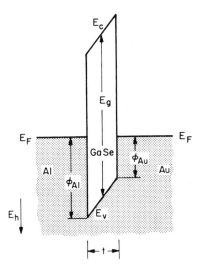

Fig. 10. Energy-band diagram for an Au–GaSe–Al structure. The energy gap of gallium selenide is 2.0 eV, ϕ_{Au} is 0.52 eV, and ϕ_{Al} is 1.05 eV (see Fig. 9).

of ordinary Scotch Magic Transparent Tape. With certain rudimentary precautions this procedure results in rather uniform samples of reasonably large area, with thicknesses in the 50- to 1000 Å range. Subsequently gold dots are evaporated through a fine mesh mask onto the freshly peeled surface.

A knowledge of the barrier energies derived from Fig. 9 and the band gap of gallium selenide (2.0 eV) allows us to construct an energy-band diagram for the sample, as shown in Fig. 10.

For samples at the upper end of the thickness range, tunneling can be neglected to first order, and a simple thermionic-current-flow process is observed. The current-voltage characteristic of a typical sample 600 Å thick is shown in Fig. 11. It is immediately apparent that the qualitative features are similar to those for the thermionic vacuum diode. The current rises exponentially with voltage until the electric field in the device approaches zero and then flattens out.

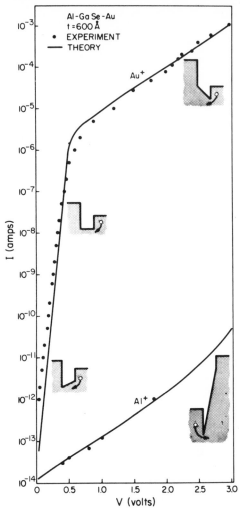

Fig. 11. Current-voltage characteristic of an Au–GaSe–Al structure 600 Å thick. Insets show the relevant portion of the energy diagram at various biases. As indicated, holes are the dominant current
carrying species.

Some quantitative differences from the vacuum diode are also apparent.
The rate of increase in the exponential portion of the curve is not quite
that expected from Eq. (3). In addition, current above the flat band
condition is not totally independent of voltage, but still increases.

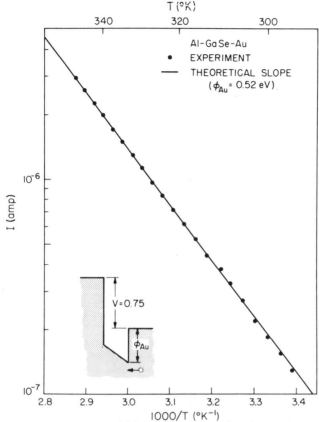

Fig. 12.   Temperature dependence of the forward current in the Au-GaSe-Al structure of Fig. 10.

The same is true for the reverse current of the device. However, the electric fields present in this device are many orders of magnitude larger than in the thermionic vacuum diode, and one would have to know the detailed dependence of the barrier energy on electric field to determine the exact shape of the current-voltage curve. The way in which this has been done to calculate the theoretical curve shown in Fig. 9 is discussed in Ref. [1].

Qualitatively, we expect the device to be operating in precisely the same way as the vacuum diode, and when biased just above the knee of the curve (e.g., +0.75 V applied to the gold electrode), we

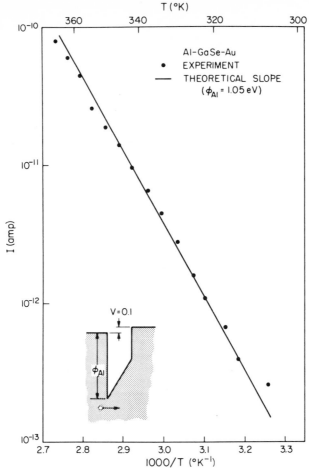

Fig. 13. Temperature dependence of the reverse current in the Au–GaSe–Al structure of Fig. 10.

would expect to measure the barrier energy of the gold electrode directly by observing the dependence of the current flow upon device temperature. The results of this experiment are shown in Fig. 12 and are essentially identical with those for the thermionic vacuum diode except that the barrier energy is only 0.52 eV instead of 1.4 eV. This of course accounts for our ability to perform the experiment at room temperature instead of 470°C, as was necessary with the vacuum diode.

A similar measurement can be performed in the reverse direction, which was not possible with vacuum diode because of the very high work function of the titanium anode. The results are shown in Fig. 13. In this case the experiment gives an energy of approximately 1 eV for the aluminum-gallium selenide barrier. Both of these values are in excellent agreement with the results of photoresponse shown in Fig. 9 and give us confidence in the simple thermionic model of current flow.

V. TUNNELING THROUGH THIN INSULATING FILMS

So far we have discussed a mechanism of current flow that involved electrons surmounting a potential barrier caused by the forbidden gap of an insulator. This is not the only mechanism by which current can flow through an insulator. When the insulating film becomes sufficiently thin, it is no longer opaque to electrons. Electrons that pass directly through the forbidden gap without surmounting the barrier are said to "tunnel" through the forbidden gap. This lack of total opaqueness of a forbidden gap of any insulator (including vacuum) is a consequence of the basic quantum mechanical nature of the electron.

The wavefunction of a free electron in vacuum, ψ, is of the form

$$\psi = \exp(ikx) , \tag{4}$$

where

$$\hbar k = \sqrt{2mE} . \tag{5}$$

Here the electron energy E is measured from the vacuum level. For E > 0 this expression represents a traveling wave in the x direction whose momentum $\hbar k$ is proportional to the square root of the electron energy. This result is familiar from elementary mechanics and is

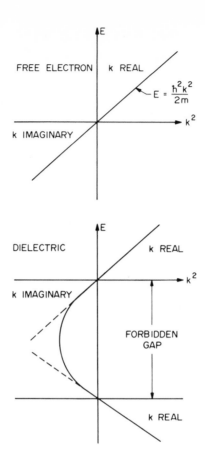

Fig. 14. Schematic representation of the dependence on electron energy E of the real and imaginary parts of the wave vector k. Top: free electron. Bottom: electron in a dielectric or semiconductor. Regions of imaginary k correspond to damped wavefunctions.

contained in the plot of E versus k^2 shown in Fig. 14. A not so familiar result is also evident in this figure. For negative values of E, the quantity k^2 is negative and hence k is imaginary. The electron no longer has propagating solutions, but rather damped ones of the form

$$\psi = \exp(-kx) , \tag{6}$$

where k is now the imaginary part of the wave vector. These "forbidden" damped solutions are not appropriate for an electron in an

extended space since they increase without limit for negative x.
However, in a finite structure they are quite appropriate and contain
an important fundamental physical fact: it is not possible to confine
an electron absolutely. No matter how high a potential barrier we
set up, the electron has a finite probability (given by the square of
the wavefunction) of being where it does not have the energy to be.

For a metal-vacuum interface like that shown in Fig. 2, it is clear
that electrons in the metal cannot stop abruptly at the surface but in
fact penetrate a short distance into the vacuum, the probability P of
finding one dying out with increasing x as

$$P \propto \exp(-2kx) , \tag{7}$$

where now x is measured from the metal surface. Under normal cir-
cumstances the distance $1/2k$ over which probability decreases by
a factor of e is less than an angstrom, and hence for practical purposes the interface is abrupt. However, when two electrodes are
very closely spaced (e. g., less than 100Å), an electron in one electrode can make a transition to the other electrode through the tail of
its damped wavefunction. The probability of such a transition is just
proportional to the probability P of finding an electron from electrode 1 in electrode 2. From Eq. (7)

$$P \propto \exp(-2kt) , \tag{8}$$

where t is the spacing of the electrodes.

For any tunneling path between a filled state in one electrode
opposite an empty state in the other, Eq. (8) gives the probability per
second that a transition will occur. The tunneling current is just the
charge on the electron times the sum of all probabilities (those from
left to right taken as positive, and those from right to left as negative). Since the number of filled states in one electrode opposite
empty states in the other is proportional to the voltage, we expect

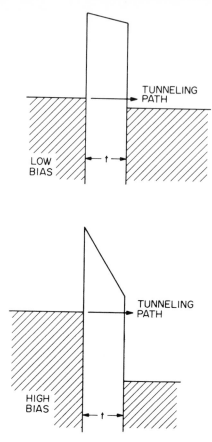

Fig. 15. Energy diagram of metal–vacuum–metal structure at low and high applied bias. Note that although the length of the tunneling path is independent of bias voltage for $V < \phi$, the tunneling path at higher bias is nearer the vacuum level and hence the average k is lower.

the current to be proportional to voltage, and for a small applied bias this is the case.

At higher voltages, however, another effect completely dominates the current–voltage characteristic. Referring to Fig. 15, we see that by applying a bias voltage we have moved the tunneling path of electrons near the left-electrode Fermi level closer to the vacuum level, where (from Eq. (5)) the value of k is lower. Hence the average k

for the entire path is lower, and the tunneling probability will increase
in a generally exponential way with bias. Although a detailed analysis
can be made (and has been many times in the literature), it is not useful in the study of real dielectrics because the dependence of k on E
in these materials never follows the simple form given in Eq. (5).
Rather, the value of k must decrease at sufficiently low energies since
it must again be zero at the valence-band edge.

The type of dependence expected is shown in Fig. 14. Since k is
in the exponent, the k versus E behavior of any material completely
dominates the current versus voltage characteristic, and essentially
no progress was made toward a quantitative understanding of thin-film
tunneling until this fact was recognized and squarely faced. The real
question then is how to determine the k versus E relationship in any
real material in order to calculate the current-voltage characteristics
of a metal-dielectric-metal sandwich formed from this material. It
will probably come as no surprise to the reader that tunneling itself
provides the most sensitive probe with which to determine the nature
of the forbidden gap in solids.

As a first step, we notice that the reciprocal of the probability
per second that an electron will tunnel from one electrode to the
other is the relaxation time of this system when a small quantity of
charge is taken from one electrode and placed on the other. In electrical engineering terms this time is just the RC time constant of the
metal-insulator-metal sandwich, R being the zero-bias resistance
resulting from the tunneling mechanism and C being the capacitance
of the structure. If we plot the log of the RC time constant as a
function of insulator thickness, we should obtain a straight line
whose slope is twice the average value of k encountered on the
tunneling path and whose intercept is given by a group of fundamental constants that can be found in any detailed treatment on the
subject.

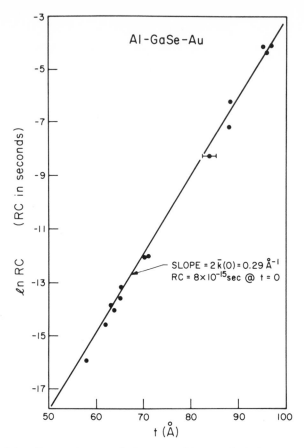

Fig. 16. Dependence of the tunneling probability for samples shown in Fig. 10 on the thickness of the gallium selenide layer. The slope of this plot is a direct measure of k.

One straightforward way to evaluate k as a function of energy for a given material is to make structures with different barrier energies, vary their thickness, and plot the data in this way. Although this approach is straightforward in principle, it is difficult in practice because of the limited number of metal–insulator interfaces suitable for study for technical reasons. However, one can choose certain metal–insulator systems and make such plots to obtain information on the average k along the tunneling path for a particular pair of metals involving particular values of the two barrier energies.

The results of this procedure applied to gallium selenide samples [5] are shown in Fig. 16. Notice that in fact an excellent straight line is obtained, indicating that tunneling is the dominant current-flow mechanism. The intercept is within a factor of 2 of that calculated from the fundamental constants measured earlier. This is the first and most severe test of any tunneling experiment.

VI. THE VOLTAGE DEPENDENCE OF THE TUNNELING CURRENT

If a voltage is applied to the tunneling sample, the Fermi level of one metal will be displaced relative to that in the other metal, and tunneling can take place from all allowed states in the negative metal into the opposite empty states in the positive metal, that is, in the entire energy range between the Fermi levels of the two metals. Thus a calculation of the tunneling current involves determining the average k for each energy between the two Fermi levels, calculating the contribution of this energy to the tunneling current, and integrating the result over the energies between the Fermi levels. (This analysis assumes that the temperature is low enough for thermally excited electrons to contribute negligibly to the current.) This rather complex procedure can be carried out only if the details of the dependence of k on energy are known. Since in general this dependence is not known, we are faced with the problem of determining k as a function of energy before proceeding. One approach is to use the thickness dependence of structures with different barrier energies and different applied biases. Another is to notice that the dependence of k on energy is necessary to determine the current–voltage relationship for any single tunneling sample. Both approaches have been successfully used.

Typical current–voltage characteristics for gallium selenide samples of various thicknesses are shown in Fig. 17. The vertical

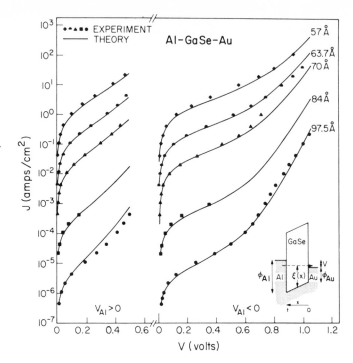

Fig. 17. Current-voltage characteristics of Au-GaSe-Al structures of various thickness. Vertical displacement of the curves is
due to change in thickness; increase in current with applied bias is
due to the effect shown in Fig. 15, that is, change in the average k.

displacement between curves for samples of different thickness is
due to the exponential dependence of the wavefunction on sample
thickness, the same effect which gave rise to the exponential plot of
Fig. 16. The current for any given sample increases in a generally
exponential way with the bias applied to the sample. As already
mentioned, the reason for this increase is that, as more bias is
applied to the sample, the tunneling paths for electrons, at one or
the other Fermi level (or perhaps both), are brought closer to the
band edges where the value of k is smaller, and hence the average
value of k for these tunneling paths is decreased and the current is
increased. The curve in the forward and reverse direction for any

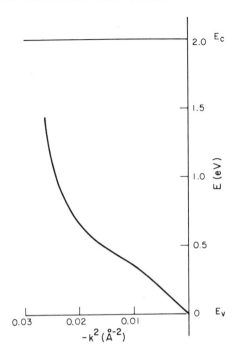

Fig. 18. Energy versus k relationship for the forbidden gap of gallium selenide.

given sample, together with the knowledge of the barrier energies, is sufficient to determine the k versus E relationship for the dielectric. Because of the integral nature of the relationship it is necessary to unwind the dependence by numerical iteration. Conceptually the underlying process is straightforward. It consists of taking a trial solution for the k versus E relationship, calculating the current as a function of voltage, noticing the difference between the calculated and experimental curve, and using this to adjust the k versus E relationship until the process converges.

The result of such a calculation done for gold-gallium selenide-aluminum structures is shown in Fig. 18. Notice that although the general form of the curve is similar to that of Fig. 14, there is

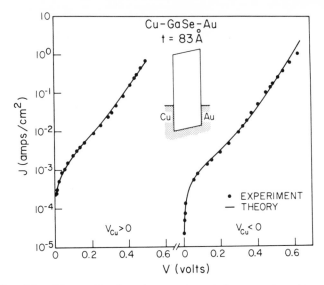

Fig. 19. Theoretical and experimental current–voltage characteristics of a Au–GaSe–Cu sample, obtained by using the k versus E
relationship of Fig. 18. This experiment conclusively demonstrates
the internal consistency of the model.

structure that would not have been predicted by any simple approximation, and therefore it is necessary to do an experiment to determine
the exact relationship. Once the relationship is known, however,
there are no longer any adjustable parameters, and the current can be
calculated as a function of voltage for any sample thickness and any
combination of electrode materials. The results of such a calculation are shown as the solid lines in Fig. 17. Notice the excellent
agreement between theory and experiment over the rather wide range
of current density and thickness involved. The results indicate a
high order of confidence that the theory is, in fact, representative of
the actual physical process.

Since the k versus E relationship was derived from one of the
curves taken on a gold–gallium selenide–aluminum structure and
currents were calculated only for structures involving the same two
metals, it might be thought that what we have done is construct a

model that is adequate for these two particular barrier energies only and not really characteristic of the dielectric itself. In order to dispel this fear we must repeat the calculation for a different set of barrier energies (which result in a different tunneling path for any given applied voltage) and compare the results with experiment. This has been done for copper-gallium selenide-gold samples and is shown in Fig. 19. Notice once more the excellent agreement between theory and experiment.

The result of this study has led us to several conclusions:

1. The simple theory of electron tunneling through a forbidden gap is adequate to explain with reasonable precision experimental data taken on well-controlled and well-characterized dielectric films.

2. Deriving the k versus E relationship for a dielectric can give new insight into the structure and nature of the forbidden gap.[*]

3. The ability to vary independently the energy and thickness of the tunneling samples allows a high degree of overdetermination in the system, and thus the self-consistency of the results can be checked in a number of important ways.

VII. HIGH-FIELD TUNNELING

Tunneling can occur in any situation where filled states and empty states at the same energy are separated by a forbidden region that is sufficiently thin. Even when a vacuum or dielectric film is itself much too thick for direct electrode-to-electrode tunneling to take place, a very large applied bias will result in a high electric field F:

---

[*]Similar experiments have been done on Schottky barriers on semiconductor materials originally by Padovani and Stratton [6] (see also Refs. [7, 8]).

C.A. MEAD

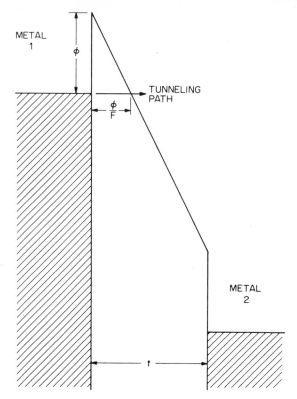

Fig. 20. Energy diagram of thick metal–insulator–metal structure under high applied bias. The tunneling path carries electrons
from the metal into states in the vacuum or dielectric conduction
band. Note that the tunneling distance is ϕ/F.

$$F = \frac{V}{t} \, . \tag{9}$$

Near the electrode the vacuum level (or conduction band) will be very
steep, as shown in Fig. 20, and tunneling can take place from the
Fermi level of the electrode to allowed states in the vacuum (or dielectric). This process was first considered in detail by Fowler and
Nordheim and bears their name.

For any electric field, electrons tunnel from the metal Fermi level
to the vacuum level, the tunneling path being independent of applied
bias. For this reason the average k does not depend on bias voltage,

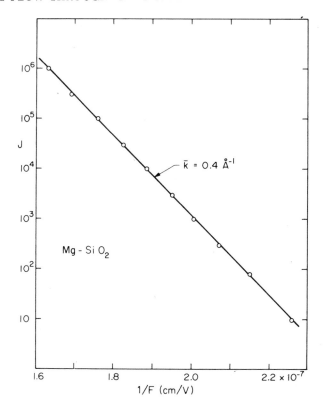

Fig. 21. Fowler-Nordheim plot of the current-voltage characteristics of silica on magnesium. Although the preexponential factors are assumed to be constant in this simplified analysis, the results are essentially identical with those obtained in a more detailed treatment [9]. The ordinate scale J is expressed relative to those values at the lowest field, F.

and the form of the current-voltage characteristics may be realized immediately from Fig. 20. The distance that an electron at the Fermi level must tunnel is $d = \phi/F$. Hence the tunneling probability is

$$P \propto \exp\left(-\frac{2k\phi}{F}\right) . \qquad (10)$$

Electrons with energies below the Fermi level must tunnel further and hence make smaller contributions to the current. For this reason the current increases essentially as Eq. (10) with some slowly varying preexponential factors.

Data taken by Lenzlinger and Snow [9] on a thin film of silica are shown in Fig. 21, where log J has been plotted as a function of 1/F as indicated by Eq. (10). Note that a straight line is obtained over five orders of magnitude in current. From the slope of this line the average k in silica is 0.4 Å^{-1} over the energy range involved.

REFERENCES

[1] T. C. McGill, S. Kurtin, L. Fishbone, and C. A. Mead, J. Appl. Phys., 41, 3831 (1970).

[2] C. A. Mead, Solid State Electronics, 9, 307 (1966).

[3] P. C. Leung, G. Anderman, W. G. Spitzer, and C. A. Mead, J. Phys. Chem. Solids, 27, 849 (1966).

[4] S. Kurtin and C. A. Mead, J. Phys. Chem. Solids, 29, 1865 (1968).

[5] S. Kurtin, T. C. McGill, and C. A. Mead, Phys. Rev., B3, 10 (1971).

[6] F. A. Padovani and R. Stratton, Phys. Rev. Letters, 16, 1202 (1963).

[7] G. H. Parker and C. A. Mead, Phys. Rev. Letters, 21, 205 (1968).

[8] M. F. Millea, M. McColl, and C. A. Mead, Phys. Rev., 177, 1164 (1969).

[9] M. Lenzlinger and E. H. Snow, J. Appl. Phys., 40, 278 (1969).

Chapter 4

ELECTRICAL DOUBLE LAYER

AT METAL OXIDE-SOLUTION INTERFACES

Syed M. Ahmed

Department of Energy, Mines, and Resources

Mineral Sciences Division, Mines Branch

Ottawa, Canada

319

I. INTRODUCTION

When two conducting phases with different electrochemical properties are brought into close contact with one another, a redistribu-
tion of charge and potential occurs at the interface until, at equilibrium,
the electrochemical potential of the charge carriers in both phases is
equal. A double layer is associated with such a nonuniform distribution of charge and of potential at the interface. The subject of
potential differences at interfaces has been discussed by a number
of authors [1-3]. Using the simplified approach of Gerischer [3], if
E_F' and E_F^0 refer to the work done in transfering an electron from the
Fermi level of the bulk-solid phase to free space in vacuum, in the
presence and in the absence of any surface electrostatic potentials,
respectively, then

$$E_F' = E_F^0 - e\Delta\varphi \tag{1}$$

and

$$E_F' = E_F^0 - e(\Delta\psi + \Delta\chi) , \tag{2}$$

where φ is the Galvani, or inner, electrical potential, ψ is the Volta, or outer, electrical potential, χ is the potential due to the presence of surface dipoles, and e is the electronic charge. It may be noted that E_F' and E_F^0 in Eqs. (1) and (2) are equivalent to the electrochemical potential $\bar{\mu}_e/N$ and the chemical potential μ_e/N of a single electron in the solid phase. Also the negative value of E_F', as defined above, is the work function of the semiconductor in the presence of an excess surface charge. When two different phases are in equilibrium with one another, much of the difference in their initial Fermi levels may be compensated by changes in the various electrostatic potentials in the double-layer regions which can extend from the interface from a few angstroms to several thousand angstroms. If the Fermi levels of the two phases are greatly different from one another, a third chemical compound may also be formed at the interface, and it, in turn, will be in separate equilibria with the other two phases. Since the electric field strength at the interface is intense (10^6 to 10^7 V/cm), the interfacial compounds are strongly polarized and may even dissociate and give rise to further charge and potential distributions in the boundary region. The above effects can also be produced or modified by the application of an electric field normal to the surface.

When in equilibrium with aqueous solutions, oxide films on metals have two main interfaces: the metal-metal oxide interface and the metal oxide-solution interface, each interface consisting of several double-layer regions. The electrical and electrochemical properties of the metal-metal oxide-solution system as a whole are determined by the double-layer structures, which in turn are greatly influenced by several physicochemical properties of the bulk phases, such as structure, composition, stoichiometry, conductivity, ion and surface solvation, and dielectric constant and polarizability. Comprehensive data on all the double-layer regions involved in the metal-metal oxide-solution system are seldom obtained simultaneously by studying this system as a whole. Different techniques are required to study different regions of the interface, and hence our overall knowledge of the system is distributed in a number of disciplines. The purpose of this chapter is to make a critical and comparative study of the results obtained from different fields that are relevant to the various double layers in the metal-metal oxide-solution system in an attempt to gain a basic understanding of interfacial phenomena on oxides.

The metal-metal oxide interface has been the subject of numerous investigations in solid-state studies of metal oxide semiconductor (MOS) devices (e. g. , see Ref. [4], p. 170, and the references therein), and this work is not dealt with here. The present review is chiefly concerned with the metal oxide-solution interface, which can possess two diffuse double layers, one in the solid phase and the other in the solution (dilute) phase, separated by a few-angstroms-thick Helmholtz region. The Helmholtz region is known to extend on both sides of the interface and is in equilibrium both with the space-charge layer in the solid phase and with the double layer in the aqueous phase. Henceforth the term "space charge" will be used to refer to the double layer in the solid phase, and the term "double layer" itself will be restricted to the solution phase. The oxide phase can behave as a metal, a semiconductor, or an insulator, and accordingly the space-charge layer can be either electronic or essentially ionic in character. Much of our knowledge of the electronic space-charge layer and the Helmholtz layer at the oxide-solution interface

has been gathered from the extensive work carried out on the interfacial properties of semiconductors and insulators (wide-band-gap
semiconductors). This work is related directly or indirectly with
studies of electrode potentials, electrolytic rectification, conduction
and charge-transfer mechanisms, electrode kinetics, and also to
several applied fields (e.g., batteries and capacitors).

Many sophisticated techniques have been developed for space-
charge investigations of solids in the above-mentioned studies. In
the case of the metal-solution interface, capacitance measurements
(and electrocapillary studies for Hg) are used for double-layer studies.
Several nonelectrochemical methods, including spectroscopic [5-9]
and ellipsometric (see Refs. [10], [11], and the references therein)
techniques, that have been used in investigating surface and interfacial properties of metals have been recently reviewed [6, 10-12].
However, with a few exceptions [7, 8, 13, 14], the double layer on
oxides has not been investigated by these techniques. The need to
investigate the aqueous double layer on solids having an oxide or
hydroxide surface, so as to obtain a better understanding of the
semiconductor- and insulator-solution interface, has been stressed
[15, 16]. Zeta-potential data, obtained by electrokinetic measurements, provide information on a part of the potential distribution at
the oxide-solution interface. However, the interpretation of the
ζ-potential data on oxides is complicated by several experimental
and theoretical uncertainties (see Section VI.K for details).

Data on the double layer on oxides have been obtained recently
by a potentiometric method of measuring the equilibrium distribution
of H^+ and OH^- (potential-determining ions for oxides) at the oxide-
solution interface as a function of pH, ionic strength, and solution
composition. This work has provided considerable information on the
reversible double layer at the oxide-solution interface, such as the
pH at the zero point of charge (zpc), the interfacial charge densities,

the change in the interfacial energies with adsorption, the nature and
the mechanism of counterion adsorption on oxides, the distribution of
potential in the double layer, and the standard free energies of ionic
adsorption.

In investigations of the space charge at polarized interfaces the
diffuse double layer is usually suppressed by using concentrated
solutions and the time constants so chosen in the solid-state measurements that the slow states, which are partly in equilibrium with
the solution phase, do not interfere in these studies. The Helmholtz
layer is assumed to be effectively unchanged in such measurements.
Any deviations of the measured space-charge properties from the calculated values are attributed to changes in the properties of the
Helmholtz layer, which can thus be investigated. In studies of the
reversible double layer on oxides the properties of the space-charge
layer may be initially assumed to be constant, and the aqueous
double layer, including the Helmholtz layer, can be investigated.
These two methods of investigation are thus complementary, and
together they provide comprehensive information on the oxide-solution
interface.

A number of reviews have been published recently on the metal-
solution [2, 17-24], semiconductor-solution [3, 15, 16, 25-30], and
insulator-solution [31-34] interfaces. These subjects are not dealt
with in detail here. However, for a comparative study of the subject
a summary of the basic concepts and the relevant experimental results
in the above fields are included in the first few sections of this
chapter. Key references are also provided for those who wish to
investigate the subject in more detail. The discreteness-of-charge
effect in the double layer has been reviewed by Levine et al. [35, 36]
and by Barlow and Macdonald [37]. This subject is also not examined
here with reference to the double layer on oxides in view of the limited experimental work carried out in this field.

The physicochemical behavior of transition-metal oxides is complex and often not explained by the simple band theory of solids.
The bulk electronic properties of transition-metal oxides also appear
to have exceptionally profound effects on their surface properties.
Hence complex behavior may be expected in the surface properties of
transition-metal oxides with semiconducting or insulator properties.
A summary of the bulk electronic, magnetic, and electrical properties
of transition-metal oxides is therefore included in the beginning of
this chapter.

Many metals either possess a surface-oxide layer or readily
acquire one when in contact with aqueous solutions. Thus a germanium surface in contact with aqueous solutions, particularly in the
anodic region, has been shown [38-41] to behave in practice as an
oxide or hydroxide surface of germanium. Hence with respect to
reactions with the aqueous electrolyte phase the germanium-solution
interface probably represents an extreme case of a semiconductor
with a thin oxide film. The general properties of the semiconductor-
solution interface have been established from the extensive work on
germanium. Accordingly, together with the work on a number of
oxides, work on the germanium-solution interface is also included
in this chapter.

The data, to be discussed, on the reversible double layer on
oxides have been obtained by using oxide powders. Some of the
solid-state studies with wide-band-gap oxides (e.g., TiO_2 [42] and
$KTaO_3$ [43]) have also been carried out with massive electrodes.
However, the experimental conditions for these studies are such that
the results are representative of the thin metal-oxide films in equilibrium with solutions. The significance and the applicability of
such work to thin metal-oxide films are discussed in a separate section.

The reversible double layer at the oxide-solution interface has
been treated in considerable detail, for to date this has been the most

neglected and least understood branch of the oxide-solution interface, and no detailed reviews of the subject exist. The significance of the ζ potential at the oxide-solution interface will also be examined in relation to studies of the reversible double layer on oxides.

II. THE BAND THEORY OF SOLIDS AND THE BULK PROPERTIES OF OXIDES

The oxide-solution interface is influenced by the bulk properties of both the solid and the liquid phases. For most practical purposes all solids can be classified as metals, semiconductors, or insulators on the basis of their electrical conductivity. Metal oxides with conductivities of 10^{-9} to 10^3 ohms^{-1} cm^{-1} are generally regarded as semiconductors. However, many insulators whose conductivities range from 10^{-10} to 10^{-22} ohms^{-1} cm^{-1} are best regarded as semiinsulators (particularly at high fields), analogous to semiconductors. However, the main distinction between these groups lies in the manner in which the electrical conductivity arises. The electrical conductivity of semiconductors increases exponentially with increasing temperature, whereas the conductivity of metals decreases almost linearly with increasing temperature.

A. Simple Oxides of Nontransition Metals

The electrical properties of most of the simple metal oxides, such as ZnO, SnO_2, Al_2O_3, and GeO_2 are adequately accounted for by the Bloch-Wilson band theory of solids [44], which postulates that a material is an insulator if its energy bands are either completely full or completely empty and is metallic in character if its energy bands are partly filled or when an empty band overlaps a filled band. In semiconductors an empty conduction band is separated from a filled valence band by a forbidden energy gap E_g. Conduction in pure (intrinsic) semiconductors arises at a given temperature if the thermal energy (kT = 0.026 eV at 300°K) is sufficient to excite enough electrons from the valence band to the allowed quantum states in the conduction band (kT > E_g). Such excitation of electrons from the conduction band leaves electron vacancies, or holes, in the valence band, so that the carrier densities due to electrons and holes in an intrinsic semiconductor are equal.

Conduction in a nontransition metal like Zn is attributed to the partly filled 4s conduction band. On the other hand, the band structure of the nontransition-metal oxides, such as ZnO or SnO_2, is represented by a filled valence band composed of the 2p levels of oxygen and an empty conduction band composed of the 4s or 5s levels of zinc or tin [45], respectively. The energy gap between the conduction and valence bands ($E_g = E_c - E_v$) is about 3 eV (~ 100 kT) for ZnO [26] and 3.7 eV for SnO_2 [46]. Hence pure ZnO and SnO_2 are essentially insulators. However, in an insulator of this kind, if an electron-donor level is introduced slightly below the conduction band by adding a suitable impurity, then thermal or photoexcitation of electrons from the donor level to the conduction band now becomes possible. The material then becomes an n-type semiconductor, with electrons as the majority current carriers (e.g., Sb-doped SnO_2 [46]). Similarly a p-type semiconductor can also be made by introducing electron-acceptor levels slightly above the valence band, the majority current carriers now being holes in the valence band. The ionization energy required to excite electrons from the donor level to the conduction band or from the valence band to the acceptor level is obviously much less than the band gap E_g of the intrinsic (pure) material. Donor or acceptor levels can be introduced in the semiconductor by several means, such as (a) addition of impurities (substitutional or interstitial), (b) nonstoichiometry of host material, (c) valency differences in the constituents of the lattice, and (d) the presence of structural defects.

Several authors [47, 48] have plotted or tabulated the band gaps of various oxides against various physical properties. Oxides with a band gap of 1 to 4 eV include Ag_2O, AgO, Cu_2O, CdO, Bi_2O_3, In_2O_3, and SnO_2; those with a band gap of 4 to 6 eV include GeO, Ge_2O_3, Nb_2O_3, SrO, BaO, and SnO; whereas Al_2O_3, MgO, CaO, and H_2O (ice) have $E_g > 6$ eV.

B. Transition-Metal Oxides

Transition-metal oxides have been the subject of several recent reviews [47, 49-54]. The observed electrical, magnetic, and optical properties of transition-metal oxides present glaring anomalies when compared with the predictions of the Bloch-Wilson band theory, which has been successfully applied to elementary semiconductors and also to many compounds of nontransition metals. Transition metals are known to form a large number of oxides with the general formulas MO (rock-salt structure), MO_2 (fluoride, rutile (tetragonal), distorted rutile, and other structures), M_2O_3 (corundum or rhombohedral), and a large number of ternary oxides $MM'O_3$, $M_xM'O_3$, $M_2M_2'O_7$, and $M_3M_5'O_{12}$. Many metals like titanium and niobium form oxides with a wide range of stoichiometry (e.g., TiO_x, with x ranging from 0.8 to 1.2). Transition-metal oxides are predominantly ionic as compared with their predominantly covalent sulfides and arsenides, and their metallic carbides and nitrides. Although most of the oxides are antiferromagnetic [53], there are many that are ferromagnetic, paramagnetic, and diamagnetic. Their electrical conductivity also shows a complex dependence on temperature. Many of these oxides at room temperature are semiconductors, but a few are metallic. Some oxides are insulators at low temperatures and behave like metals at higher temperatures, the transition from insulator to metal being rather abrupt. The electrical, magnetic, and optical properties of these oxides show such diversity that the simple band theory fails to predict, even qualitatively, their behavior.

There have been several partly successul attempts at modifying the band theory. For instance, metals of the first transition series possess, in addition to their s and p electrons, one to nine electrons in the d levels which form a narrow d band and are responsible for most of the characteristic properties of the transition metals and of their oxides. The electrical conductivity of transition metals is attributed to the overlapping of the wide ($\sim$ 10 eV) 4s bands with the characteristically narrow d band ($\sim$ 2.8 eV for Ni) of the transition metals (Fig. 1). However, in transition-metal oxides the 4s band of the metal and the 2p band of oxygen are separated by nearly 10 eV, with the narrow d band passing through the Fermi level in the middle of

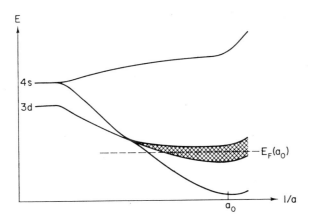

Fig. 1. Energy-band structure of a transition metal of the iron group shown as a function of the inverse lattice parameter a, a_0 being the equilibrium value of the lattice parameter. Reprinted from Ref. [49], p. 2, by courtesy of Academic Press, Inc.

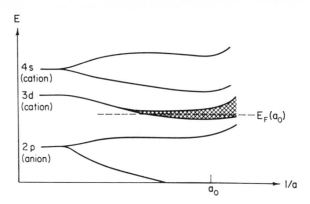

Fig. 2. Energy-band structure of a transition-metal oxide of the iron group shown as a function of the inverse lattice parameter a, a_0 being the equilibrium value of the lattice parameter. Reprinted from Ref. [49], p. 2, by courtesy of Academic Press, Inc.

the s-p band gap (Fig. 2). The s band will be completely empty at $0°K$, and, unlike in the case of the transition metals, there is no overlapping between the s and d bands. Thus the electrical, magnetic, and optical properties of transition-metal oxides are still determined by the electrons in the d band. Because of the narrow width of the d band and the absence of overlap between the outer electron orbitals, the 3d energy levels should be considered essentially localized on the cations. Each cation experiences a strong electrostatic field due to a regular arrangement (tetrahedron or distorted octahedron) of O^{2-} anions, and hence the crystal field splitting of the d band should also be taken into account in calculating the band structure.

The ligand-field theory has been applied [51-54] successfully to many optical problems of transition-metal complexes, such as $Ni(H_2O)_6^{2+}$, whose optical spectrum closely resembles that of NiO. This approach has also been applied [50-54] to account for the solid-state properties of the transition-metal oxides, but with only limited success; for instance, the metallic conductivity of TiO and NbO can

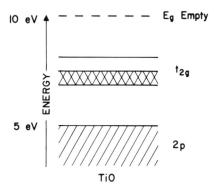

Fig. 3. A partly filled d band of TiO (d^2) formed by the overlap of the cationic t$_{2g}$ orbitals, shown schematically. Based on Ref. [47], p. 510.

be accounted for by the partly filled d band formed by the overlap of the cationic t$_{2g}$ orbitals while the E$_g$ levels remain empty, as shown in Fig. 3. However, from similar considerations, pure MnO, CoO, and NiO should be metallic, whereas actually MnO is one of the best known insulators and CoO and NiO in the pure state are poor conductors. The band theory also fails to account for the solid-state properties of the sesquioxides and other oxides.

Goodenough and Rathenau [55, 56] have examined various conditions under which the localized-electron theories, as compared with the one-electron model of the band theory or a combination of the two approaches, could be applicable to the transition-metal oxides in order to account for the observed properties. The essentially one-electron approach of the elementary band theory has also been criticized [49] for neglecting the effects of the surrounding polar field on the motion of the single electron in a crystal. The theories that take into account such electronic and polarization effects in the conduction mechanism, and also the more recent work of Goodenough on the bonding properties of transition-metal compounds, have been reviewed recently by Adler [49] and by Rao and Rao [50].

Finally, the mechanism of hole-hopping (thermally activated polarons) from Ni^{3+} sites to Ni^{2+} sites (created by Li^{+} or nonstoichiometry) that was originally proposed by De Boer and Verwey [57] to account for the conductivity of NiO has also been shown to be invalid by later experimental work [58, 59]. In addition to several other effects left unexplained, the hole-hopping mechanism predicts a carrier mobility that increases with temperature; this is contradictory to experimental observations. Further details are given in two recent reviews [49, 50] that also provide a compilation of all the experimental data collected on the bulk properties of various transition-metal oxides.

From the above considerations it is clear that the surface properties of transition-metal oxides could be extremely complex and hard to predict. Often approaches using localized-electron theories, such as the ligand-field theory, may be more appropriate in dealing with the double-layer properties of the transition-metal oxides than theories based on the band model alone.

III. THE DOUBLE LAYER AT THE SOLID-SOLUTION INTERFACE

The structure of the double layer at the solid-solution interface depends on whether the solid is a metal, a semiconductor, or an insulator. Because of the high concentration of free carriers in metals ($\sim 10^{22}$ cm^{-3}), practically no space charge can occur in the metal phase. On electrical polarization of an oxide-free metal-solution interface, for example, the characteristic length for the variation of interfacial properties in the metal phase is only 0.5 Å [32]. The dielectric constant and the capacity of the metallic phase are considered to be infinite [60]. On the other hand, in comparison with metals and insulators, a semiconductor phase has an intermediate level of carrier concentration ($\sim 10^{14}$ cm^{-3} in intrinsic Ge at room temperature). For small changes in the Galvani potentials φ in the solution and in the solid phases, it may be shown [27, 31, 61] that

$$\frac{\psi_{sol}}{\varphi_{sd}} = \frac{L_{sol} \epsilon_{sd}}{L_{sd} \epsilon_{sol}} \approx \left(\frac{C_{sd} \epsilon_{sd}}{C_{sol} \epsilon_{sol}} \right)^{\frac{1}{2}} , \qquad (3)$$

where the subscripts "sol" and "sd" refer to the solution and the solid phases, L is the Debye length, ϵ is the dielectric constant, C_{sd} and C_{sol} refer to the carrier concentration in the solid and the ion concentration in solution, respectively, and ψ_{sol} is the electrostatic potential in the solution phase relative to that in the bulk. The Debye length in a semiconductor with $\epsilon_{sd} = 16$ can vary from 10^{-3} to 10^{-7} cm, with C_{sd} varying from 10^{13} to 10^{18} cm^{-3}. In a 1:1 valent electrolyte solution L_{sol} varies [62] from 10^{-4} to 10^{-7} cm, with the electrolyte concentration varying from 10^{-7} to 10^{-1} mole/l. Hence the Debye length in semiconductors with moderate carrier concentrations is comparable to the Debye length in dilute electrolyte solution. However, the Debye length is negligible in concentrated solutions. The total potential drop at the solid-solution interface is the sum of the Galvani potentials in different regions of the interface, including the space-charge region. Hence in dilute as well as in concentrated solutions the space charge and surface states in semiconductors play a decisive role in all interfacial phenomena and electrode kinetics. The various aspects of the space charge in semiconductors and in insulators are presented in the next sections, after a brief introduction to the general properties of the double layer on metals.

A. General Properties of the Double Layer on Metals

1. General Remarks

The Gouy-Chapman model of the double layer on mercury and on other oxide-free metal surfaces, as modified by Stern and Grahame

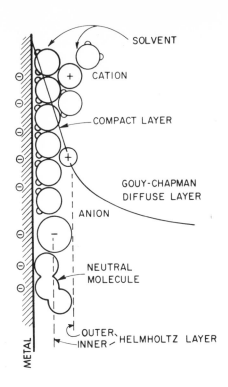

Fig. 4. A general schematic representation of the double layer
on metals showing the adsorption of solvent molecules, anions, and
cations. Reprinted from Ref. [2], p. 26, by courtesy of the Ronald
Press Company.

[60], is shown schematically in Fig. 4 [2]. The plane of the locus of

electric charge of the adsorbed hydrated ions is generally known as

the outer Helmholtz plane (OHP, ~ 3 Å), or the Gouy plane. Other

variations of the Helmholtz-layer structure with respect to the mode

of surface and ion hydration have been considered [63]. The Helm-

holtz layer is generally occupied by oriented water (or other solvent)

molecules under a high electric field. Many anions (and some cations

to a lesser degree) are said to be specifically adsorbed on the metal

surface at the inner Helmholtz plane (IHP) by displacing the water

molecules on the surface. It may be noted that Bockris and co-workers

[20, 24] use the term "contact adsorption" in preference to "specific

adsorption. " The experimental methods for double-layer investigations consist of electrocapillary techniques for mercury [60, 64],
capacitance measurements by the dc charging and ac bridge techniques [65-67], and the recent pulse techniques [68-78] using oscil-
lographic or recording methods. Ramaley and Enke [70] have briefly
reviewed the instrumentation required for capacitance measurements
and have described two methods in detail. The pulse techniques
applied to the mercury-solution or metal-solution interface have been
discussed by Hackerman and co-workers [72-74] and Barradas and
Valeriote [76-78].

Many anions show a strong tendency for specific adsorption.
Cations (e.g., alkali and alkaline-earth ions), being in general more
strongly hydrated and less polarizable than anions, stay at the OHP.
However, Cs^+, Tl^+ [18, 79-81], polyvalent cations (Al^{3+}, La^{3+}), and
tetraalkylammonium ions [18] show some tendency for specific adsorption on mercury. In the absence of specific adsorption of ions
on metal surfaces the IHP loses its significance and the Helmholtz
layer is occupied by a layer of oriented water or other solvent molecules. The measured differential capacitance of the interface, given
by

$$\frac{1}{C_{total}} = \frac{1}{C_{Helmholtz}} + \frac{1}{C_{Gouy}} \quad , \tag{4}$$

generally shows a minimum at the pzc in dilute solutions (Figs. 5
and 6). In the cathodic region the capacity is almost independent of
the nature and concentration of cations (except Tl^+ and Cs^+) and is
about 16 $\mu F/cm^2$ (Fig. 6) at a surface charge $q = 13$ $\mu C/cm^2$ and a
potential of about 0.7 V relative to the pzc. The water molecules in
the Helmholtz layer are subjected to a high electric field ($d\psi/dx \sim$
10^8 V/cm) and undergo dielectric saturation, with the dielectric constant varying from about 7 inside the Helmholtz layer [2, 63] to about

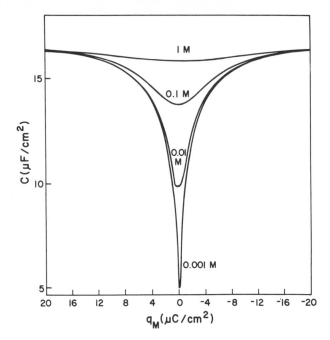

Fig. 5. Theoretical differential-capacitance curves of the double layer on solids, in solutions of 1:1 valent salts, in the absence of specific adsorption. Reprinted from Ref. [18], p. 642, by courtesy of the American Chemical Society.

30 to 40 in the vicinity of the OHP and 78.5 in the diffuse double layer. The differential capacitance of the double layer on the anodic side is quite sensitive to the nature and concentration of anions, the temperature, and the bias potential. The V-shaped capacitance curves in the presence of many anions show a characteristic hump on the anodic side, followed by a steep rise in capacitance at high anodic potentials due to increased specific adsorption of anions. The capacitance hump has been the subject of many investigations and has been discussed recently in detail by Barlow [19]. The hump is usually attributed to two opposing effects, the specific adsorption of anions and dielectric saturation. No such hump is found to occur in the case of anions showing intense specific adsorption (e.g., I^-), which would mask all other effects.

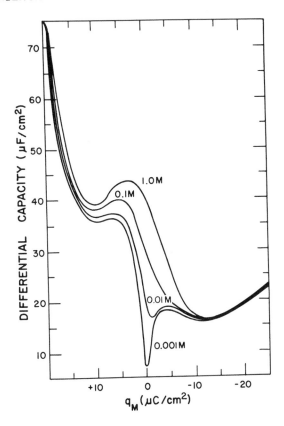

Fig. 6. The differential capacitance of the double layer on mercury in potassium chloride solutions, shown as a function of the surface–charge density q_M. Reprinted from Ref. [18], p. 643, by courtesy of the American Chemical Society.

The pzc is a characteristic electrochemical property of each metal and is related to its electronic work function [18, 82–84] (see Ref. [85] for recent compilations of the pzc data). Hence the pzc varies with different crystallographic faces and also depends on the adsorbed solvent dipoles and the nature and amount of specifically adsorbed ions. In the absence of specific adsorption the pzc coincides with the electrocapillary maximum for the Hg–solution interface and in general with the capacitance minimum for any metal–solution interface.

The pzc and the corresponding double-layer properties of the interface
shift toward positive or negative potentials, depending on whether
cations or anions are specifically adsorbed at a rate $> RT/nF$ (Esin
and Markov effect [17, 18, 86, 87]). When both anions and cations
are specifically adsorbed, the sign of the shift of potential may
be reversed as the electrolyte concentration changes.

In contrast to the precise electrocapillary methods for studying
the double layer on mercury, the interfacial tension at a solid metal-
solution interface can be studied only indirectly by measuring such
properties as contact angles and friction coefficients. The coefficient
of friction between two solids in contact with an electrolyte solution
varies with the applied potential across the solid-solution interface
and has been shown [82] to be maximal at the pzc. Electrocapillary
curves of solid metal electrodes have also been obtained recently by
measuring the differential change in the length of a metal ribbon [88]
or in the bending of a thin metal electrode [89] as a function of the
applied potential. Several uncertainties arise in these measurements
from such factors as interfacial tension at the solid-gas phase,
hysteresis effects, surface areas and surface roughness, and surface
heterogeneity.

Eyring and co-workers [90] have described a frictional method of
determining the pzc of a number of freshly exposed metal surfaces.
From the shifts in the pzc of these metal surfaces these authors have
also determined the relative adsorbability of halide ions on various
metals. However, no simple relationship could be found between the
adsorbabilities of halide ions on metals and the covalent metal-
halogen bond strengths.

Direct measurements of the double-layer capacitance have been
made for a number of metals [18, 70-72, 91-95]. These data have
provided some information on the adsorption densities and the nature

of adsorption of ions and of organic molecules on metals [18, 96]. In aqueous solutions most metals either possess an oxide layer or readily acquire one on anodic polarization (e. g. , Zn [94, 95], Sn [92], and Pt [18, 97, 98]). The presence of such an oxide layer on metals makes the results of the double-layer investigations doubtful and complicated to interpret.

Electrolytic adsorption of H^+ on platinum has also been studied in detail by a number of workers [22, 87, 98, 99]. Platinum in this respect is reported to behave entirely differently from mercury. The pzc of platinum has been reported to be both pH independent [22, 98] and also pH dependent [100], with a linear relationship between the pzc and pH. As the behavior of platinum electrodes depends on their pretreatment, it may be suggested that the quantities measured in these investigations were not the same.

Detailed analyses of the double layer on metals, and on mercury in particular, dealing with the thermodynamics, structure and the distribution of charge and potential at the interface, and adsorption isotherms [23, 101, 102] have been presented in several papers [2, 17-20]. The present state of the subject has been summarized by Conway and Gordon [103] and discussed in detail by Barlow [19].

2. The Double Layer on Metals Compared with the Double Layer on Oxides

Models for the double layer on metals have been developed strictly for polarizable interfaces in the absence of potential-determining ions in solution (e. g. , Hg^{2+} for Hg, H^+ and OH^- for oxides). Variations in extremely minute concentrations of the potential-determining ions can give rise to a reversible double layer on oxides by adsorption and surface-dissociation processes, such as

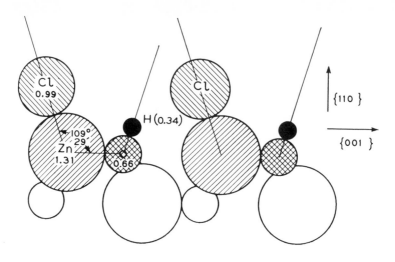

Fig. 7. The structure of the 110 surface of ZnO in contact with
a hydrochloric acid solution. The chemisorbed H^+ are shown to be
closer to the oxide surface than the other ions. The distances (in
angstroms) were obtained [104] from Pauling's tetrahedral covalent
radii. Reprinted from Ref. [104], p. 131, by courtesy of the Bell
Telephone Laboratories.

$$\begin{array}{ccccc}
\underset{OH}{\overset{I^-}{>}}M^+ & \underset{\Longrightarrow}{\overset{H^+I^-}{\longleftarrow}} & >M\underset{OH}{\overset{OH}{<}} & \underset{\longrightarrow}{\overset{K^+OH^-}{}} & >M\underset{OH}{\overset{O^-K^+}{<}}
\end{array} \qquad . \qquad (5)$$

Basic dissociation   Acidic dissociation

Taking the centers of either the surface oxygen or the surface metal

atoms as the reference plane, the chemisorbed H^+ ions (radius = 0.34Å)

lie much closer (Fig. 7) to the surface than to the OHP or the IHP, and

hence the chemisorbed H^+ ions may be said to constitute another plane

of charge centers within the inner Helmholtz layer. The adsorption

densities of H^+ and OH^- on the surface, and consequently the surface

potential and the potential drop in the Helmholtz region, all vary with

pH (a reversible double layer). An oxide-solution interface, there-

fore, under an external bias will have the combined features of both a

polarizable and a reversible double layer.

Another complication in the analysis of the double layer on oxides arises from the fact that in the reversible double layer the reference plane that separates the surface charge from the counterionic charge is established naturally by the dissociation process that occurs at the interface. Thus, as shown in Eq. (5), the surface charge for cationic adsorption on oxides lies on the oxygen sites of the surface, whereas the surface charge for anionic adsorption lies on the metal sites (M) of the oxide surface. In comparison with the double layer on metals an additional modification of the double layer on oxides arises from the semiconducting and insulating nature of bulk oxides and also from the possibility of proton injection into the space-charge layer of a semiconductor under a high applied electric field. Any double-layer model for the oxide-solution interface should also account for the pH dependence of the zpc on the bulk properties of oxides.

B. The Semiconductor Oxide-Solution Interface

At a semiconductor-electrolyte solution (sc-l) interface, in addition to the presence in the aqueous phase of an ionic, diffuse double layer (in dilute solutions) and the compact Helmholtz layer, a diffuse space-charge layer occurs in the solid phase. The Helmholtz layer is known to extend in the semiconductor phase also, due to the presence of surface states [105-108]. The Helmholtz region of the interface is also known to play a complex role in all charge-transfer processes and in the potential distribution and polarization phenomena occurring at the semiconductor-solution interface. The total, Galvani-potential (φ) difference at a semiconductor-solution interface has been written as a sum of its components g [1, 15], so that

$$\Delta\varphi^{sc-l} = g^{sc-l}_{ion, free} + g^{sc-l}_{ion, bound} + g^{sc-l}_{dipole} \ . \qquad (6)$$

The first term on the right-hand side of Eq. (6) refers to the contribution from the excess charge in the space-charge layer; the second term arises from the bound charge of the semiconductor surface, which may be due to ionized surface states or surface groups; and the third term has the usual meaning. The nature of the diffuse double layer in the aqueous phase for semiconductors and for insulators remains the same as that for metals and will not be discussed here further. Compared to the metal-solution interface, the distinguishing features of the semiconductor-solution interface are the presence of a space-charge region and a Helmholtz region consisting of surface states and surface groups. These aspects of the semiconductor-solution interface, with particular reference to oxides, are discussed in the sections that follow.

1. The Fermi Level of Solids and Interfacial Equilibria

The probability f_n of an available quantum state E in a solid being occupied by an electron or by a hole ($f_p = 1 - f_n$) is given by the Fermi-Dirac distribution law. This distribution law is conventionally written for a system obeying the Pauli exclusion principle as

$$f_n = \frac{1}{1 + \exp\left[(E - E_F)/kT\right]} \tag{7}$$

and

$$1 - f_n = f_p = \frac{1}{1 + \exp\left[(E_F - E)/kT\right]} \; , \tag{8}$$

where k and T have the usual meaning and E_F is the Fermi level. The significance of E_F may be readily realized by writing the above expressions in the following form, which is more familiar to chemists [28]:

$$E_F = E + kT \ln \frac{f_n}{1 - f_n} \; . \tag{9}$$

When $f_n = 1 - f_n = f_p$, the probability of an energy level's being occupied or being empty is $\frac{1}{2}$ and the corresponding value of E is the Fermi level. Also $E_F = E_i \sim E_g/2$, E_i being the Fermi level of the intrinsic semiconductor. Hence E_i, in effect, can be considered as an internal, standard electrochemical potential of electrons with reference to which the distribution of electrons under other conditions of occupancy can be expressed. When $|E - E_F| \gg kT$, the Fermi-Dirac distribution law reduces to the Maxwell-Boltzmann distribution law (e.g., within 2% error when $E - E_F = 4kT$). Within this approximation classical thermodynamics can be used in most practical problems concerning interfacial equilibria, and Eqs. (7) and (8) can be written as

$$f_n = \exp\left[-(E - E_F)/kT\right] \tag{10}$$

and

$$f_p = \exp\left[(E - E_F)/kT\right] \tag{11}$$

or

$$f_n = A \exp\left(-E/kT\right) \quad \text{for} \quad (E - E_F) \gg kT \tag{12}$$

and

$$f_p = B \exp\left(E/kT\right) \quad \text{for} \quad (E - E_F) \ll kT \ , \tag{13}$$

A and B being normalizing constants. Hence the carrier concentrations in the bulk phase of a nondegenerate semiconductor are also related to the Fermi level by the standard Maxwell-Boltzmann relationship. For an intrinsic semiconductor, for example, the electron (n_b^i) and hole concentrations (p_b^i) in the bulk phase are given as

$$n_b^i = N_c \exp\left[-(E_c - E_F)/kT\right] \tag{14}$$

and

$$p_b^i = N_v \exp\left[-(E_F - E_v)/kT\right] \ , \tag{15}$$

where N_c and N_v refer to the available density of states in the conduction and valence bands of energy levels, E_c and E_v, respectively.
The carrier concentrations in extrinsic semiconductors are also related
by standard equations [16, 26, 109-112] to the different energy levels in
the band model. It has also been shown from these equations that the
Fermi level of an intrinsic semiconductor almost coincides with the
mid-gap position. A thermodynamic state of equilibrium between two
phases represents an optimum balance between the conditions of minimum free energy and maximum entropy of the system in the distribution
of particles in the two phases. Thus the condition for two phases to
be in thermal equilibrium is that the Fermi level is the same in both
the phases, so that all levels at a given energy have equal probability
of being occupied. Initially, if the Fermi levels of the two systems
are not the same, charge will flow from the system of higher Fermi
level to the other until their Fermi levels are equalized.

2. Surface States and Surface Groups at Oxide-Solution Interfaces

a. Surface States. As a result of the termination of lattice periodicity
at a crystal surface, the energy levels at the surface (or interface)
are not the same as those in the bulk. These surface energy levels
were first shown by Tamm [113], and later by Shockley [114], to be
localized states where the electrons are confined to the surface atoms
and whose energies lie in the forbidden zone [105-108]. Two kinds
of surface states have been distinguished theoretically: Tamm and
Shockley states. The type of states that occur for a given surface
depends on the symmetry of the lattice termination and the interatomic
distances. Details concerning the origin of the surface states, their
three-dimensional models, and calculations of their energy-band
structure have been discussed in several papers [107, 108, 110, 115-
120], including surface states at the semiconductor-electrolyte
interface [120-124]. The surface states may also originate from

(a) lattice imperfections, chemical inhomogeneity, and geometrical irregularities; (b) adsorbed impurities; (c) electrical polarization; and (d) contact with another phase with a different work function. The surface states are known to be either fast or slow, depending on their relaxation times. However, the fast surface states are known to be too few to affect the aqueous double layer on semiconductors.

The occupation of surface states by electrons is also given by the Fermi-Dirac distribution law, which is written for a simple case under nondegenerate conditions as

$$f_n(E_t) = \frac{1}{1 + \exp\left[(E_t - E_F)/kT\right]} \tag{16}$$

and

$$f_p(E_t) = \frac{1}{1 + \exp\left[(E_F - E_t)/kT\right]} \cdot \tag{17}$$

The possible multiplicity of the surface levels is taken into account by replacing E_t in Eqs. (16) and (17) by an effective energy of state E_t^f [120]. In the presence of an electric field at the interface the energy bands at the surface are normally bent up (p-type space charge) or bent down (n-type space charge) relative to E_F by a term $u_s = \pm e\varphi_s/kT$. Hence $E_F = E_i + u_s kT$ (Figs. 8 and 9), φ_s being the inner potential at the surface. Under such equilibrium conditions the exponential terms in Eqs. (16) and (17) are replaced by $[(E_t - E_i)/KT - u_s]$ and $[u_s - (E_t - E_i)/kT]$, respectively, and may be written in a combined form as

$$u_s = \frac{E_t - E_i}{kT} + \ln \frac{f_n(E_t)}{f_p(E_t)} \cdot \tag{18}$$

If N_t (cm^{-2}) refers to the total density of the surface states (which are assumed to be identical) and n_t (cm^{-2}) and $p_t = (N_t - n_t)$ (cm^{-2}) are the densities of the occupied and unoccupied surface states, then

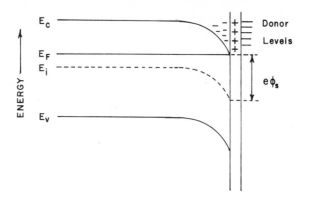

Fig. 8. Schematic energy-band diagram showing an n-type space charge formed at a semiconductor surface by electron transfer from the surface levels.

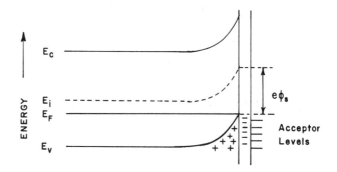

Fig. 9. Schematic energy-band diagram showing a p-type space charge formed at a semiconductor surface by hole transfer from the surface levels.

$$u_s = \frac{E_t - E_i}{kT} + \ln \frac{n_t}{p_t} . \tag{19}$$

When $n_t = p_t = N_t/2$, $u_s - (E_t - E_i)/kT = 0$. At this point $E_F = E_t$; that is, the Fermi level passes through the surface energy level E_t, and a maximum change occurs in the occupation statistics of surface states. A surface may have a series of independent energy levels, such as above, when the total density of electrons captured by all the surface states is given by the sum of their individual occupancies. The fractional occupancy of the surface states under varying conditions of E_t has been discussed elsewhere [110] in detail, with numerical solutions.

b. Surface Groups at the Oxide-Solution Interface. Consider an oxide crystal (e.g., ZnO, SnO_2, SiO_2) subjected to a random fracture in an aqueous medium, thereby creating a series of surface energy levels by virtue of the residual chemical bonds on the surface. A statistically equal distribution of the oxygen and metal atoms on the two surfaces is expected to occur, with overall electrical neutrality for each surface. One such fracture is shown schematically in two dimensions in Fig. 10, which also indicates the partially ionic and partially covalent character of the surface metal-oxygen bonds. A maximum of about 10^{15} unit charges per square centimeter ($\sim 100 \ \mu C/cm^2$) can occur on the surface. The surface is also shown to have donor and acceptor

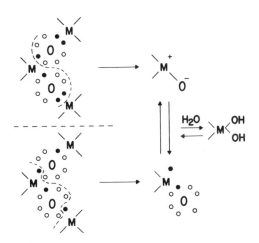

Fig. 10. Schematic representation of the formation of unsaturated chemical bonds from the fracture of an oxide crystal and the formation of surface hydroxyl groups.

properties. In the case of a semiconductor oxide, when the energy difference $E_t - E_F$ (Eqs. (16) and (17)) falls within the range of thermal energy, electrons can be excited from the donorlike surface levels to the conduction band (Fig. 8) or from the valence band to the acceptor-like surface levels (Fig. 9), as the case may be [120]. When the energy difference $|(E_t - E_F)| >> kT$, the number of surface energy levels acting as donor or acceptor levels will be so small as to be negligible, as in the case of an insulator oxide like quartz. In such a case no space charge could exist in the solid phase, and ideally any potential drop in the surface region will be confined to an inter-atomic distance only.

From the foregoing considerations, therefore, the surface states can be identified as those surface energy levels that give rise to, and are in electronic and thermal equilibrium with, the space charge. The following loose distinction between the surface states and the surface groups has been made by Garrett and Brattain [108] and by Gerischer [25]. The charge in the surface states is described [108] as that part of the "space charge which, while not forming part of the extended space charge region, is yet in good electrical contact with the con-duction or valence band." The remaining part of the surface charge (which changes slowly with an activation energy) may be said to con-stitute the ionizable surface groups that are in ionic equilibrium with the aqueous phase. The surface states also participate in the recom-bination process and serve as trapping centers [107, 108]. The slow surface states and the surface groups differ considerably from the space-charge region in their relaxation times, and the latter can therefore be investigated experimentally by short-duration pulse tech-niques if the density of the fast surface states is low.

The density of measured surface states in semiconductors has been found to depend on whether the surfaces are clean or contami-nated and also on the nature of impurities adsorbed on the semiconductor surface. Clean germanium and silicon surfaces in vacuum are found to have about 10^{15} surface states per square centimeter, corresponding to a maximum of about 10^{15} unsaturated bonds per square centimeter that can occur on any surface. However, oxidation of germanium surfaces either anodically or chemically and chemisorption of oxygen greatly reduce the number of fast surface states (relaxation time $< 10^{-3}$ sec) and give rise to slow surface states. Thus a maximum of about 10^{11} fast surface states per square centimeter and more than 10^{13} slow surface states per square centimeter have been measured on chemically etched germanium with an oxide layer, using various methods [30, 118, 122]. The unsaturated bonds on the surfaces of oxides are also readily neutralized with H^+ and OH^- from water and give rise to surface groups. Depending on the solution pH, these surface groups may be either neutral (at the zpc) or charged due to dissociation, thus giving rise to a reversible double layer. From the principle of overall electrical neutrality of the interface,

$$q_{sc} + q_{ss} + q_s + q_{el} = 0 ,$$ (20)

where q_{sc}, q_{ss}, q_s, and q_{el} refer to the densities per square centimeter of the excess charge in the space-charge region, in the surface states, on the oxide surface (due to surface groups), and in the aqueous-double-layer region, respectively. Furthermore, in the absence of an applied electric field, the sum of the Galvani potential differences in different interfacial regions should also be equal to zero. It should be emphasized, however, that the space-charge region in the semiconductors under consideration is essentially electronic in nature compared with the aqueous double layer, which is composed of mobile counterions as the charge carriers. The effect of an applied electric field on the surface potential will be examined briefly after considering the subject of space charge in some detail.

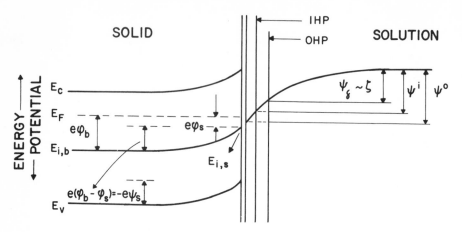

Fig. 11. Schematic representation of the energy-band structure of a p-type space-charge region in an n-type semiconductor and the distribution of potential in the double-layer region.

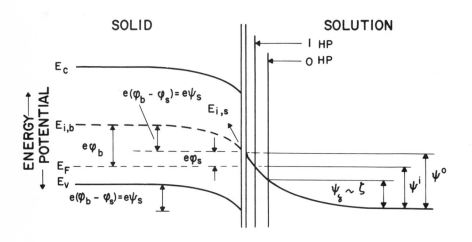

Fig. 12. Schematic representation of the energy-band structure of an n-type space-charge region in a p-type semiconductor and the distribution of potential in the double-layer region.

3. Charge and Potential Distribution at the Semiconductor-Solution Interface

The variation in potential in the space-charge region and in the aqueous double layer at a semiconductor-solution interface is shown schematically in Figs. 11 and 12. The space-charge regions shown in these figures are depletion layers. In accumulation and depletion layers the bulk majority carrier is still in excess in the space-charge region, but it increases or decreases respectively relative to the bulk. The bands are shown bent up for a p-type and bent down for an n-type space charge (the convention for the sign of the potential, in solid-state physics, as shown in Figs. 11 and 12, is opposite to that used in electrochemistry [104]). The intrinsic Fermi level E_i is parallel to the band edges and has different positions at different points relative to the Fermi level in the space-charge layer. In the case of an inversion layer (minority carrier in relative excess), E_i crosses E_F near the surface. These and other junctions that are possible at the semiconductor-electrolyte interface have been considered by a number of authors [25, 125]. The ZnO-solution interface, for example, as discussed by Dewald [27, 104], is shown in Fig. 11. The space-charge layer at the semiconductor-electrolyte interface can originate, even without an applied bias, as a result of the surface states, which together with the ionizable surface OH (or OH_3O) groups constitute the Helmholtz region.

The difference between E_F and E_i at different positions along the x coordinate in the space-charge region represents the degree of band bending or the potential barrier at that position. If $E_{i,b}$, $E_{i,x}$, and $E_{i,s}$ refer to the intrinsic Fermi levels in the bulk phase, at any point x in the space-charge region, and at a point just inside the surface, then the potential barriers at the corresponding points are related to the differences in the inner potentials φ_b, φ_x, and φ_s as

$$e\varphi_b = (E_F - E_{i,b}); \quad e\varphi_x = (E_F - E_{i,x}); \quad e\varphi_s = (E_F - E_{i,s}) \quad (21)$$

for the p-type space-charge region in Fig. 11. In the n-type space-charge region (Fig. 12) the φ terms in Eq. (21) will have signs opposite to those shown above. Other useful terms, $e(\varphi_b - \varphi_x)$ and $e(\varphi_b - \varphi_s)$, follow from Eq. (21). Also, with reference to Fig. 11,

$$\varphi_b - \varphi_s = \psi_b - \psi_s = -\psi_s , \quad (22)$$

where ψ is the electrostatic potential and ψ_b refers to the bulk potential, which by convention is taken as zero. The electrostatic potential ψ_s, also known as the surface potential, has the opposite sign for the situation in Fig. 12. The dependence of the field strength $(d\varphi/dx)$ on different parameters of the bulk and of the space-charge regions is obtained in a manner similar to that for the diffuse double layer by solving the Poisson-Boltzmann equation [108]. From the one-dimensional Poisson equation,

$$\frac{\partial^2 \varphi_x}{\partial x^2} = -\frac{4\pi}{\epsilon} \rho(x) , \quad (23)$$

where ϵ is the dielectric constant and $\rho(x)$ is the net charge density (C/cm^3) due to both mobile and static charges, so that

$$\rho(x) = e(-n_x + p_x + N_D - N_A) . \quad (24)$$

For a nondegenerate semiconductor the electron and hole concentrations are given by the Boltzmann equation, so that

$$n_x = n^0 \exp [e(\varphi_x - \varphi_b)/kT] , \quad (25)$$

$$p_x = p^0 \exp [-e(\varphi_x - \varphi_b)/kT] . \quad (26)$$

By substituting for $\rho(x)$ in Eq. (23) from Eqs. (24), (25), and (26) and integrating we get the general expression for the field strength:

$$\frac{d\varphi}{dx} = \pm \left\{ \frac{8\pi kT}{\epsilon} \left[-(N_D - N_A)y + p^0(e^{-Y} - 1) + n^0(e^{-Y} - 1) \right] \right\}^{\frac{1}{2}} , \qquad (27)$$

where

$$y = \frac{e(\varphi_x - \varphi_b)}{kT} . \qquad (28)$$

By substituting φ_s for φ_x in Eq. (27) the field strength at the surface, E_s, can be obtained and written as

$$E_s = \pm \left(\frac{8\pi kTn_i}{\epsilon} \right)^{\frac{1}{2}} F(Y, \lambda) \qquad (29)$$

or

$$E_s = \pm \frac{kT}{eL} F(Y, \lambda) , \qquad (30)$$

where, since $n^0 p^0 = n_i^2$,

$$\lambda = \left(\frac{p^0}{n^0} \right)^{\frac{1}{2}} = \frac{p^0}{n_i} = \frac{n_i}{n^0} , \qquad (31a)$$

$$Y = \frac{e(\varphi_s - \varphi_b)}{kT} , \qquad (31b)$$

$$F(Y, \lambda) = [\lambda(e^{-Y} - 1) + \lambda^{-1}(e^{Y} - 1) + (\lambda - \lambda^{-1})Y]^{\frac{1}{2}} , \qquad (31c)$$

and

$$L = \left(\frac{\epsilon kT}{8\pi n_i e^2} \right)^{\frac{1}{2}} \qquad (31d)$$

is the Debye length in an intrinsic semiconductor and is analogous to the reciprocal of the Debye-Hückel constant in electrolytes. The space-charge density (net charge per unit area) in the semiconductor is given by

$$q_{sc} = -\frac{\epsilon E_s}{4\pi} = \mp \frac{\epsilon kT}{4\pi eL} F(Y, \lambda) . \qquad (32)$$

The first term on the right-hand side of Eqs. (29) and (30) will have a positive or negative sign depending on whether $Y > 0$ or $Y < 0$, that is, $\varphi_s > \varphi_b$ or $\varphi_s < \varphi_b$ (bands bending downward or upward), and according to Eq. (32) the excess space charge will have an opposite sign to these terms. When $Y = 0$, the potential distribution and the concentration of charge carriers are uniform from the bulk of the semi-conductor up to the surface, and the electrode potential at which $Y = 0$ is known as the flat-band potential. These equations for q_{sc} and the field strength can be further simplified depending on the relative concentrations of the majority and minority carriers that control the shape of the potential barrier. Numerical solutions to these equations have been presented and the shape of the potential barrier has been discussed in various papers [108, 110, 126].

The charge densities q_{ss} due to surface states have been calcu-lated [16, 25, 110, 120] by using the appropriate expressions for the Fermi level and also measured experimentally [119, 121–124]. From the principle of electroneutrality and from Eq. (20), the sum of the excess charge in different interfacial regions — namely, the space charge, the surface states, the surface groups, and the aqueous phase (the countercharge) — should be zero.

Quantitative information on the space-charge layer and some qualitative information on the behavior of the Helmholtz layer can be obtained by measuring several solid-state properties of the interface. The most important and useful of these properties are (a) the surface conductance, (b) the surface photopotentials, (c) the electrode poten-tials and i-V characteristics, and (d) the differential capacitance. The differential capacitance (henceforth referred to as "capacitance") is by far the most informative property of the interface.

As conductivity is proportional to the concentration of free carriers, the surface conductivity in the presence of an applied elec-tric field arises from the difference in the concentration of the free carriers in the space-charge layer and the free carriers in the bulk. This solid-state surface conductivity is different from the surface conductivity encountered in electrokinetic work, which is discussed in a subsequent section. The surface photopotential arises on illumination of the semiconductor surface with light of energy greater than the band gap when electrons can be excited from the valence band to the conduction band. The excess (nonequilibrium) free carriers resulting from the absorption of light in the space-charge region change the surface potential ($\varphi_b - \varphi_s$) and hence the electrode potential. The nonequilibrium charge carriers produced by a light pulse may also recombine with surface levels at the semiconductor-solution interface. Hence another useful measurement, particularly for determining the characteristics of surface states, is the surface recombination velocity. Its measurement is based on the difference between the lifetime of charge carriers in the surface and in the bulk region of the semiconductor.

The theory correlating the interfacial properties with the bulk- and with the space-charge properties of semiconductors has been dealt with in several papers [104, 108, 126-128] and reviews [3, 5, 15, 16, 26, 27, 110]. The experimental methods of investigating these properties have also been described and the results discussed [41, 104, 122, 128-138] by a number of authors. In the following section only some basic principles concerning the capacitance and electrode potentials of the semiconductor-electrolyte interface and the nature of information obtainable from their measurements will be outlined. Some experimental results that are of direct interest to us in this chapter will also be presented.

4. The Semiconductor-Electrode Potentials

Changes in the measured electrode potentials ΔV_E of a cell such as

$$\text{metal} \, | \, \text{semiconductor} \, | \, \begin{array}{c} \text{electrolyte} \\ \text{solution} \end{array} \, | \, \text{reference electrode} \, | \, \text{metal}$$

may be written [134] as the sum of the changes in the Galvani potentials at different interfaces, so that

$$\Delta V_E = \Delta \varphi^{M-sc} + \Delta \varphi^{sc-sol} + \Delta \varphi^{sol-M} , \tag{33}$$

where M, sc, and sol refer to the metal, semiconductor, and the solution phases, respectively. If the variations in the liquid-junction potentials are insignificant, then $\Delta \varphi^{sol-M} = 0$. Splitting φ^{sc-sol} into its components, we have

$$\Delta V_E = \Delta \varphi^{M-sc} + \Delta (\varphi_b - \varphi_s)^{sc} + \Delta (\varphi_s - \varphi_{sol})^{sol} , \tag{34}$$

where the subscripts b, s, and sol refer to the bulk semiconductor (sc), the surface, and the solution, respectively. If V_H and χ refer to the potential of the Helmholtz layer and potential due to surface dipoles, respectively, then in sufficiently concentrated solutions, where the Guoy layer is suppressed,

$$\Delta (\varphi_s - \varphi_{sol})^{sol} = \Delta V_H + \Delta \chi . \tag{35}$$

Brattain and Boddy [134] have distinguished further between the χ potential due to the orientation of the solvent molecules $(\chi^{sol(ijk)})$ and the χ potential due to the different arrangements of the surface atoms and of the dangling bonds at different crystallographic planes $(\chi^{M(ijk)})$. In Eq. (35), therefore,

$$\Delta \chi = - \Delta \chi^{M(ijk)} + \Delta \chi^{sol(ijk)} . \tag{36}$$

For a given composition of the bulk and of the space-charge region of a semiconductor it is seen from Eq. (22) and Fig. 11 that $\varphi_b - \varphi_s = -\psi_s$, and, relating φ^{M-sc} to the bulk-carrier concentration, Eq. (34) is written as

$$\Delta V_E = \frac{kT}{e} \Delta \ln \lambda - \Delta \psi_s + \Delta V_H - \Delta \chi^{M(ijk)} + \Delta \chi^{sol(ijk)} . \qquad (37)$$

According to the sign convention generally used, V_E becomes increasingly positive as the electrode becomes more anodic relative to the solution and ψ_s becomes more negative (Figs. 11 and 12) as the free positive charge increases in the semiconductor. Equation (37) includes possible changes in the potential drop in the Helmholtz layer (V_H), such as changes due to the chemical dissociation of the surface OH groups depending on pH [41]. Boddy [15] has discussed in detail the variation in the electrode potential with different parameters. Rearranging Eq. (37), we have

$$\Delta(\psi_s - \frac{kT}{e} \ln \lambda) = -\Delta V_E + \Delta V_H - \Delta \chi^{M(ijk)} + \Delta \chi^{sol(ijk)} , \qquad (38)$$

where λ is known from the measurement of the bulk semiconductor properties (electrical conductivity and Hall-effect measurements). Hence a plot of the left-hand side of Eq. (38) against the measured electrode potentials V_E, in a solution of a given composition, should be linear with a slope of minus unity if V_H and χ potentials are constant.

It was found that the electrode-potential plots for germanium are linear with a unit slope, at potentials that are not too anodic (Fig. 13), only if the measurements are made under conditions of fast polarization. Under steady-state conditions of slow polarization (Fig. 14) these plots for germanium are linear at low electrode potentials, but the slope is far from unity. This deviation of the slope from unity is attributed to changes in the surface dipole moments with relaxation times longer than those of the space-charge region. These results also indicate that space-charge parameters (in the absence of fast states), such as ψ_s, can be obtained from capacitance, surface conductance, and photovoltage measurements only by using high-frequency

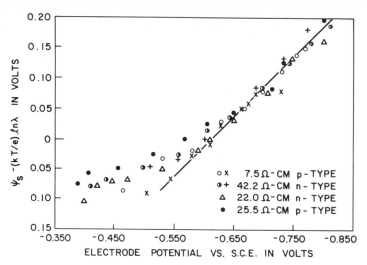

Fig. 13.   Values of $\psi_S - (kT/e) \ln \lambda$ plotted against electrode potentials for germanium with different bulk properties.   Data obtained by fast electrode polarization.   Reprinted from Ref. [137], p. 579, by courtesy of the Electrochemical Society, Inc.

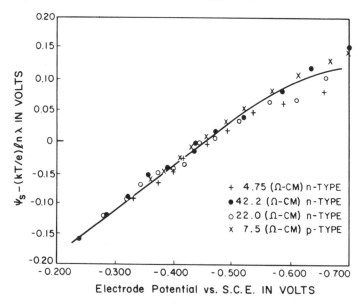

Fig. 14.   Values of $\psi_S - (kT/e) \ln \lambda$ plotted against electrode potentials for germanium with different bulk properties.   Data obtained under steady-state conditions of electrode polarization.   Reprinted from Ref. [137], p. 579, by courtesy of the Electrochemical Society, Inc.

ac or short-duration dc pulse techniques. Furthermore, parallel
shifts and/or deviations in the plots of the left-hand side of Eq. (38)
against electrode potentials, and also in the plots of other space-
charge properties versus electrode potentials, have been found to
occur with variations in (a) current densities, (b) pH of the solution,
and (c) crystallographic orientation of the surface. These shifts in
the interfacial properties of the semiconductor-solution interface may
be attributed to shifts in the pzc resulting from changes in V_H, χ^M,
or χ^{sol}, as the case may be.

The electrode potentials, as well as the flat-band potentials, of
ZnO, antimony-doped SnO_2 [46, 139], and lithiated NiO [140] have
also been found to change with pH at a rate of approximately
0.059 V/pH unit. These variations in electrode potentials with pH
may be attributed to changes in V_H due to reversible double-layer
effects [134, 141]. Differences in the electrode potentials of germanium with different crystallographic planes have been attributed to
differences in the Ge/OH ratios for each plane [15, 134]. It will be
shown subsequently that any difference in the metal/oxygen ratio of
an oxide surface also gives rise to differences in the zpc of the
aqueous double layer.

The linear plots of electrode potentials, shown in Figs. 13 and
14, also show parallel shifts on the potential scale with varying current densities (0.18 V for 10 $\mu A/cm^2$) when these current densities are
not above 10 $\mu A/cm^2$ (not shown in Fig. 14). These changes in electrode potentials with current densities were attributed to concentration
polarization by Brattain and Boddy [137]. However, Mayamlin and
Pleskov [142] have attributed these shifts in electrode potentials to
the interphase potential differences existing initially in different sets
of measurements.

Similar effects resulting from variations in pH, crystallographic
orientation of the surface, current densities, and the time effects in

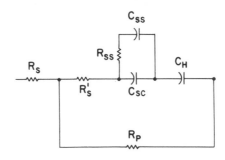

Fig. 15. Equivalent circuit showing various contributions to the measured capacity of a semiconductor-solution interface. R_p is the cell resistance to the Faradaic current flow; R_s, R_s', and R_{ss} refer to the resistances of bulk semiconductor, space-charge region, and surface states, respectively; C_H, C_{sc}, and C_{ss} refer to the capacities of the Helmholtz layer, space-charge region, and surface states, respectively. Reprinted from Ref. [15], p. 222, by courtesy of the Elsevier Publishing Company.

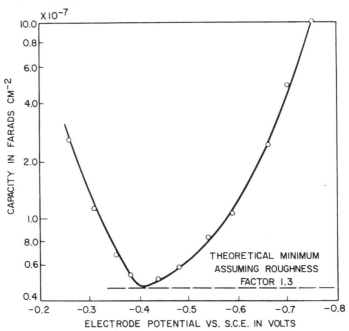

Fig. 16. Space-charge capacitance as a function of electrode potential of an n-type germanium electrode, in a phosphate buffered 0.1 M K_2SO_4 solution of pH 7.45. Data obtained by the fast-polarization method. Reprinted from Ref. [137], p. 578, by courtesy of the Electrochemical Society, Inc.

polarization are also observed in measurements of other interfacial properties: capacitance, surface conductance, and photovoltages. This variation in V_H may be compared with the variations in V_H obtained from studies of the reversible double layer on oxides.

5. Interfacial Capacitance

In solutions sufficiently concentrated to suppress the diffuse double layer the measured capacitance ($C_m = dq/d\varphi$) of the semiconductor-electrolyte interface is the sum of the Helmholtz-layer capacitance C_H in series with the parallel capacitances of the space charge, C_{sc}, and of the surface states, C_{ss}:

$$\frac{1}{C_m} = \frac{1}{C_H} + \frac{1}{C_{sc} + C_{ss}} . \tag{39}$$

The equivalent circuit in Fig. 15 [15] shows various contributions to the measured capacity. The space-charge capacitance C_{sc} can be calculated by differentiating the expression for the space-charge density (Eq. (32)) with respect to the potential drop $\varphi_b - \varphi_s$ in the space-charge region. At the flat-band potential, C_{sc} has a minimum value, which depends on the degree of doping for a given semiconductor. For a moderately doped semiconductor the space-charge capacitance is much smaller (Fig. 16) than C_H. The minimum in the space-charge capacity for intrinsic germanium, for example, is 2.04×10^{-8} F/cm^2. If C_{ss} is also very much smaller than C_H, then $C_m \approx C_{sc} + C_{ss}$. Furthermore, under conditions of a depletion layer, when Y is sufficiently negative (Fig. 11), the space-charge capacitance may be related to the space-charge potential barrier [$\varphi_{sc} = (\varphi_b - \varphi_s)$] by the Mott-Schottky relationship:

$$\frac{1}{C^2} = \frac{8\pi}{\epsilon\, en^0} \left(-\varphi_{sc} - \frac{kT}{e} \right) . \tag{40}$$

According to Eq. (40) a plot of $1/C^2$ against the electrode potential should be linear for conditions of a depletion layer in the space-charge region and $\varphi_{sc} = kT/e$ when $1/C^2 = 0$. The electrode potential at the point where $1/C^2 = 0$ differs from the flat-band potential by kT/e, and hence the latter potential, which is of similar significance as the pzc, can be calculated. These theoretical predictions were first confirmed for ZnO electrodes by Dewald [27, 104]. The slope of the linear plots of $1/C^2$ against the electrode potential also agreed with the theoretical value within 2%. After bringing ZnO to equilibrium with electrolyte solutions for several hours, no time effects were observed on the measured space-charge capacity of ZnO. This indicates the absence of surface states in ZnO. More recently, antimony-doped SnO_2 [45, 46], lithiated NiO [140], TiO_2 [42], and $KTaO_3$ [43] electrodes have also been shown, under appropriate conditions, to satisfy the Mott–Schottky relationship, and the corresponding plots have been used to examine the space-charge region.

The most extensive investigations of the interfacial properties of low-band-gap semiconductors have been carried out on germanium and silicon. In general the interfacial capacitance, conductance, and photovoltage all show a minimum at the flat-band potential $(\varphi_b - \varphi_s = 0)$, at which the carrier densities are uniform from the bulk up to the surface. The minima, and hence the entire curve of the plot of the interfacial properties against electrode potentials, shift (cf. Eqs. (37), (38)) to more anodic or cathodic potentials (relative to the intrinsic values) for the more n- or more p-type semiconductors. The curves of capacitance against electrode potential also shift with current densities (Fig. 17) for reasons already mentioned.

In general the measured photovoltages and the minima in the measured capacitance and surface conductance of germanium were found to agree with the calculated values only if the measurements were made by fast electrode polarization (Fig. 16). However, the

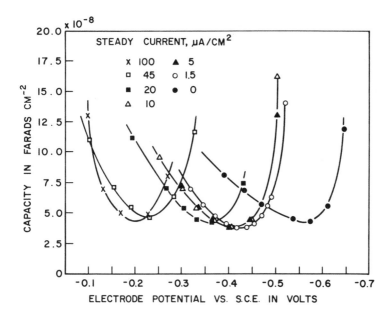

Fig. 17. Effect of varying current density on the space-charge capacitance of a p-type germanium electrode in an unbuffered 0.1 M K_2SO_4 solution. Data obtained under steady-state conditions. Reprinted from Ref. [137], p. 580, by courtesy of the Electrochemical Society, Inc.

measured surface conductance of germanium, at high anodic potentials and at high pH, was found to be in poor agreement with theory. With slow electrode polarization changes in surface dipoles (Eq. (38)) occur and the experimental values are greater than the theoretical ones. Apart from these time-dependent variations, regular displacements of the interfacial capacitance minima occur from changes in the following variables: (a) current densities (Fig. 17), (b) crystallographic plane, (c) the pH of the solution, (d) the ionic strength of the solution, and (e) the concentration of ions that are specifically adsorbed (e.g., I^- on Ge). The effects of the solution composition on the interfacial properties of semiconducting oxides are considered briefly

below, and a detailed discussion of the above properties may be found
in several other publications [15, 16, 41, 134, 137, 141].

6. The Variation of Interfacial Properties of Semiconducting Oxides with the Solution Composition

Several features of the interfacial properties of semiconducting-
oxide electrodes have not been clearly understood. These include
the nature of surface states and the Helmholtz layer, the effects of
the solution composition and of the surface charge on the space-
charge properties, and the distribution of potential in the double-
layer region. Information on these properties has been obtained by
studying electrode potentials [41, 134, 137], flat-band potentials, and
other space-charge properties, such as capacitance [15, 41, 121, 128,
130, 131, 133-135, 137, 141, 143], surface conductance [131, 136], and
the surface recombination velocity [122, 124] as a function of pH [15,
41, 128, 134, 141], ionic strength [15], and the type of anions or
cations present in the solution [15, 128, 130].

The space-charge properties of semiconductor electrodes can be
determined from capacitance measurements only when the slow surface states are prevented from responding to the measurements and
the density of fast states is also reduced substantially ($< 10^9$ cm^{-2}).
The former requirement is accomplished by using short-duration-pulse
techniques, and the latter is met by preanodizing the germanium electrodes for a few minutes. However, cathodic treatment of germanium
electrodes is known to restore the surface states, by converting the
Ge-oxide or Ge-hydroxide surface into a hydride surface [124, 138].
Our interest in germanium in this chapter is confined to the interfacial
properties of germanium with a thin oxide or hydroxide surface, either
present normally or obtained by anodic treatment or by treatment with
H_2O_2 [129].

The preanodization of silicon, to minimize fast surface states, forms a passivating and insoluble silica film that can be dissolved out at a high pH (above 12 [132]) or in a hydrofluoric acid solution [121]. The capacitance of oxide-free silicon electrodes can also be measured in the cathodic region. The Si/SiO_2 film/electrolyte system has always been found to give a higher capacitance than the oxide-free Si/electrolyte system [132].

The flat-band potential of germanium [15, 41, 128, 134] as well as of passivated silicon surfaces [15, 132], has been found to vary with pH by about 0.059 V/pH unit. Similar shifts with pH in the electrode potentials and other interfacial properties of ZnO [104], lithiated NiO [140], and antimony-doped SnO_2 [46, 139] have also been established, and hence such behavior appears to be a general characteristic of semiconducting-oxide electrodes. In contrast to germanium and silicon, the ZnO-solution interface has been found to be essentially free from fast surface states ($< 10^9$ cm^{-2}) [27, 104].

The flat-band potential of TiO_2-doped α-Fe_2O_3 (n-type), as obtained from the Mott-Schottky plots of $1/C_{sc}^2$ versus electrode potential, has been reported by Yoneyama and Tamura [144] to vary at a rate of 0.35 V/pH unit. These measurements were, however, limited to a small pH range of 0.75 to 1.5. These authors have attributed the dependence of the flat-band potential on pH to the specific adsorption of protons on the oxide surface. Compared with other oxide electrodes [15, 46, 104, 132, 140], this variation in the electrode potential (0.35 V/pH unit) is too large to be due only to the chemisorption of protons on the doped Fe_2O_3, even if the surface potential ψ_s in the space-charge region is assumed to be affected by the penetration of protons in the oxide phase. This discrepancy may be due to any of the following causes: (a) the extrapolation of the Mott-Schottky plots at different pH to $1/C_{sc}^2 = 0$ is doubtful (see Figs. 7

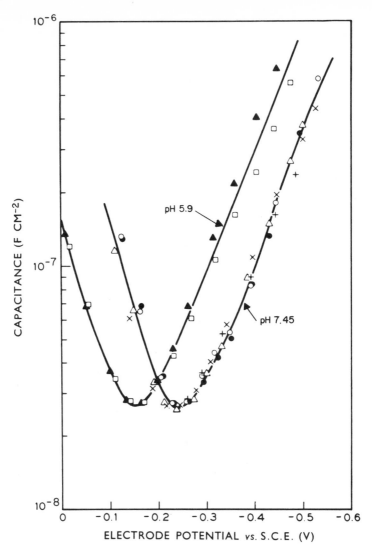

Fig. 18. The space-charge capacitance of a germanium electrode, (111) plane, at pH 5.9 and 7.45, in the presence of various
anions: ●, K_2SO_4, $M/10$, KH_2PO_4, $M/20$; o, $NaClO_4$, $M/10$,
NaH_2PO_4, $M/20$; △, KNO_3, $M/10$, KH_2PO_4, $M/20$; ✕, KH_2PO_4, $M/5$;
+, K_2SO_4, $M/10$, KH_2PO_4, $M/50$; ▲, K_2SO_4, phosphate buffered;
□, K_2SO_4, phthalate buffered. Reprinted from Ref. [15], p. 230, by
courtesy of the Elsevier Publishing Company.

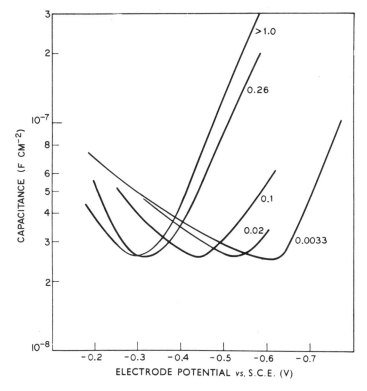

Fig. 19. Effect of ionic strength on the interfacial capacitance
of a p-type germanium electrode, (110) plane. Solutions with an
ionic strength below 0.1 M contained NaH_2PO_4/NaOH buffer, at
pH 7.5.   Solutions with an ionic strength of 0.1 to 3.0 M contained
$NaClO_4$ also. Reprinted from Ref. [15], p. 237, by courtesy of the
Elsevier Publishing Company.

and 8 in Ref. [144]), (b) the range of pH tested is too limited to gener-

alize the results, and (c) the effect of dissolved metal ions may also

have to be taken into account.

The flat-band potential of germanium (the potential of the minima

in C_{sc}), and hence the entire capacitance-versus-potential curve, as

shown in Fig. 18, shifts to more cathodic potentials by increasing the

solution pH; the same curves shift in the opposite direction with

increasing ionic strength, as shown in Fig. 19.

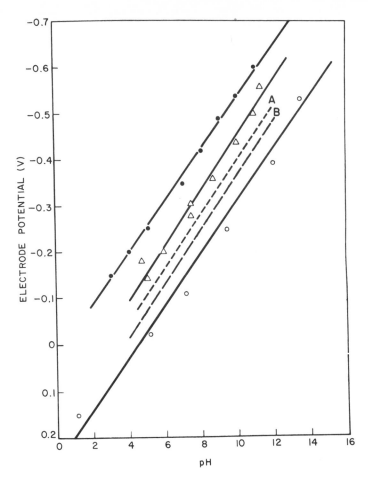

Fig. 20. Effect of pH, for the intrinsic germanium, on the potentials of ●, the surface conductance minimum (versus S.C.E., (111)
plane); o, the capacitance minimum (versus N.H.E., (111) plane);
△, $\psi_s - (kT/e) \ln \lambda = 0$ (versus N.H.E., (100) plane); A and B, same
as in △, but for (110) and (111) planes, respectively. Reprinted from
Ref. [15], p. 233, by courtesy of the Elsevier Publishing Company.

The pH dependence of the measured C_{sc} of germanium has been
extensively investigated by a number of workers [15, 41, 128, 129, 132,
134]. In Fig. 20 the variation in the electrode potential and in the
flat-band potential of intrinsic germanium, as measured by various

methods, is shown as a function of pH. This displacement of the flat-band potential with pH, at a rate of 0.059 V/pH unit is attributed by Gerischer [3, 141] and Brattain and Boddy [15, 41, 128, 134] to the change in the electrode potential resulting from a potential drop in the Helmholtz region, ΔV_H, which itself occurs because of a charge separation resulting from an acidic dissociation of the surface hydroxyl groups and also varies with pH. The effect of increasing ionic strength on the flat-band potential of germanium can also be reasonably accounted for on the basis of the properties of the Helmholtz region derived from studies of the reversible double layer on oxides (Section VI).

It has been further found [128, 130] that many anions (SO_4^{2-}, ClO_4^-, PO_4^{3-}, and NiO_3^-) and cations (K^+, Li^+, Na^+, and Cs^+) at pH 7.45 have no specific effect (surface inactive) on the capacitance and on the flat-band potential of germanium, as shown in Figs. 18 and 21, respectively. These conclusions are valid only at constant ionic strength and perhaps also at the experimental pH ~ 7.45. The anions need not necessarily be surface inactive in solutions of pH < 6.

On the other hand, cations like Cu^{2+} [123], Au^{3+}, and Ag^+ [122], even at very low concentrations (~ 10^{-6} mole/l), can give rise to fast surface states (~ 10^{11} cm^{-2}) in germanium and hence affect its space-charge properties. Several other ions, such as Fe^{2+}, Fe^{3+}, Ni^{2+}, and Co^{2+} [145], at concentrations of about 10^{-2} mole/l, were also found to have similar effects on the space-charge properties of germanium. The increase in the interfacial capacitance of germanium after the addition of Cu^{2+} to test solutions, up to a concentration of 10^{-7}mole/l, was not noticeable below pH 5, but a strong increase in the additional capacitance occurred above pH 6 [143]. An increase in the Cu^{2+} concentration from 10^{-7} to 10^{-5} mole/l also did not affect the capacitance of germanium at pH < 5. A similar pH dependence of germanium capacitance was also found in Ag^+ and Au^{3+} solutions

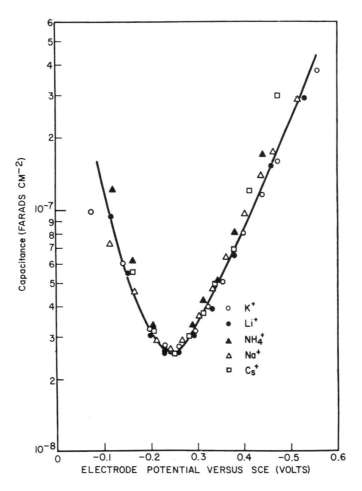

Fig. 21. Space-charge capacitance of a germanium electrode in M/6 solutions of various cations. Reprinted from Ref. [128], p. 55, by courtesy of the North-Holland Publishing Company.

(10^{-7} and 10^{-6} mole/l, respectively). From direct analytical methods Cu^{2+} and Au^{3+} have been found to be strongly adsorbed on germanium [143, 145]. The adsorption of Cu^{2+} on germanium increased with increasing Cu^{2+} concentration in solution and with increasing pH from 4.9 to 7.8 (see Fig. 4 in Ref. [143]). However, Memming [143] reports that the adsorption of Cu^{2+} on germanium is independent of pH between pH 4 and 9, which is probably due to the precipitation of $Cu(OH)_2$ in alkaline solutions. Adsorption of Cu^{2+} on Ge-oxide surfaces has been found to occur by Sparnaay [145] by the replacement of H^+ from Ge-oxide surfaces. The origin of surface states in germanium by the adsorption of these metal ions is not clearly understood. However, from the foregoing data it does appear that these phenomena are closely associated with the specific adsorption of the metal ions on germanium or on negatively charged germanium surfaces with a thin oxide film. Specific adsorption of cations on oxides as a function of pH is dealt with in detail in Section VI. I.

Unlike many other anions, iodide ions, at a constant ionic strength and pH of the solution, have been found to displace the flat-band potential of germanium to more cathodic potentials (Fig. 22), and this displacement of the flat-band potential is attributed to changes in V_H caused by the replacement of OH^- on germanium by I^- in a manner yet to be clearly understood [15, 128, 130]. At a high pH (~ 10.4) the iodide ions had very little effect on the capacitance, this being attributed [15, 128, 130] to increased competition with OH^- for adsorption. Replacement of OH^- on oxidized germanium by I^- should increase the pH of a KI solution on reacting with the germanium powder [146]. In recent unpublished measurements by this author, no such increase in pH occurred and this behavior needs further investigation.

The foregoing mechanisms, proposed to account for the variations in the interfacial properties of germanium with the solution composition, are in general agreement with the behavior of the Helmholtz layer derived from studies of the reversible double layer on oxides. In fact, studies of the reversible double layer on oxides (Section VI) lead to a more comprehensive understanding of the oxide-solution interface as a whole. The Ge-Ge oxide-solution interface acquires an effective zero surface charge in the double layer by chemisorption of H^+ and OH^- on the surface at a characteristic pH (~ 2.0 [147]),

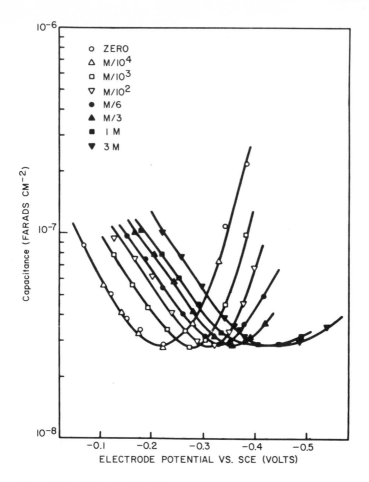

Fig. 22. Effect of iodide-ion concentration on the space-charge capacitance of a p-type germanium electrode ((100) plane). Reprinted from Ref. [128], p. 57, by courtesy of North-Holland Publishing Company.

known as the zero point of charge (zpc). In the pH range above the zpc the Ge-oxide surfaces acquire a negative surface charge due to an acidic dissociation of the surface hydroxyl groups. This surface charge is neutralized by an excess accumulation of counterions (cations) in the double-layer region, where cations may also be adsorbed in a partly dehydrated form (specific adsorption) on the oxide surface.

The total change in the Galvani potential at the interface is -0.059 (pH $-$ pH$_{zpc}$) at 25°C, whereas the distribution of potential itself in different double-layer regions depends, for a given pH, on the ionic strength of the solution and also on the degree of specific adsorption of counterions. In concentrated solutions any potential drop in the double layer occurs almost entirely in the Helmholtz layer. Hence the shifts in the flat-band potentials of the semiconducting-oxide electrodes with pH, in concentrated solutions, are attributed entirely to changes in V_H.

Similarly it can also be shown that at a constant pH any change in potential in the Helmholtz layer of a negatively charged oxide surface, with increasing ionic strength of the solution, occurs in a direction opposite to that caused by increasing the pH. This is in qualitative agreement with the observation that the direction of the shift in the flat-band potential of germanium with increasing pH (Figs. 18 and 20) is opposite to that observed on increasing the ionic strength of the solution (Fig. 19). A detailed and quantitative evaluation of the distribution of potential in the double layer on quartz surfaces as a function of pH and ionic strength is given in Section VI. I. 4. b. The double-layer properties of negatively charged surfaces of all oxides, in the pH range above the zpc [146], have been shown to be similar when examined under conditions of equal surface coverage. Since the zpc of both Ge-oxide (~ pH 2.0 [147]) and of quartz (~ pH 1.5) have about the same value, the double-layer properties of these two oxides in the pH range above the zpc may also be assumed to be similar in many respects, except for the higher metallic character, solubility, and reactivity (e. g. , with I$^-$) of the oxide or hydroxide surface of germanium compared with the corresponding properties of quartz.

The shifts in the flat-band potential of germanium surfaces with pH and with ionic strength are generally not attributed to a change in

ψ_s (i.e., bending of the bands) in the space-charge region, except perhaps in the presence of heavy-metal ions like Cu^{2+}, Ag^+, and Au^{3+}, as discussed before. The negative surface charge on an oxide or a hydroxide surface of germanium that originates from an acidic dissociation of the surface hydroxyl groups, at a pH greater than the zpc, is more readily neutralized in the solution phase itself than affecting the space-charge region, because of (a) a smaller Debye length in the solution phase than in the solid phase and (b) specific adsorption potentials arising from the chemical affinity between the charged surface and the adsorbate ions or molecules, including H_2O, H^+, and OH^-, in solution. This conclusion is substantiated by experimental observations (Figs. 18 and 22) that the minimum C_{sc} value for germanium at any pH, ionic strength, or I^- concentration can be restored to the original value by applying a suitable potential so as to compensate for the potential drop in V_H.

Variations have also been observed by Brouwer in the surface conductance and hence in the apparent flat-band potentials of oxidized germanium surfaces, with changes in the pH of solutions containing H_2O_2. This variation with pH in the surface conductance of germanium electrodes with an oxide layer has been attributed by Brouwer [129] to variations in the surface potential ψ_s in the space-charge region, where the charge transfer is said to occur through the oxide phase by singly charged interstitial oxygen defects O^-. However, this variation in the surface conductance with pH can also be reasonably accounted for, as before, by considering changes in the V_H of the oxide-solution interface; such changes of the Galvani potential in the double-layer region have not been taken into account by Brouwer [129].

The foregoing conclusion that variation in the solution pH has only a second-order effect, if any, on the space-charge parameters of germanium need not necessarily be true for other oxides, particularly for transition-metal ones. It will be shown in the subsequent sections that for certain transition-metal oxides not only proton adsorption can occur on the oxide surface in excess of that present at the zpc but also protons can penetrate into the oxide phase and thus affect the space-charge properties.

C. The Insulator Oxide-Solution Interface

The insulator-solution interface has received much attention in the past few years in connection with the theoretical and practical problems concerning electrolytic rectification by anodic oxide films, and with regard to problems relating to space-charge-limited currents [148]. Although many of the modern ideas concerning the insulator-solution interface are based on experiments carried out on anthracene (an organic insulator [33]), extensive work has also been done on the oxide films of valve metals and on massive electrodes of ZnO [27, 104, 149-153], SnO_2 [46, 139], NiO [140], $KTaO_3$, TiO_2, and Ta_2O_5 [42, 43]. Several reviews have appeared recently on the mechanism of electrolytic rectification [4] and on the insulator-solution interface [31-34]. The bulk concentration of free carrier in an insulator is small, and hence no space charge can result from surface states or from an applied electric field alone. However, if the insulator surface is in ohmic contact with an electrolyte, say of a redox type,

donor $\rightleftarrows$ acceptor + carrier ,

that can inject electrons into the conduction band or holes into the valence band, the contact can serve as a source of charge carriers. These carriers can then flow under an applied electric field. Such a dynamic exchange of carriers at equilibrium can give rise to an "exchange current" and thus to an induced space charge. In general, though strongly reducing electrolytes can inject electrons into the insulator surface with an applied electric field, strongly oxidizing electrolytes can inject holes into (or extract electrons from) the surface. Thus an n- or p-type space charge can be induced into the insulator surface while in contact with a suitable electrolyte solution. The injection of holes, or of protons from acidic solutions, has been reported to occur with an applied electric field in different types of insulators, such as anthracene [33] and metal-oxide films [4, 154-157]. Photoinjection of electrons and holes at the insulator-electrolyte interface is also possible and has been observed in organic insulators [31, 33] as well as in oxides [31, 42, 43, 150, 158, 159]. The space charge and the corresponding Debye length in an insulator, unlike those in a semiconductor, are entirely controlled by the concentration of the injected carriers. In the following survey we shall be concerned with only such aspects of the studies as provide information on the double-layer properties of the oxide-solution interface.

The phenomenon of electrolytic rectification by oxide-covered valve metals (e.g., Ta, Ti, Zr, Nb, Al) is well known [4, 160-162]. In aqueous electrolytes the cathodic currents of oxide-covered valve metals are larger than the anodic ones (metal oxide positive) by several orders of magnitude. Electrolytic rectification also occurs in aprotic solvents (liquid SO_2) in which the discreteness-of-charge effect was found to be much more pronounced than it is in an aqueous medium. Thus with Ta/Ta_2O_5 electrodes in liquid SO_2 containing different alkali ions the cathodic currents (in the high-current-density range) were observed to be greater by several orders of magnitude ($\sim 10^5$) for Li^+ and Na^+ [4, 163] than they are in the presence of larger cations, K^+ and Rb^+. This dependence in nonaqueous media of the cathodic current on the nature of cations indicates varying degrees of chemisorption of cations on oxides. However, no evidence of penetration of cations (Na^+, Li^+) into the oxide phase has been experimentally observed.

There is strong experimental support, particularly at high fields, for the adsorption and penetration of protons into the oxide phase from aqueous solutions. Adsorption and penetration of protons on SiO_2 films on silicon has also been reported [164, 165]. The physico-chemical properties of oxide films have been studied [4, 75, 154-157, 163] by measurements of the space-charge and bulk capacitances, the i-V characteristics, and film conductance. For some oxide films spectroscopic studies [166-169] and adsorption measurements with D_2O have also been carried out [156, 164]. Such studies of the SiO_2 films on silicon have shown that the hydration of oxide surfaces and proton transfer in the oxide layer are of primary importance in determining the surface properties of SiO_2 films in general and their conductivity in particular.

Anderson and Parks [164] have shown from spectroscopic, conductivity, and radiotracer measurements that the conductivity due to silanol groups is essentially ionic, protons being the predominant charge carriers. The protons dissociate from the silanol groups with variable dissociation energies, depending on the dielectric constant of the oxide phase. Similar results have been obtained for a number of other oxides by Vermilyea [154-157] and by Isaacs and Leach [75].

Vermilyea found [155, 156] that, as the electrode potential is decreased in the cathodic or nearly cathodic potential range, the interfacial capacitance increases sharply (~ 100 $\mu F/cm^2$) for anodic oxide films (Fig. 23) on Nb, Ta, W, Ti, and Sb but does not increase in the case of anodic films on Al, Zr, Si, Be, Ga, and Bi. The increase in capacitance in the first series of oxides is attributed to the protonic space charge (through proton injection) in the solid phase, whereas the behavior of the latter group of oxides is attributed by Schmidt [4] to the lower mobility of protons in these oxide layers.

In similar studies of the oxide films on valve metals Isaacs and Leach [75] found that the surface capacitances of oxide-covered Al,

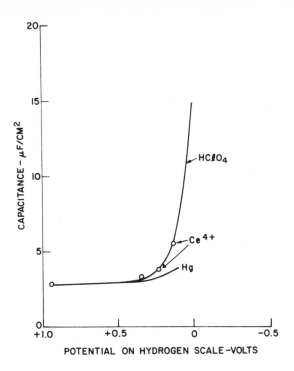

Fig. 23. The capacitance-voltage behavior of a 210-Å-thick Nb_2O_5 film in 0.1 M $HClO_4$ solution and in contact with mercury. Potential scale has been adjusted for the curves to coincide. The Ce^{4+} ions are shown to have no effect on the capacitance of the oxide film in $HClO_4$ solutions. Reprinted from Ref. [157], p. 180, by courtesy of the Electrochemical Society, Inc.

Ta, and Zr electrodes vary with the H content of the oxide but are always less than the double-layer capacitance. However, the surface capacitances of oxide-covered Ti, Nb, V, and U electrodes were found to be greatly in excess of the double-layer ones and have been attributed to the reversible reactions of H^+ with the metal oxide. Differences between the individual metal oxides in their affinity for proton penetration or proton conductivity have also been demonstrated [4]: thin anodic Al_2O_3 windows, when sandwiched between two electrolyte contacts and biased, were found to pass only negligible currents. On the other hand, in a hydrogen atmosphere Ta_2O_5 films with platinum contacts showed appreciably higher conductivity under a cathodic bias than under an anodic one.

According to Vermilyea [154-157] the injected protons form a space charge in the oxide film (under cathodic bias) and give rise to electronic states that allow conduction in the same way as Li adsorbed on oxides in liquid SO_2 increases the electrical conductivity of the oxide film, as found by Schmidt and co-workers [4, 163]. A more detailed account of the behavior of anodic oxide films may be found in recent reviews on this subject [4, 170, 171].

The semiconducting nature of anodic oxide films has been considered by several workers, and the rectification properties of these films have been attributed to p-i-n [159] or p-n [158, 172] junctions existing within the oxide films. This work has been reviewed by Schmidt [4] and by Huber [172].

Boddy and co-workers [42, 43] have studied the interfacial properties of several semiconductor-oxide electrodes with wide band gaps by measuring (a) current-voltage characteristics, (b) interfacial or space-charge capacitance as a function of applied voltage with superimposed short-duration current pulses, (c) photocurrents in the ultraviolet range, and (d) transient currents produced in response to a rectangular anodic-voltage pulse. These measurements, which were carried out on 1-mm-thick oxide crystals, are claimed to be representative of the general behavior of anodic oxide films, because the interfacial properties of the semiconductors studied are essentially controlled by the narrow barrier width imposed by the space-charge region. At 3 eV below the conduction-band edge and at a field strength of 4×10^6 V/cm the barrier width was calculated to be only 75Å (Fig. 24) compared with the space-charge depth of several thousand angstroms. Under these conditions the high field, and hence

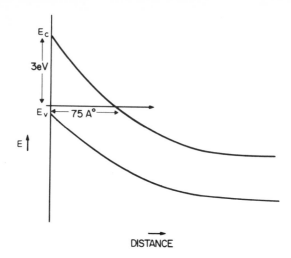

Fig. 24. The barrier width or effective thickness of the space-charge layer for electron tunneling at 3 eV below the conduction-band edge at a field of 4×10^6 V/cm, in a semiconducting TiO_2 electrode [42].

the surface activity, is confined to a relatively narrow surface region, while the bulk of the oxide crystal, being an insulator, has a negligible field and remains inert. Both n-type TiO_2 [42] and n-type $KTaO_3$ [43] exhibit electrolytic rectification (cathodic current >> anodic current) similar to the anodic oxide films of valve metals. Similar behavior has also been reported [42, 43] for Ta_2O_5, $SrTiO_3$, and ZnO. Hence it has been suggested that the semiconducting properties of anodic oxide films could perhaps be a significant factor in their electrolytic rectification properties.

Although the anodic leakage current is generally smaller than the cathodic current by several orders of magnitude, it is significant in several respects. For example, TiO_2 [42] and $KTaO_3$ [43] were found to evolve oxygen under anodic bias at high current densities, and ions like iodide, borohydride, ferrocyanide, and hydrogen, which are more reducing than water, had no effect on the anodic current flow. In a similar investigation Kiess [150] found that a sudden increase in the anodic current of ZnO occurs at a sufficiently high field, and the choice of electrolyte had no effect on this current.

The passage of anodic current at the oxide-solution interface requires either injection of electrons into the surface region or consumption of holes. The mechanism of electron transfer in electrochemical processes is a vast subject and has been discussed in detail with particular reference to the semiconductor-solution interface in several papers [3, 25, 27, 151–153] and reviews [173, 174]. According to Boddy [42, 43], the above observations rule out the mechanism of direct electron transfer from the solution species to the conduction band of insulators. Such a mechanism had been proposed by a number of workers [3, 25, 27, 151] and is supported by experimental work with low-band-gap elementary semiconductors (Ge and Si) and with organic insulators [33]. According to this mechanism, the electron or hole transfer from the solution states to the conduction or to the valence band becomes possible if the occupied energy levels of ions (in solution) overlap the empty levels in the conduction or valence band, respectively. In such a process of electron transfer, for instance, the anodic current should increase with increasing concentration of the reducing species in solution, which was not found to occur with TiO_2, $KTaO_3$, and ZnO.

However, electron injection into the conduction band of ZnO by several reducing agents with high oxidation potentials (e. g., aquo-complexes of Cr^{3+}, Eu^{2+}, V^{2+}, Ti^{3+}, ethylenediamine cobalt^{2+}) and by certain organic radicals has also been reported by Freund and co-workers [149, 152]. A high oxidation potential, however, is not the only requirement for the reducing agent to inject electrons into the conduction band, since borohydride, for example, also with a high oxidation potential ($E^0 = 0.5$ V), is inactive in this reaction [149].

It was further found that compared to the i-V plots, which depend
on the carrier density, the plots of i versus the surface field E_s were
linear and more meaningful. The surface field $F_s = eN_D/C$, where e
is the electronic charge, N_D is the donor density, and C is the interfacial capacitance per unit area. From an analysis of the i-V and the
i-F_s plots and the behavior of the photocurrents Boddy [42, 43] attributed the charge-transfer process at the interface to the tunneling of
electrons into the empty levels of the conduction band from the occupied surface (or interface) levels lying either within the forbidden
gap or at the top of the valence-band edge of the oxide surfaces. A
similar tunneling mechanism of charge transfer at high fields has
been proposed recently for SnO_2 electrodes by Elliot et al. [46] and
for ZnO by Kiess [150]. Even in the absence of an applied voltage
there is, in general, a "built-in field" at the semiconductor-solution
interface due to the alignment of the electrochemical potentials of
electrons in the solid and in the solution, causing a space charge in
the oxide phase. This space-charge potential at zero applied voltage
is determined by the electron exchange between the surface states
and the bulk semiconductor (cf. the distinction between surface states
and surface groups) by a tunneling process through the surface barrier.
On anodic or cathodic polarization this built-in surface barrier is
modified. The narrowness of the surface barrier at the semiconductor
surface (~ 70Å for $KTaO_3$ [43], Fig. 24) is particularly favorable for
electron tunneling. As already described, the overlapping 2p states
of oxygen in the oxide constitute the valence band, and electron
tunneling can occur from these oxygen 2p states at the oxide surface
into the conduction band when the surface field is sufficiently high.
The following sequence of the electrochemical reactions has been proposed [42, 43]:

$$\text{(surface oxide)} \quad \text{—O}^- \xrightarrow{\quad \text{tunneling} \quad} \text{—O} + e \;,$$

followed by

$$2(-O) \longrightarrow 2(-V) + O_2 ,$$

where $-V$ is a surface oxygen vacancy (i. e. , metal excess), which can act as an electron donor,

$$-V \longrightarrow -V^+ + e .$$

The surface level due to $-V^+$ is replenished by reacting with H_2O (or OH^-), so that the leakage current is continuous:

$$-V^+ + H_2O \longrightarrow -O^- + 2H^+ .$$

Cathodic reduction of several oxidizing agents on the ZnO electrode has been recently investigated in considerable detail by Morrison and Freund [153], who found that for the ions (ferricyanide and permanganate) to be reduced, the energy level of the oxidizing ion in solution, as estimated from its redox potential, should be slightly below that of the conduction band. As in the case of other oxide semiconductors, variation in the solution pH was found to affect the Helmholtz-layer potential at the ZnO-solution interface and hence also the cathodic currents.

These mechanisms still do not present a microscopic picture of the interface, particularly with respect to the specific adsorption of anions on oxides. The electron-tunneling mechanism, for example, does not explain why the same process should not occur with other ions (e.g., halide ions). In order that the anodic currents of oxide electrodes be affected by the solution composition, the anions would have to replace the surface OH groups or the surface water of hydration by specific adsorption. Whether or not such a replacement of surface OH groups by anions occurs depends on the hydration-dehydration energies of both the anions and of the surface. Such interactions have been considered by Gerischer [25]. The same problem will be encountered in the work on the reversible double layer.

To sum up the behavior of the oxide-solution interface, protons
can be injected into the oxide films of certain metals under a cathodic
bias and thus can give rise to a space charge and to a current flow
of appreciable magnitude. However, it appears that, under an anodic
bias, electrons from the 2p levels of surface oxygen (replenished by
H_2O or OH^-) can tunnel through the surface barrier into empty levels
in the conduction band, this process giving rise to leakage currents
and to oxygen evolution. The adsorbed H^+ (or H_3O^+) and OH^- on
oxide surfaces together constitute the Helmholtz layer, which controls
several properties of the oxide-solution interface. It will be seen
later that any pH variation in solution also changes the H^+ and OH^-
adsorption densities on the oxide surfaces and hence the potential
drop in the aqueous part of the interface, particularly in the Helmholtz region. The Galvani potentials in the aqueous part of the
reversible double layer on oxides can be either compensated or
superimposed on by anodic or cathodic polarization of the interface.
Hence the pH variations in cathodic and anodic polarization studies
only serve to shift the pzc by a value of 0.059 V/pH unit. Ions, if
specifically adsorbed at the experimental pH, could affect interfacial
properties so that the shifts in the pzc do not equal 0.059 V/pH unit.
In the absence of specific adsorption of ions, H^+ (or H_3O^+) and OH^-
ions are expected to control exclusively the observed properties of
the oxide-solution interface. Thus in the case of TiO_2 and $KTaO_3$ the
experiments [42, 43] were performed at both extremes of pH and in
neutral, phosohate-buffered solutions, and the same results, unaf-
fected by solution composition, were obtained. None of the experimental studies so far described has provided quantitative information
on the surface-charge densities and the potential distribution in the
aqueous double layer on oxides.

IV. INTERFACIAL PROPERTIES OF METALS WITH OXIDE FILMS
AND METAL OXIDES

The interfacial properties of the system

metal | metal-oxide film | gas or electrolyte solution

depend to a large extent on whether the film behaves like a metal, a
semiconductor, or an insulator. If the oxide film is a semiconductor,
the interfacial properties of the system also depend on the film thickness. Extensive investigations have been carried out on metal surfaces with oxide films, in connection with studies of adsorption,
catalysis, electrode kinetics, passivation and corrosion of metals,
and electrolytic rectification. However, there have been relatively
few experimental or theoretical investigations directly relating to the
double-layer characteristics of thin metal-oxide films.

Considerable information on the interfacial properties of metal
systems with oxide films can be obtained from double-layer studies
carried out with massive oxide electrodes or with metal-oxide powders.
It has already been shown that the presence of a space charge in a
semiconducting oxide or an induced space charge in an insulator oxide
greatly alters the interfacial properties of the material. It was also
shown that the properties of the insulator-oxide surfaces that were
investigated [42, 43, 150] with thick oxide electrodes can also represent the properties of thin oxide films. Whereas the space-charge
region in a semiconducting oxide may extend up to several thousand
angstroms from the surface, the thickness of anodic films, for example,
ranges from 10 to 10^4 Å. In systems where the substrate is a metal
and the oxide is a semiconductor the oxide-film thickness becomes
critical in determining the interfacial properties. By solving the

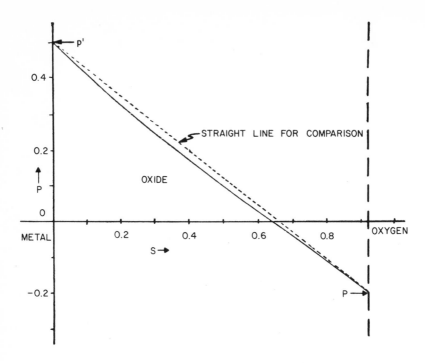

Fig. 25. Diagram showing a nearly uniform distribution of potential in a thin oxide film on a metal. Reprinted from Ref. [175], p. 638, by courtesy of the American Institute of Physics.

Poisson–Boltzmann equation, Butler [175] has calculated the field and the total potential drop across a homogeneous, semiconducting-oxide film in the system

$$\text{metal or semiconductor} \,|\, \text{oxide} \,|\, \text{oxygen}$$

as a function of the oxide-film thickness. The general form of the variation in potential with distance in oxide films is shown for three ranges of film thickness in Figs. 25, 26, 27, and 28. These variations in the film potential have been calculated for an arbitrary choice of constants relating to the bulk properties of the semiconducting oxides and have been expressed, in Figs. 25 through 28, in terms of the dimensionless parameters p and S. The parameter p is the total

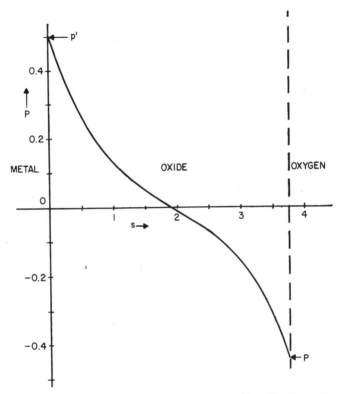

Fig. 26. Diagram showing a nonuniform distribution of potential due to space charge in a moderately thick oxide film on a metal. Reprinted from Ref. [175], p. 638, by courtesy of the American Institute of Physics.

potential drop in the oxide film and is equal to $p' - P$, where $p' = e\varphi'/kT$ and $P = e\Phi/kT$, φ' and Φ being the electrostatic potentials at the metal-oxide interface and at the oxide surface, respectively, relative to the neutral point in the oxide bulk. The parameter S is expressed in dimensionless units as the ratio of the film thickness to the Debye length. In Figs. 29 and 30 the film thickness is expressed both in terms of S and in angstrom units for the particular case of Cu_2O. For any given values of bulk properties the field is shown to be nearly uniform in relatively thin oxide films (Fig. 25),

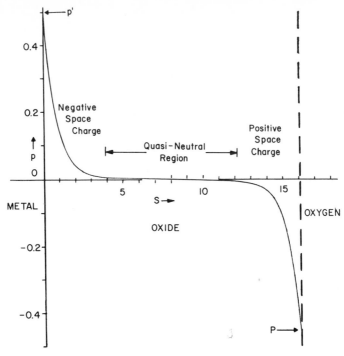

Fig. 27. Diagram showing a nonuniform distribution of potential due to space charge in a thick oxide film on a metal. Reprinted from Ref. [175], p. 639, by courtesy of the American Institute of Physics.

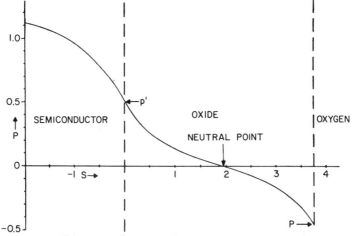

Fig. 28. Diagram showing the distribution of potential in a moderately thick oxide film, with a space charge, on a semiconductor. Reprinted from Ref. [175], p. 641, by courtesy of the American Institute of Physics.

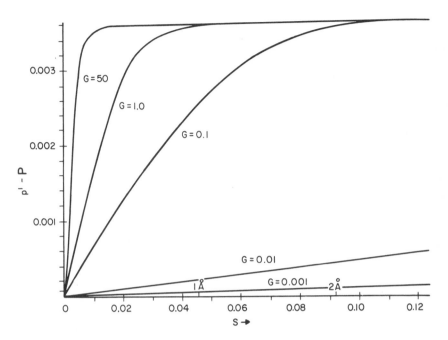

Fig. 29. Diagram showing the variation in the potential drop (p'-P, Figs. 25-28) across a Cu₂O film at 1000°C as a function of film thickness for various values of G. The film thickness is expressed both in angstroms and as S (see text for abbreviations used) [175].

but the space charge is shown to affect the field in thicker films (Figs. 26 to 28). In practice very thick films (Fig. 27) behave essentially as bulk oxides.

In a practical case of Cu-Cu₂O, for example, the order of magnitude of the film thickness (in angstroms) and the effect of bulk properties, G, on the field strength within the film may be seen in Figs. 29 and 30. The dimensionless parameter G in Figs. 29 and 30 is approximately given by

$$G = \left(\frac{4\pi e^2 n_{B^0}}{\epsilon kT} \right)^{\frac{1}{2}} \frac{n(O^-/ads)}{n(M^{2+}\square/x)} , \qquad (41)$$

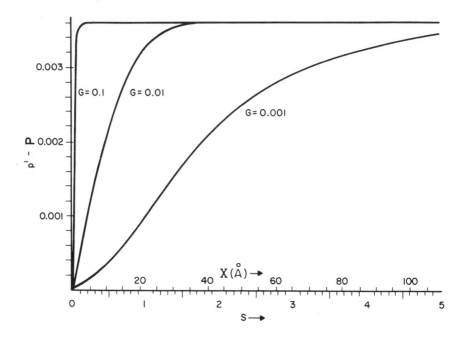

Fig. 30.   Variation in the potential drop (p'-P, Figs. 25-28) across a Cu_2O film at 1000°C as a function of film thickness for various values of G.   The film thickness is expressed both in angstroms and as S.   Reprinted from Ref. [175], p. 640, by courtesy of the American Institute of Physics.

where $n_{B^-}{}^0$ is the concentration of the negatively charged species at a point in the oxide film where the electrostatic potential is zero and ϵ is the dielectric constant of the oxide phase.   The parameter G is seen to be a product of the reciprocal Debye length (the first term in parentheses) and the ratio of the concentrations of the adsorbed O^- to the cation vacancies ($M^{2+}\square$) on the surface.

   If the substrate for the oxide film is a semiconductor with a space-charge region at the semiconductor-oxide interface, the potential p' at this interface will be dependent on the film thickness and also on the potential drop across the oxide film.   The distribution of potential at different interfacial regions of moderately thick oxide film

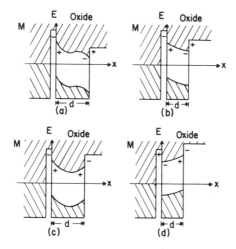

Fig. 31. Schematic diagrams showing different types of energy-band structure of semiconducting oxide films (thickness, d) on metals (M). The nature of surface charge is also indicated. Reprinted in a simplified form from Ref. [176], p. 620 (English transl.), by courtesy of the Consultants Bureau.

on a semiconductor is shown in Fig. 28. A detailed discussion on the effect of different variables on the field strength in the film is given by Butler [175]. In brief, the variation of field within the oxide layer depends on the film thickness relative to the Debye length and on the surface potential, which is also affected by the presence of surface states.

Four types of energy diagrams are shown in Fig. 31 [176], corresponding to different types of semiconducting-oxide films (on metals) whose thickness is less than or approximately equal to the screening distance of the oxide ($\propto$ the Debye length). Figures 31a and 31c correspond to relatively thick films, and Figs. 31b and 31d correspond to thin films. In relating the catalytic activity of oxide films on metals to their semiconductor properties, Vol'kenshtein et al. [176] also studied the relationships between the film thickness and the work functions of metals and oxides. In these studies, in order that

the oxide film may be considered as an independent phase, the film
thickness should be greater than 10^{-6} cm and be less than the screening distance, $\sim 10^{-4}$ to 10^{-5} cm.

Now consider a system

metal | metal-oxide film | electrolyte solution

where the oxide is a semiconductor whose surface is in equilibrium
with a redox reaction $A^{2+} \rightarrow A^{3+} + e^-$. The space charge arising
in thick oxide films on electrical polarization of such a system tends
to screen the surface charge on the metal at one side of the film from
the double-layer charge at the other side of the film. However, if the
film thickness is relatively small, then the total, excess space charge
in the oxide film will be substantially smaller than the excess charges
on the metal surface and in the aqueous double layer on the oxide film.
In such a case the space charge in the oxide film will not be enough
to screen the electrostatic interactions between the metal phase and
the solution phase. In electrode kinetic studies, and also for a
double-layer analysis of the above system, it is important to know
the critical film thickness required to screen effectively the surface
charge of the metal from the double-layer charge on the oxide. This
problem has been analyzed theoretically by Kuznetsov and Dogonadze
[177]. By assuming the oxide film to be a homogeneous intrinsic
semiconductor without surface levels and by solving the Poisson-
Boltzmann equation applied to this system, these authors have
obtained expressions for the variation in field strength across the
oxide film. The energy diagram of this system under anodic polarization is shown in Fig. 32, where E'_c and E'_v refer to the positions of
the conduction- and the valence-band edges at the point of contact
between film and electrolyte. The extent of band bending, $E_F - E'_v$
and $E'_c - E_F$, is equivalent to the surface potential (relative to the

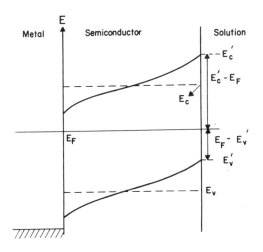

Fig. 32. The energy-band structure of a semiconductor oxide film on a metal. Reprinted in a simplified form from Ref. [177], p. 2043, by courtesy of the Consultants Bureau.

bulk solid) and has been shown [177] to be related to the properties of the metal and of the oxide film as

$$E_F - E'_V = W_B - e(\varphi^0_M - \varphi^0_C) + W^0_S - W^0_M \qquad (42)$$

and

$$E'_C - E_F = W_B + e(\varphi^0_M - \varphi^0_C) - W^0_S + W^0_M , \qquad (43)$$

where W^0_S and W^0_M are the work functions of the semiconductor and of metal, respectively, under electrically isolated conditions, W_B is the energy difference between the Fermi level and the band edge $(E_C - E_F)$, φ^0_M is the equilibrium inner potential of metal, and φ^0_C is the equilibrium inner potential of the film at the point of contact with the electrolyte. Both φ^0_M and φ^0_C are referred to the inner potential of the solution (φ_{sol}), which is assumed to be zero. The other terms in Fig. 32 are self-explanatory. The charge distribution in the film as a function of distance x (film thickness) is given by

$$n_x = n_0 \exp\left(\frac{W_s^0 - W_M^0}{kT}\right) \exp\left(\frac{e\varphi_x - e\varphi_M}{kT}\right) \tag{44}$$

and

$$p_x = p_0 \exp\left(-\frac{W_s^0 - W_M^0}{kT}\right) \exp\left(-\frac{e\varphi_x - e\varphi_M}{kT}\right), \tag{45}$$

where n_0 and p_0 are the concentrations of carriers in an isolated
intrinsic semiconductor. The differences between the present thin
film system and the space-charge distribution in a massive semiconductor phase may be noted by comparing Eqs. (44) and (45) with
Eqs. (25) and (26). It is seen that (a) the inner potential φ_M of the
metal phase in the present system is equivalent to φ_b in Eqs. (25)
and (26), and (b) the difference between the work functions of the
metal and of the semiconductor also contributes exponentially to the
effective charge densities in the system metal|oxide film|solution.

The general expressions for the field strength in an oxide film of
thickness d have been obtained [177] by combining Eqs. (44) and (45)
with the Poisson equation and by integrating the resultant Poisson-
Boltzmann relationship. From the approximate solutions of these
expressions the limiting conditions for the linear variation in the
field strength in the film have been derived. Under conditions of
small potential differences ($e\varphi_M/kT \leq 1$) it is shown that the sufficient condition for a linear potential drop in the film is that $\kappa d \ll 1$,
where κ is the reciprocal of the Debye length L and d is the film
thickness. Under these limiting conditions the following relationships are shown to be valid:

$$\varphi_c = \varphi_M \left(1 + \frac{d}{\delta_H} \frac{\epsilon_H}{\epsilon}\right)^{-1} \tag{46}$$

and

$$\frac{\eta_C}{\eta} = \left(1 + \frac{d}{\delta_H} \frac{\epsilon_H}{\epsilon}\right)^{-1}, \tag{47}$$

where φ_C is the inner potential at the point of contact between film
and electrolyte, δ_H is the Helmholtz-layer thickness, and ϵ_H and ϵ
are the dielectric constants of the Helmholtz layer and of the semiconductor oxide. The potential η_C, effective at the film-solution
interface when the applied potential is η, is given by Eq. (47). The
conditions for a linear potential drop in films with large potential
($e\varphi_M/kT \geq 1$) have also been discussed [177] and have been shown
to depend on the exponential factors of Eqs. (44) and (45) in addition
to the parameters in Eq. (46). For a given set of values, such as
$\varphi_M = 0.4$ V, $W_s^0 - W_M^0 = 0.2$ V, and $e\varphi_C \sim 0.5kT$, the condition for
the linear potential drop is that the film be less than 240 Å thick,
when Eqs. (46) and (47) are still valid.

Studies of the reversible double layer on oxides have been mostly
carried out with oxide precipitates or oxide powders as half-cells in
the absence of any applied electric field. If the oxide is an insulator
(e.g., SiO_2), it is reasonable to assume that these studies of the
reversible double layer on oxide powders are also representative of
the double layer on insulating oxide films. Mechanical defects in
the oxide films may, however, lead to direct metal-solution interactions even if the oxide film is of an insulating type. Furthermore,
if the oxide film is fairly soluble in aqueous solutions, it may also
be assumed that it is the metal surface with metal-hydroxide groups
that will be in equilibrium with the solution phase (e.g., anodized
Ge surface in equilibrium with aqueous solutions).

If the oxide film on a metal is a semiconductor, then the film
thickness, depending on its Debye length, may also influence the
properties of the reversible double layer on the oxide film. Hence

properties of the reversible double layer obtained by using semi-conducting metal-oxide powders should be representative of such metal-oxide films whose thickness is much greater than the Debye length of the oxide. If the thickness of the semiconducting-oxide film is less than its Debye length, the proximity of the metal phase to the solution phase may be expected to affect the reversible double layer on the oxide film in several ways.

In the presence of an indifferent, supporting electrolyte the reversible double layer on oxides is determined mainly by the zpc of the oxide. The zpc, to be discussed in detail in Section VI, is the pH at which the difference of the Galvani potential ($\Delta\varphi^0$) between the oxide surface and the bulk solution is zero. Hence, at a pH other than the zpc, $\Delta\varphi^0 = \Delta\chi + \Delta\psi^0$, where ψ^0 is the electrostatic potential of the oxide surface, relative to the bulk solution, that results from charge separation at the interface due to the ionization of the surface hydroxyl groups, and $\Delta\chi$ refers to the potential drop that occurs in the hydration layer of the oxide surface mainly due to dipole orientation (see Sections VI. F and VI. I. 4).

The zpc of the oxide is usually determined by the following solid-phase properties: surface stoichiometry, the coordination number of the cation, the ratio of the bonds exposed to the solution and the bonds coordinated to the oxide phase, and the degree of surface hydration of the oxide. Compared with these major influences, the effect of the Fermi level of the oxide phase on the zpc is relatively insignificant; for example, the zpc of both n- and p-type Ge with a thin oxide or hydroxide surface occurs at the same pH (~ 2.0).

However, a semiconducting-oxide film whose thickness is much smaller than its Debye length is too thin to prevent interactions between the solution and metal phases. In such a case the stoichiometry of the oxide film and hence the zpc (χ potentials in particular)

may also depend on the film thickness. Some work has been carried
out on the reversible double layer on metal powders with thin oxide
films [178, 179]. The zpc values of the metal powders with an oxide
film are significantly different from those of the corresponding metal-
oxide powders [146]. However, comprehensive studies are lacking
of the reversible double layer on oxides, with known solid-state
properties, that could establish the effects of film thickness and
other variables on the reversible double layer. Since the distribution
of potential at the oxide-solution interface is also not well understood, the ζ potential of oxides is often wrongly assumed to represent the entire potential difference at the oxide-solution interface.

V. STUDIES OF THE AQUEOUS DOUBLE LAYER
ON POLARIZABLE OXIDE ELECTRODES

Studies of the aqueous double layer on metal oxides by direct
capacitance measurements are restricted by the low conductivity of
the material, and such studies are scarce. However, it has been
pointed out [13] that the double layer on a few polarizable oxide
electrodes (e.g., oxides of Pb, Mn, Cu, Bi, and Ag) that have sufficiently high conductivity can be investigated by direct capacitance
measurements. Lead dioxide is similar to bismuth in its metallic
character and is reported to exhibit an electronic conductivity that
decreases with increasing temperature. Lead dioxide electrodes
have been investigated by a number of workers [13, 180-183]. The
polarizable range of PbO_2 electrodes, as determined [13, 181] from
the i-V plots, was pH dependent and was found to be 0.7 to 1.2 V
(SCE) at pH 12, 1.2 to 1.7 V (SCE) at pH 1.8, and 0.4 to 1.7 V (SCE)
at pH 5.7. The α and β forms of PbO_2 showed slight differences
in their polarizable range.

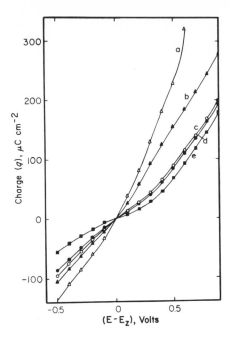

Fig. 33. Variation in surface charge q with potential of α-PbO$_2$ in KNO$_3$ solutions (mole/l): (a) 0.307, (b) 0.115, (c) 0.018, (d) 0.0121, (e) 0.0043. Measured at 23°C and 120 Hz. Reprinted from Ref. [13], p. 113, with the abscissa modified on author's advice, by courtesy of the Elsevier Publishing Company.

The double layer capacitance of PbO$_2$ electrodes in the polarizable range was measured by Kabanov et al. [180] in HClO$_4$ and H$_2$SO$_4$ solutions. The pzc of PbO$_2$, reported by Kabanov et al., is 1.8 V (NHE) in SO$_4^{2-}$ or ClO$_4^{-}$ solutions (pH ~ 1). This value is in serious disagreement with a pzc of 0.8 to 0.9 V (SCE) for PbO$_2$ in NO$_3^{-}$ or SO$_4^{2-}$ solutions (pH 5.7) as reported by Carr, Hampson, and Taylor [13, 181]. From the magnitude of the reported pzc [180] it appears that the potential measured by Kabanov et al. was perhaps the E_{rev} (pH dependent) of the PbO$_2$–PbSO$_4$ electrode [184] (E^0 = 1.68). Evidence for the specific adsorption of SO$_4^{2-}$ on PbO$_2$ was obtained from a slow shift of the pzc with time toward more cathodic potentials [180] and also from studies of the oxygen overvoltage on PbO_2 [180, 182].

These studies have been extended recently by Hampson and co-workers [13, 181, 183]. The surface charge q obtained for PbO_2 in KNO_3 solutions by the integration of the C versus E curves, are shown plotted against the electrode potential in Fig. 33. The C versus E (SCE) plots of β-PbO_2 (not shown in Fig. 33) were flatter than those of α-PbO_2. In both cases a considerable frequency dispersion of capacitance was also experienced, which was probably due to surface heterogeneity. As the electrode surface area was not known, a roughness factor of 5 to 10 had to be assumed to bring the measured C and q values (Fig. 33) to the theoretically expected order of magnitude. The capacitance minima (assumed for pzc) were independent of KNO_3 concentration, indicating that K^+ and NO_3^- are not specifically adsorbed on PbO_2 at the experimental pH of 5.7.

The absence of specific adsorption of K^+ and NO_3^- on PbO_2 was also checked by examining the Esin and Markov effect for the entire range of q values from the standard plots shown in Fig. 34. In the absence of specific adsorption of counterions it is shown from the Gouy-Chapman theory that the slope $(\partial E/\partial \mu_{salt})_q$ in Fig. 34 should be equal to 1, 0, or -1, depending on whether q is highly positive, zero, or highly negative. This requirement is shown to be approximately satisfied for PbO_2 (Fig. 34) at low values of q. If the pzc does not change with the electrolyte concentration, then deviations of the Esin and Markov plots from the predicted behavior (as seen in Fig. 34) at high values of q have been attributed by Delahay [185] to defects in the Gouy-Chapman theory and also to minor contributions of the diffuse double layer to the interfacial properties at conditions of high charge densities.

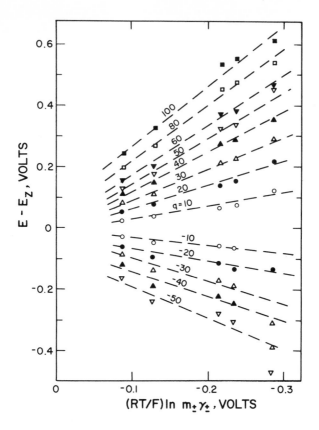

Fig. 34. Plot showing the Esin and Markov effect for an α-PbO$_2$ electrode. Reprinted from Ref. [13], p. 115, by courtesy of Elsevier Publishing Company.

From similar studies of the double layer on PbO$_2$ in SO$_4^{2-}$, ClO$_4^-$ [181], and PO$_4^{3-}$ [183] solutions, specific adsorption of SO$_4^{2-}$, HSO$_4^-$, ClO$_4^-$, and PO$_4^{3-}$ has been found to occur on PbO$_2$. In ClO$_4^-$ no sharp minimum was found to occur even in dilute solutions. The effect of changing pH from 12 to 1.8 on the interfacial capacitance of PbO$_2$ in a solution of 0.034 mole/l of K$_2$SO$_4$ is shown in Fig. 35 [181]. The increase in the magnitude of C with decreasing pH in the acidic range has been attributed to a pseudocapacitance arising from several interfacial reactions, including the reduction of SO$_{4,ads}^{2-}$ to HSO$_{4,ads}^-$

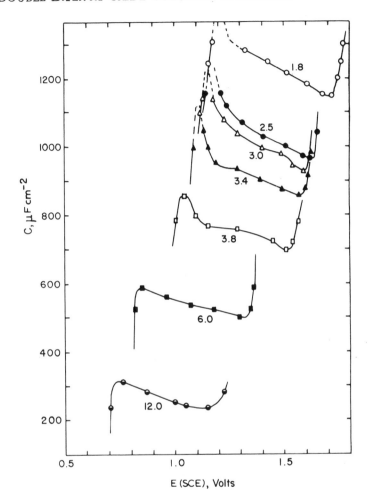

Fig. 35. Effect of pH on the plots of differential capacitance versus potential for electrodeposited α-PbO$_2$ in K$_2$SO$_4$ solution (0.0344 mole/l) at 23°C, measured at 120 Hz. The numbers on the curves correspond to pH values. Reprinted from Ref. [181], p. 204, by courtesy of the Elsevier Publishing Company.

by H$^+$ adsorption on the electrode surface. Above pH 6, adsorption of OH$^-$ on the electrode surface and a displacement of the SO$_{4,ads}^{2-}$ by OH$^-$ has been suggested.

The shifts in the C versus E curves of Fig. 35 to more anodic potentials with decreasing pH may be accounted for by the adsorption of H^+ on the electrode and may also be due to changes in the E_{rev} of the PbO_2–$PbSO_4$ electrode,

$$PbSO_4 + 2H_2O = PbO_2 + 4H^+ + SO_4^{2-} + 2e^- \, ,$$

which is pH dependent [182, 184]. However, in the presence of SO_4^{2-}, the electrode potentials of PbO_2 have been shown to vary by about 60 mV/pH unit only in acidic solutions with pH < 4 [182]. The electrode potentials of PbO_2 in sulfate solutions over the entire pH range are determined by a series of potential-determining reactions, which have been expressed as pH-potential diagrams by Rüetschi and Angstadt [182]. Further details of the capacitance behavior of PbO_2 electrodes are discussed by Hampson and co-workers [13, 181, 183].

The presence of oxide films on metals has been known to result in abnormal values of the double-layer capacitance on metals [18]. A pH-dependent capacitance has also been reported for some metals (e. g., Pt [100], Sn [92], Zn [94]) and could be due to an oxide or hydroxide film that can also be acquired readily in the anodic region.

VI. THE REVERSIBLE DOUBLE LAYER AT THE OXIDE-SOLUTION INTERFACE

A. General Considerations

Hydrogen and hydroxyl ions generally function as potential-determining species at the oxide-solution interface. In most cases the standard free energies of adsorption of H^+ on oxides are several orders of magnitude more negative than those of the indifferent electrolyte ions. The potential-determining nature of H^+ and OH^- ions for oxides may be attributed to several of their unique electrochemical properties.

Hydrogen ions are protons in an ideal, nonsolvated state. In comparison with other ions they are exceptionally small and require a high ionization energy (13.6 eV, or 313.5 kcal/mole) for formation from atomic hydrogen. Although protons have a radius of about 10^{-13} cm, their effective radius in the solvated state is about 0.34 Å [104, 186]. In addition, hydrogen ions show an extremely high ionic mobility in solutions due to a chain or tunneling mechanism and also exhibit a great affinity for electron donors [186, 187]. Hence hydrogen ions readily polarize surrounding molecules and almost always exist in a solvated form in aqueous media. The hydrogen ion has a permanent hydration number of 1 (H_3O^+), a primary hydration number of 3 per H_3O^+ (in a tetrahedral hydrogen-bonded structure), and a loosely held secondary hydration shell of nine water molecules. However, in charge-transfer reactions the hydrogen ion can still take part as a nucleus. Henceforth the terms "H^+" or "hydrogen ion" will be used to refer to the hydroxonium, or H_3O^+, ion in aqueous solutions.

Several authors [26, 188] have pointed out that there are a number of similarities between the electron-hole equilibria in a semiconductor phase and the OH^- and H^+ equilibria in electrolyte solutions. Thus the concentrations of charge carriers in a semiconductor at room temperature are comparable to the H^+ and OH^- concentrations in electrolyte solutions. Also, the constancy of the charge-carrier product in a semiconductor is similar to the constancy of the concentration product (K_w) of H^+ and OH^- in water. The dissociation energy of water, as calculated from the temperature dependence of K_w, is 13.5 kcal/mole. This dissociation energy is comparable to a corresponding value of 15.7 kcal/mole for intrinsic germanium obtained from the temperature dependence of the electron-hole equilibrium constant K_i.

The electron donor-acceptor properties of H_3O, H_3O^+, OH, and OH^- are evident from such discharge reactions as

$$H_3O^+ + e^- \rightleftharpoons H_3O$$

and

$$OH^- \rightleftharpoons OH + e^- \,,$$

which occur predominantly on electrode surfaces. Whereas the electrochemical potential of electrons in a semiconductor is expressed in terms of the Fermi level, the electrochemical potential of H^+ or OH^- in aqueous media is expressed in terms of pH. Hence pure water at pH 7 is comparable to an intrinsic semiconductor, whereas basic and acidic solutions are analogous to n- and p-type semiconductors. Although the bulk dielectric constant of water is high, the dielectric constant of water in the Helmholtz layer is usually considered to be much lower (~ 6) and is often comparable to the dielectric constant of a solid semiconductor phase.

The charge-transfer processes in aqueous solutions chiefly involve electrostatic forces. The subjects of proton solvation, water structure, and proton-transfer kinetics in both homogeneous and heterogeneous systems have been discussed in a number of publications [24, 174, 186, 189, 190]. The mechanisms of proton transfer from H_3O^+ in the aqueous phase to an electrode surface has been discussed [24, 189, 190] with the aid of potential-energy diagrams that can be constructed from the Morse equation. The transfer of protons from H_3O^+ to metal electrodes in the presence of an applied electric field is controlled by the activation energy and, thermodynamically, is an irreversible process. A reversible process of proton transfer is characterized by a low activation energy. The double-layer studies of the oxide-solution interface to be discussed subsequently are carried out in the absence of an applied electric field. In these studies the process of proton transfer at the oxide-solution interface is reversible, and the method is based on the potential-determining nature of H^+ and OH^- for oxides.

B. Oxide-Solution Equilibria and Double-Layer Studies
on Oxides (Method A)

Attempts to obtain information on the double-layer properties of oxides from direct acid-base titrations of oxide surfaces (or pH measurements) were first made in 1962 by Parks and De Bruyn [191] on precipitated Fe_2O_3, and also by Herczyńska and Prószyńska [178, 179, 192] on Al_2O_3, Fe_2O_3, NiO, and oxide-bearing surfaces of iron, nickel, and zinc. The method used by Parks and De Bruyn was based on the same well-established principles as those employed previously for studies of the AgI system [193, 194]. The theory dealing with solid-solution equilibria in general and the double layer on silver halides in particular has been considered in detail in the early work of Grimley and Mott [195], Grimley [196], Verwey and Overbeek [62], and more recently by Lyklema and co-workers [197, 198] and Honig and co-workers [199-201]. However, the oxide-solution system has proved to be much more complex than the silver halide system because of the multiple oxidation states of the different metal ions involved and the variety of bulk properties encountered with metal oxides.

When a sparingly soluble oxide is added to an electrolyte solution, the following reactions will occur [202, 203]:

1. A primary equilibrium between the surface atoms of the solid and the potential-determining ions in solution, leading to formation of metal aquocomplexes on the surface and dissociation of the surface groups. This primary oxide-solution equilibrium is known [203-206] to be complete within a few minutes after the addition of the oxide to the solution.

2. Dissolution of the oxide leading to a series of secondary reactions, such as hydrolysis, complex formation and dissociation of the dissolved complexes, adsorption of the dissolved complexes on the

surface (possibly in multilayers), precipitation of the complexes in colloidal form, and several other nucleation processes [207, 208].

The secondary reactions (type 2) depend on the solubility of the oxide and the dissociation constants of various metal aquocomplexes. The slow equilibria of the hydroxy complexes of iron in electrolyte solutions have been investigated in detail by Biedermann and Chow [207] and by Van der Giessen [208]. The processes of oxide dissolution and the secondary reactions of type 2, both of which result in changes in pH, are known to continue for days or weeks.

Studies of the reversible double layer on oxides are based on acid-base titrations of oxide surfaces of known surfaces areas, as a function of pH and ionic strength. The amount of acid or base consumed in such titrations is obviously equal to the sum of the acid or the base bound to the oxide surface and that bound, in various forms, to the dissolved aquocomplexes in solution. Hence solubility corrections, when necessary, should be made in order to obtain unambiguous information on the surface properties of oxides. Even if the amount of metal-hydroxy complexes in solution at a given time is small, because of the usually low solubility products of metal complexes (e.g., Fe complexes), the total pH change due to the cumulative secondary effects of type 2 could be considerable and sufficiently large to mask the primary surface effects. If, in the experimental method used, these side or secondary effects are not effectively prevented, corrections for them are difficult and doubtful, as pointed out by the author in earlier work on quartz, zirconia, and thoria [203].

Hence in one experimental approach [146, 202, 203, 206] the secondary reactions were almost totally prevented (with minor solubility corrections) by using a batchwise method in which the oxide sample was not allowed to contact the test solution for more than 10 min for each adsorption-density measurement. Fresh samples of oxides and solutions were used for each point obtained on the plots of adsorption density versus pH. This procedure is comparable to the use of fresh mercury surfaces in electrocapillary measurements and provides information on the primary equilibrium between the oxide surface and H^+ and OH^- ions. This method will henceforth be referred to as method A.

The necessity of not only minimizing the secondary effects (including ion-exchange reactions [209]) but also correcting for oxide dissolution when necessary (e. g., ZnO [210, 211]) has been realized in the more recent work of De Bruyn and co-workers. These authors, as a modification to their previous prolonged- and continuous-titration method [191], used a fast-titration technique in studying the double layer on Fe_2O_3 [205], TiO_2 [212, 213] and ZnO [209-211]. In the case of ZnO [210] solubility corrections were also made. In the fast-titration method the entire adsorption isotherm is obtained in one single and continuous titration, which still lasts for more than 1 hour. Hence the solution will be saturated with the dissolved aquocomplexes of metals in this method (method B), in contrast to method A.

It is hard to compare the results obtained from the two experimental approaches described here. It is also doubtful that the results obtained with experimental procedures A and B represent identical phenomena at the oxide-solution interface. Hence as far as possible these two experimental approaches will be presented and discussed separately. However, measurements conducted by method A relating to the primary oxide-solution equilibrium (type 1) are performed in solutions that are practically free of dissolved complexes and would correspond to conditions prevailing in streaming-potential measurements where a fresh sample of solution will be in contact with the oxide. On the other hand, the experiments performed by method B in solutions that are practically saturated with their dissolved complexes perhaps correspond more closely to conditions in colloidal systems.

C. Double-Layer Studies from Acid-Base Titrations of Oxide Surfaces in Solutions Saturated with Dissolved Metal Aquocomplexes (Method B)

The recent work of De Bruyn and co-workers on TiO_2 [212, 213] and ZnO [209-211] may be taken as representative of method B, noted in the preceding section. The basic assumptions and thermodynamics underlying this method of studying the oxide-solution interface, as presented by Bérubé and De Bruyn [212] for TiO_2 and Blok and De Bruyn [210] for ZnO, are outlined here.

The electrochemical potential $\bar{\mu}$ of an ion in any system at equilibrium should be constant throughout that system. Hence

$$(\bar{\mu}_{M^{z+}})_{sd} = (\bar{\mu}_{M^{z+}})_{sur} = (\bar{\mu}_{M^{z+}})_{sol} , \tag{48}$$

where sd, sur, and sol refer to the solid phase, to the surface, and to the solution phase, respectively, and the electrochemical potential is defined, as usual, as

$$\bar{\mu}_i = \mu_i^0 + RT \ln a_i + z_i F\varphi , \qquad \bullet \tag{49}$$

where a_i is the activity and μ_i^0 is the standard chemical potential of an ion i. The potential φ was defined by Bérubé and De Bruyn as the macropotential (of i), which was assumed to be a function of position — an assumption that has been criticized by Ball [214]. From Eqs. (48) and (49),

$$d(\varphi_{sd} - \varphi_{sol}) = \frac{RT}{zF} d \ln (a_{M^{z+}})_{sol} . \tag{50}$$

By taking into account various reactions of ion hydrolysis and of the dissolved metal complexes, the changes in the activities $a_{M^{z+}}$ in solution can be related to changes in a_{H^+} or a_{OH^-} of the bulk solution, so that

$$d(\varphi_{sd} - \varphi_{sol}) = \frac{RT}{F} \, d \ln a_{H^+} \, , \tag{51}$$

$$d(\varphi_{sd} - \varphi_{sol}) = - \frac{RT}{F} \, d \ln a_{OH^-} \, . \tag{52}$$

By writing (cf. Ref. [212]

$$\varphi_{sd} - \varphi_{sol} = (\varphi_{sd} - \varphi_{sur}) + (\varphi_{sur} - \varphi_{sol}) \, , \tag{53}$$

we obtain

$$d(\varphi_{sd} - \varphi_{sur}) + d(\varphi_{sur} - \varphi_{sol}) = \frac{RT}{F} \, d \ln a_{H^+} \, . \tag{54}$$

By assuming $d(\varphi_{sd} - \varphi_{sur}) = 0$ for a given oxide,

$$d(\varphi_{sur} - \varphi_{sol}) = \frac{RT}{F} \, d \ln a_{H^+} \, . \tag{55}$$

Equation (55) does not imply that $\varphi_{sd} - \varphi_{sur} = 0$, but that the change in charge density and Galvani potentials in the electronic and ionic space-charge region of the solid phase is assumed to be zero. This assumption is justified on the basis of low ionic mobilities in the oxide phase at room temperature in addition to the fact that the surface responds rapidly to the potential-determining ions in the oxide-solution system. However, it may not be valid when the positions of the energy bands of the solid phase are altered by the application of an external electric field. It may also not be valid when chemisorption of protons on oxide surfaces affects the space charge [154-157, 206] or when the dissolution of oxide changes its stoichiometry.

In deriving Eqs. (54) and (55) the chemical potential of the dissolved lattice ions, such as Zn^{2+}, was related to the chemical

potentials of a number of ionic species in solution, which in turn
were related to a_{H^+} and a_{OH^-}. The pH-dependent equilibria [209,
215] encountered in aqueous ZnO suspensions, for example, are as
follows:

$$ZnO + H_2O \rightleftharpoons Zn^{2+} + 2OH^- \qquad (pK = 16.4) ,$$

$$\rightleftharpoons ZnOH^+ + OH^- \qquad (pK = 12.0) ,$$

$$\rightleftharpoons ZnO_2H^- + H^+ \qquad (pK = 17.0) , \qquad (56)$$

$$\rightleftharpoons ZnO_2^{2-} + 2H^+ \qquad (pK = 29.7) .$$

The pH-independent solubility of $Zn(OH)_2$ is 10^{-5} mole/l [216].
The equilibrium-solubility data of ZnO are shown in Fig. 36. In

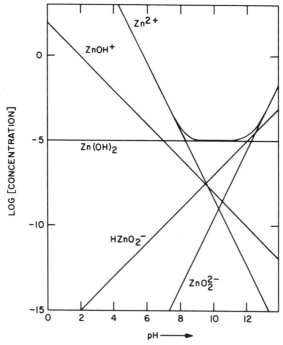

Fig. 36. Concentration of various zinc hydroxy complexes in
equilibria with ZnO surfaces, as a function of pH. Reprinted from
Ref. [210], p. 519, by courtesy of Academic Press, Inc.

principle, all the ionic species in Eqs. (56) are regarded as potential
determining in order to account for the changes in the surface-charge
density q of ZnO as a result of the adsorption of these ionic species.
Hence [217]

$$\Delta q = F[\Delta\Gamma^*_{H^+} - \Delta\Gamma^*_{OH^-} + 2\Delta\Gamma^*_{Zn^{2+}} - 2\Delta\Gamma^*_{ZnO_2^{2-}} + \Delta\Gamma^*_{ZnOH^+} - \Delta\Gamma^*_{HZnO_2^-}$$

(57)

and

$$\Delta q = F \Delta\Gamma_{H^+} .$$

(58)

where $\Delta\Gamma^*$ refers to the real changes in the surface excess of the
potential-determining ions and $\Delta\Gamma_{H^+}$ is the equivalent experimental
change in the amount of H^+ adsorbed per unit area of oxide.   The
minimum solubility of ZnO, from Fig. 36, is seen to occur at pH 9.7.
Similarly the minimum solubility of the iron oxide/hydroxide system
has been shown [191] to occur at a pH of about 8.5. According to
the postulate of Parks and De Bruyn [191], the zero point of charge
(zpc) of oxide surfaces occurs at the isoelectric point, which is also
the pH of minimum oxide/hydroxide solubility (however, see Sections
VI. F and VI. I). By relating the changes in the electrochemical potentials of various ionic species in solution to the activities of H^+ and
OH^- ions in solution, the change in the surface potential ψ^0 of the
oxide, relative to the zpc, is given from Eq. (55) as

$$d\psi^0 = \frac{RT}{F} d \ln \frac{a_{H^+}}{(a_{H^+})_{zpc}} ,$$

(59)

$$d\psi^0 = - \frac{RT}{F} d \ln \frac{a_{OH^-}}{(a_{OH^-})_{zpc}} .$$

(60)

Since $\varphi^0 = \psi^0 + \chi$, the use of Eqs. (59) and (60) implies that the
χ potential at the oxide-solution interface does not change in the
double-layer studies. However, it will be seen in Sections VI. F and

VI.I.5 that changes in the χ potential of some oxides are considerable. The zpc is formally defined [191] as the pH at which the surface
excess of H^+ and OH^- ions on oxides is equal; that is, $\Gamma^*_{H^+} = \Gamma^*_{OH^-}$
at the zpc. The term "zero point of charge" has been used by Over-
beek and co-workers [62, 194] in the field of colloids, by several
others in studies of the double layer on AgI [193, 197, 198], and also
by Parks and De Bruyn in their early work on Fe_2O_3 [191]. However,
Bérubé and De Bruyn [212] have recently used the term "potential of
zero charge (pzc)" in preference to zpc. Because of the possible
confusion with the pzc of electrified interfaces, the term "zpc" will
be used throughout this work. A more detailed discussion of the zpc
and the results of work on ZnO and TiO_2 will follow.

D. The Primary Oxide-Solution Equilibrium
and the Surface Charge

The origin of surface charge during the primary oxide-solution
equilibrium has been attributed [146, 202, 206, 218] to the formation of
neutral or charged ($\pm$) aquocomplexes of metals on the oxide surface,
as shown schematically in the following equation:

$$[\geq M(H_2O)_2(OH)]^+ \xrightarrow[A]{H^+, NO_3^-}$$

$$[\geq M(H_2O)(OH)_2] \underset{H^+}{\overset{OH^-}{\rightleftarrows}} \left[\begin{array}{c} \geq M \\ H_2O \end{array} \begin{array}{c} O^- \\ | \\ OH \end{array} \right] ,$$

$$[\geq M(H_2O)_2^{2+} Cl^-]^+ \xleftarrow[H^+, Cl^-]{B}$$

$$\text{etc.} \quad (61)$$

$$\text{anodic } q^+ \xleftarrow[pH < zpc]{E^+} \left[\begin{array}{c} \text{neutral complex} \\ zpc \\ q, \psi^0 = 0 \end{array} \right] \xrightarrow[pH > zpc]{E^-} q^- \text{ cathodic.}$$

According to the mechanism shown in Eqs. (61), the surface acquires
an effective zero charge at the zpc through the formation of a neutral

aquocomplex, such as $[\overset{=}{\ } M(H_2O)(OH)_2]$, where M refers to the metal ion (e.g., Sn^{4+}, Ti^{4+}, or Fe^{3+}) in a six coordination state. In this surface complex only three of the six coordination sites and two out of four (Sn^{4+}) or three (Fe^{3+}) valence states of M are assumed to be exposed to the solution.

1. The Negative Surface Charge of Oxides

The negative surface charge at the oxide-solution interface orig-inates from an acidic dissociation of the surface hydroxyl groups (at pH > zpc) and increases with increasing pH as well as with increas-ing electrolyte concentration. Cations may be adsorbed [146, 203, 218] on the oxide layer either in a hydrated form or in a form partly dehydrated in a direction normal to the surface and possibly also in the lateral direction. The latter type of ionic adsorption of cations in a partly dehydrated form will henceforth be referred to as specific adsorption, by analogy with similar adsorption occurring on metal surfaces. However, specific adsorption of cations on the oxide layer is not to be confused with the general lack of specific adsorption of cations on oxide-free metals and on mercury surfaces. The specific adsorption of K^+ on oxides could as well be considered as the specific adsorption of anionic species such as $(OK)^-$ on the cationic sites of oxide surfaces. The evidence for the specific adsorption of cations on oxides, the adsorption isotherms of cations, and the standard free energies of adsorption of cations on oxides will be presented in the discussion of the experimental results.

2. The Positive Surface Charge of Oxides

Most oxide surfaces (except quartz) acquire a positive surface charge in solutions of pH less than the zpc of the oxide. Although the positive-surface-charge densities q^+ of oxides increase with decreasing pH, the variation in q^+ with electrolyte concentration, that is, the slope $(\partial q^+/\partial \mu_{salt})_{pH}$, was found to depend strongly on the nature of the anion as well as of the oxide [146]. The behavior of oxide surfaces in the anodic region (pH < zpc), as shown in Eqs. (61), is best explained [146, 202, 206, 218] on the basis of proton addition to the neutral aquocomplex, together with or without the replacement of the surface hydroxyl groups and possibly also the sur-face H_2O ligands by the anions.

In mechanism A (Eqs. (61)) the anions do not replace the surface hydroxyl groups and stay as counterions outside the primary hydra-tion shell of the surface. The oxide electrode in the above case is reversible to H^+ ions. Such electrodes have been known as elec-trodes of the second kind, whose half-cell potential in the present case is given by [184, 219]

$$E = E_{0_M} + \frac{RT}{zF} \ln \frac{K_s}{(K_w)^z} + \frac{RT}{F} \ln a_{H_3O^+} , \qquad (62)$$

$$E = E_0' + \frac{RT}{F} \ln a_{H_3O^+} , \qquad (63)$$

where K_s is the solubility product of the metal hydroxide involved in the equilibrium. The rest of the terms in Eq. (62) have their usual meaning.

In mechanism A the positive-surface-charge densities q^+ are expected to depend primarily on a_{H^+}, and not on the anion concen-tration directly. Hence on increasing the anion (salt) concentration, at a given pH below the zpc, any increase in q^+ will be determined essentially by the effect of the increased ionic strength on the charge density of the diffuse double layer and on a_{H^+} itself. From the Poisson-Boltzmann equation the surface-charge density due to the net excess charge in the diffuse double layer for a symmetrical z:z valent electrolyte, in the absence of specific adsorption, is given by [60, 220]

$$q = \left(\frac{2kTc\bar{\epsilon}}{\pi}\right)^{\frac{1}{2}} \sinh \frac{ze(\psi_\delta - \psi_{sol})}{2kT} , \qquad (64)$$

$$q = 11.72\sqrt{c} \ \sinh(19.46 \ z\psi), \ \mu C/cm^2 , \qquad (65)$$

where c is the electrolyte concentration in moles per liter, $\bar{\epsilon}$ is the mean dielectric constant (78.5 assumed in Eq. (65)), and $\psi = \psi_\delta - \psi_{sol}$ is the potential difference between the bulk of the solution and the plane of the closest approach (the OHP) of the ion toward the surface. The sign of the diffuse-double-layer charge density (η^d) is opposite to that of ψ, whereas the surface-charge density q has the same sign as ψ. Numerical values of q for the mercury-F^- system have been tabulated by Russell [221] as a function of electrode potential and electrolyte concentration and should also be applicable to other surfaces in the absence of specific adsorption. It can be shown from Russell's tables [221] that the increase in q^+ values, based on mechanism A, on increasing the ionic strength of the solution from 0.001 M to 1 M of a 1:1 valent indifferent electrolyte is only 1 to 2 $\mu C/cm^2$ for an electrode potential of 100 mV relative to the zpc. This variation in q^+ is normally within the experimental error ($\pm 1 \ \mu C$) of studying the reversible double layer on oxides. Hence, if mechanism A is applicable, the q^+ values are expected to depend mainly on a_{H^+} and to be independent of the salt concentration within 1 to 2 $\mu C/cm^2$. Such behavior was found to occur in the case of SnO_2 in NO_3^- [202, 218], Al_2O_3 [146, 218], and TiO_2 [202, 218] in NO_3^-, ClO_4^-, and Cl^-, and possibly Fe_2O_3 in NO_3^- and ClO_4^- solutions [206, 218]. These results are discussed further in a subsequent section.

Recent investigations of $Al(OH_2)_6^{3+}$ complexes by nuclear magnetic resonance (NMR) spectroscopy have shown that this complex is stable in Cl^-, NO_3^-, or ClO_4^- solutions, with no detectable replacement of H_2O by Cl^-, and that proton-exchange reactions are predominant

between the aquocomplex and the acid solutions [222, 223]. These
results are in agreement with mechanism A.

In mechanism B (Eqs. (61)), in addition to a proton-transfer
process, the surface hydroxyl groups can undergo a basic dissociation, with anions replacing the surface OH groups. Such a basic
dissociation process in acid-base titrations liberates more OH^- into
the bulk solution, compared with that in mechanism A, and depends
markedly on the anion concentration of the solution. The adsorption
of anions on oxides in mechanism B, by the basic dissociation of the
surface OH groups, is similar to the acidic dissociation of the surface
groups in the cathodic region. Hence the apparent q^+ values in mechanism B would depend not only on pH but also markedly on the anion
concentration. Surface reactions consistent with this mechanism have
been found for specular hematite in Cl^- [206, 218], ZrO_2 and ThO_2 in
NO_3^- [203], Cl^-, and ClO_4^- [146, 218] solutions.

If the specific interaction of anions with the metal atoms of the
oxide surface is very strong, a saturation limit may be reached in
anionic adsorption at a low anion concentration, such as 10^{-3} mole/l,
so that a further increase in the anion concentration would show little
increase in the q^+ values. Such a reaction was found to occur in the
case of SnO_2 in Cl^- [202, 218]. There is also a possibility of anions
replacing the water molecules coordinated to the metal atoms of the
oxide surface. These two possibilities will be referred to as mechanism C in future discussion.

On the basis of the known solution chemistry of iron complexes,
Atkinson et al. [224] have eliminated the possibility of Cl^- and NO_3^-
entering the first coordination shell of Fe^{3+} at pH 3 to 5 and thus
being specifically adsorbed on Fe_2O_3. However, it should be pointed
out that the electrochemical conditions at the interface are drastically
different from those in the solution phase. For example, the dielectric
constant in the region of the oxide-solution interface is comparable to

that of a nonaqueous medium. Replacement by anions of water or hydroxyl groups in the first coordination shell of metal complexes (e.g., Cl^- in Sn complexes) in aqueous solutions of organic liquids (even in slightly acidic solutions), has been well established in recent studies with NMR spectroscopy (see Refs. [225] through [227] and the references therein). Direct measurement of the double-layer capacitance of PbO_2 electrodes, as already discussed, has also indicated specific adsorption of ClO_4^-, HSO_4^-, and SO_4^{2-} on PbO_2 even in slightly acidic solutions. Specific adsorption of Cl^- on SnO_2 electrodes has also been detected from i-V measurements [139].

The foregoing interactions at the oxide-solution interface appear to be characteristic of all the transition-metal oxides and other amphoteric oxides. In the case of quartz or oxidized-silicon surfaces neither complex formation with H^+ nor the basic dissociation of the surface hydroxyl groups occurs (except probably in the presence of F^-). Hence no excess positive surface charge and consequently no excess of anionic, double-layer charge is expected to result at the quartz-solution interface [203] (mechanism D). Certain similarities of the double layer on oxidized germanium and silicon surfaces, including an almost identical pH for their zpc, were pointed out in Section III. B. 6. Except in the presence of I^-, mechanism D also appears to hold good for the double layer on germanium with an oxide or hydroxide surface. In the presence of I^- mechanism B or D has been suggested [15, 128, 130]. However, in the present studies I^- was not found to replace the OH^- on Ge surfaces (III, B, b).

E. Thermodynamic Considerations

The surface-charge density (q/cm^2) at an oxide-solution interface may be written, for the general case, as

$$q = F\left[(zC_{M^{z+}} - \Gamma_{OH^-}) - (C_{O^-} - \Gamma_{H^+})\right] , \qquad (66)$$

where Γ_{H+} and Γ_{OH-} are the surface excess of H^+ and OH^- in M (cm^{-2}) and C_{Mz+} and C_{O-} are the concentrations of the cationic and oxygen sites on the oxide surface. For a stoichiometric surface, $zC_{Mz+} = C_{O-}$; hence the net or effective surface charge is given as

$$q = F[\Gamma_{H+} - \Gamma_{OH-}] , \qquad (67)$$

so that q is positive, zero, or negative, depending on whether Γ_{H+} is greater than, equal to, or less than Γ_{OH-}. For the thermodynamic treatment the surface structure at the zpc, where q is zero, may be taken as the reference plane. At a given pH the potential E (or φ^0) of the reference plane relative to the bulk solution is given from Eq. (59), at 25°C, as

$$E = -0.059 \, (pH - pH_{zpc}) . \qquad (68)$$

The term on the right-hand side of Eq. (68) is commonly equated with the Volta potential difference ψ^0 of the oxide surface relative to the bulk solution (cf. Eq. (59)). However, the right-hand side in Eq. (68) actually involves changes in the χ potential (Sections VI. F and VI. I. 4. b) as well as ψ^0, and a general term E is therefore used in Eq. (68). The quantity E is negative or positive, depending on whether the experimental pH of the solution in equilibrium with the oxide is greater than or less than the zpc. The Gibbs adsorption equation, applied to this system, in the presence of a symmetrical 1:1 valent salt MX, may be written as

$$d\gamma = -(\Gamma_{H+} - \Gamma_{OH-}) \, d\bar{\mu}_{H+/OH-} - \Gamma_{M+} \, d\bar{\mu}_{MX} - \Gamma_{X-} \, d\bar{\mu}_{MX} , \qquad (69)$$

where γ is the interfacial energy. At a constant ionic strength and in the absence of specific adsorption of the electrolyte,

$$d\gamma = -(q \, dE)_{\bar{\mu}_{MX}} \quad \text{or} \quad \left(\frac{d\gamma}{dE}\right)_{\bar{\mu}_{MX}} = -q , \qquad (70)$$

which is the well-known Lipmann equation. Also,

$$-\left(\frac{\partial^2\gamma}{\partial E^2}\right)_{\mu_{MX}} = \left(\frac{\partial q}{\partial E}\right)_{\mu_{MX}} = C .$$ (71)

The differential capacitance given by Eq. (71) is the rate of change of the surface charge with the half-cell potential and is a thermodynamically derived parameter in the present case. It is obtained from a graphical differentiation of the q versus E plots. The significance of this C compared with the measured capacitance is not certain and will be discussed later in some detail with reference to the experimental results. By expanding Eq. (71), the capacitance is also given by [214]

$$C = \frac{-F^2 d(\Gamma_{H^+} - \Gamma_{OH^-})}{2.3\, RT\, dpH} .$$ (72)

Although the fundamental significance of C obtained by using Eqs. (71) and (72) is yet to be established, inference can be drawn as to the nature of adsorption of counterions on oxide surfaces by following variations in C with E (Section VI, I, 5).

If γ_0 (absolute value is not known) represents the interfacial energy at the zpc, then at any other cell potential the total change in the interfacial free energy relative to γ_0 can be obtained by integrating Eq. (70) as follows:

$$\gamma - \gamma_0 = \left(\int_{E_{zpc}}^{E} -q\, dE\right)_{\mu_{MX} = -\infty} - \left(\int_{-\infty}^{\mu} \Gamma_{M^+, X^-}\, d\mu_{MX}\right)_{E} .$$ (73)

The interfacial energy γ_0 should be a maximum at the zpc when q is zero (Eq. (70)). The interfacial energy is lowered by an amount $\gamma_0 - \gamma$ [62] on both anodic and cathodic surfaces and this decrease in the

free energy of the double layer is obtained by the graphical integration of the q versus E plots [146, 203, 218].

At a constant pH, or constant half-cell potential, the change in
the interfacial energy is due to the effect of increasing ionic strength
of the solution on the adsorption of M^+ or X^- on the surface, so that

$$-\left(\frac{\partial \gamma}{\partial \mu_{MX}}\right)_{E^-} = \Gamma_{K^+} \quad \text{and} \quad \left(\frac{\partial q^-}{\partial \mu_{MX}}\right)_{E^-} = \left(\frac{\partial K^+}{\partial E^-}\right)_{\mu_{MX}^-} \tag{74}$$

at cathodic surfaces, and

$$-\left(\frac{\partial \gamma}{\partial \mu_{MX}}\right)_{E^+} = \Gamma_{X^-} \quad \text{and} \quad \left(\frac{\partial q^+}{\partial \mu_{MX}}\right)_{E^+} = \left(\frac{\partial \Gamma_{X^-}}{\partial E^+}\right)_{\mu_{MX}^-} \tag{75}$$

at anodic surfaces.

It is to be noted that it is the excess surface charge that is obtained in studies of the double layer on oxides by potentiometric
titrations, so that from the principle of electroneutrality

$$q^{\pm} = -F(\Gamma^{\mp, i} + \Gamma^{\mp, d}) = -(\eta^{\mp, i} + \eta^{\mp, d}) \,, \tag{76}$$

where Γ^i (or η^i) and Γ^d (or η^d) refer to the relative ionic (or charge)
excess with $z = 1$ in the inner and in the diffuse double layers,
respectively. This relative excess Γ^d, for instance, is the excess
charge in the diffuse double layer relative to the bulk concentration.
The validity of Eq. (76) can be checked by comparing the q^- values of
quartz obtained by potentiometric titrations [203] with the Γ_{Na^+} on
quartz obtained by Li and De Bruyn [228] using ^{22}Na as a tracer.
These two values are in good agreement (~ 9 $\mu C/cm^2$ at pH 10 in
0.001 M KNO$_3$ or NaCl). The excess charge density of the double layer
due to the adsorption of Ca^{2+} [229] and Ba^{2+} [230] on quartz, measured
by using radioactive tracers, also agrees with the q^- values of quartz
from potentiometric titrations. The values of Γ^d are small compared

with the Γ^i values that are usually encountered on oxide surfaces, and hence Γ^d can be estimated without significant error from diffuse-double-layer theory. Hence, from Eq. (76), the density of specifically adsorbed ions on oxides has been obtained [146] as a function of pH and the activity a_{MX}. From these adsorption data, the standard free energies of specific adsorption of ions on oxides have also been calculated by using a suitable adsorption isotherm [146]. The results of these calculations in relation to the distribution of potential at the oxide-solution interface will be discussed in Section VI.I.4. Adsorption isotherms based on several double-layer models have also been proposed by Herczyńska and Prószyńska [178, 179], Atkinson et al. [224], and Blok and De Bruyn [211].

F. The Zero Point of Charge of the Oxide-Solution Interface

The pH at which an oxide surface acquires a zero effective charge as a result of H^+ and OH^- adsorption is generally known as the zero point of charge (zpc). There have been very few attempts [231-233] to correlate the zpc of oxides with their bulk properties.

The zpc of some precipitated oxides, such as Fe_2O_3 [191, 205] and ZnO [209-211], in solutions that were saturated with the corresponding dissolved metal complexes, has been shown to occur at their iso-electric point, which can be defined as the pH at which the concentrations of the positively and negatively charged metal complexes in solution are equal. However, the coincidence of the zpc with the isoelectric point is not true as a general rule [202, 206] for oxides of low solubility, particularly in fresh electrolyte solutions not saturated with metal complexes (see Table 1). The pH of the zero streaming potential of oxides does not always coincide with the isoelectric point of the corresponding metal complexes in solution. Furthermore, the isoelectric point should be independent of the solid phase, whereas,

in practice, the zpc of oxides has been found to vary with the stoichiometry [233, 234], surface hydration and heat treatment [212, 235-237]
of the oxides, and possibly with their bulk electronic properties also.
For example, the zpc of alumina can vary from pH ~ 9.2 to pH < 6
[236, 237], depending on the heat treatment and on the degree of surface hydration. The zpc of rutile has also been shown to vary with
the heat treatment of the oxide [212, 235]. Similarly the zpc of man-
ganese oxide [233] and uranium oxide [234] is also known to vary
with the oxide stoichiometry.

    A major obstacle in correlating the zpc of an oxide with its bulk
or surface properties has been the lack of experimental data on oxides
with known solid-state properties. In addition to the various experimental uncertainties reviewed by Parks [231], the significance of the
zpc is also obscure and its definition vague.

    The zpc of an oxide has been defined [191] as the pH at which
the surface excess of H^+ and OH^- is equal; that is, $\Gamma^*_{H^+} = \Gamma^*_{OH^-}$ at
the zpc. Also implied in this definition is that the charge and poten-
tial distribution from the surface to the bulk solution is uniform
$(d\psi/dx = 0)$. This definition is based on the assumption that the
oxide surface is always stoichiometric, so that the ratio
$(\Gamma^*_{H^+}/\Gamma^*_{OH^-})_{sur}$ at the zpc is also unity in order that the interface be
electrically neutral. However, the oxide surface need not necessarily
be stoichiometric. Even for a stoichiometric oxide surface, such as
quartz, a major anomaly concerning the zpc is the structure of the
interface in relation to the activities of H^+ and OH^-, and the dielectric constant in the interfacial region. From streaming potential
measurement the zpc of quartz was found to occur [228, 238] at
pH ~ 1.5. Hence the ratio of the activities of H^+ and OH^- in the
bulk solution is $(a_{H^+}/a_{OH^-})_{sol} = 10^{-1.5}/10^{-12.5} = 10^{11}$. When the
pH of the bulk solution is 1.5, two conditions are satisfied simul-

taneously: (a) the interface charge is zero and (b) the Galvani poten-

tial $(\psi^0 + \chi)$ is also zero.

Thus, if C_{H+} and C_{OH-} refer to the concentrations of H^+ and

OH^-, then from condition a the ratio $(C^*_{H+}/C^*_{OH-})_{sur}$ at the inter-

face at zpc should still be unity; and from condition (b) the ratio

$(a_{H+}/a_{OH-})_{sol} \approx (C_{H+}/C_{OH-})_{sol}$ in the bulk region is equal to 10^{11}.

The only way this large difference between the ratio of $(a_{H+}/a_{OH-})_{sol}$

in the bulk phase and the ratio of $(C^*_{H+}/C^*_{OH-})_{sur}$ at the surface can

be accounted for is by assuming major differences in the activity

coefficients (or of some equivalent property) of H^+ and OH^- in the

interface region. Thus at the zpc

$$(\bar{\mu}_{H+})_{sur} = (\bar{\mu}_{H+})_{sol} , \tag{77}$$

where "sur" and "sol" refer, as before, to the surface and to the

solution phases, respectively. Because the Galvani potential at the

oxide–solution interface is zero at the zpc,

$$(\mu_{H+})_{sur} = (\mu_{H+})_{sol} \tag{78}$$

and taking adsorption of OH^- also into consideration,

$$RT \ln \left(\frac{C^*_{H+}}{C^*_{OH-}} \right)_{sur} (f_\pm)_{sur} = K + RT \ln \left(\frac{C_{H+}}{C_{OH-}} \right)_{sol} (f_\pm)_{sol} , \tag{79}$$

where $f_\pm$ is the ratio of the mean activity coefficients of H^+ and OH^-

in the solution or in the interface region and K is the sum of μ^0_{H+} and

μ^0_{OH-} for the surface and the solution. This summation will be close

to zero if one assumes $(K_w)_{sur} = (K_w)_{sol}$ [242]. As $(C^*_{H+}/C^*_{OH-})_{sur} = 1$

and $(C_{H+}/C_{OH-})_{sol} = 10^{11}$,

$$\log \frac{(f_\pm)_{sur}}{(f_\pm)_{sol}} = 11 . \tag{80}$$

The activity coefficients of H^+ and OH^- in the interface region may not have practical significance, but they do represent the degree of electrostatic interactions in the Helmholtz region and can be correlated with the dielectric constant and χ potential at the interface. From the theory of electrolytes, for a given composition of the solution, the log of the activity coefficient is inversely proportional [239] to the dielectric constant of the medium. Hence, from Eq. (80),

$$\log \frac{(f_{\pm})_{sur}}{(f_{\pm})_{sol}} = \frac{\epsilon_{sol}}{\epsilon_{sur}} = 11 \tag{81}$$

for the present example of the quartz-solution interface. Thus from Eq. (81) $\epsilon_{sur} = \epsilon_{sol}/11 = 7$, which is in good agreement with the known dielectric constant of water molecules in the Helmholtz region. Hence the above deviation of the activity coefficients of H^+ and OH^- in the interface region from that of the bulk solution may be attributed to a parallel deviation of the dielectric constant in the two regions as a result of a rearrangement of the adsorbed-water dipoles at the interface. The equivalent potential drop associated with these changes in the dielectric constant at the interface relative to the bulk solution can be identified as the χ potential.

The foregoing conclusions concerning the zpc of oxides and the variation in the dielectric constant at the interface can also be arrived at from the theory of the relative strengths of acids and bases [240, 241]. The departure of the zpc of oxides from pH 7 to a lower or a higher pH may be explained from the difference in the acid- and base-dissociation constants of the surface OH groups of oxides. The dependence of the acid- and base-dissociation constants on the dielectric constant of the medium is then given more precisely by the Born equation [241], which also takes into account the ionic sizes of the species involved.

It is normally difficult to distinguish between the χ potential and the Volta potential difference ψ^0 of the surface relative to the bulk

solution. Whereas ψ^0 in the present work arises from the ionization of
the surface OH groups, the χ potential is due to the dipole orientation of the adsorbed water on the surface. From an analysis of the
surface-charge densities and C versus pH plots of the oxide-
solution interface, a distinction has been made between the pH at
which $\psi^0 = 0$ and the zpc at which both ψ^0 and χ are zero; that is,
$\varphi^0 = (\psi^0 + \chi) = 0$ [146]. Let the pH at which $\psi^0 = 0$ but at which $\chi \neq 0$
be denoted as $pH_{\psi^0=0}$. Then the potential difference between $pH_{\psi^0=0}$
and the pH_{zpc} may be attributed to the χ potential and is given by

$$V_\chi = -\frac{2.3\,RT}{F}\,(pH_{\psi^0=0} - pH_{zpc})\,. \tag{82}$$

For a stoichiometric oxide, such as quartz, $pH_{\psi^0=0}$ has been identified as pH 7 [146] (see Section VI. I. 4). The variation in the dielectric constant at the quartz-solution interface, ϵ_{sur}, as estimated
approximately from Eqs. (80) and (81), for dilute solutions between
pH 7 and the zpc (pH 1.5) is shown in Fig. 37. Both ψ^0 and the χ
potential, and hence the zpc, will be determined by the stoichiometry

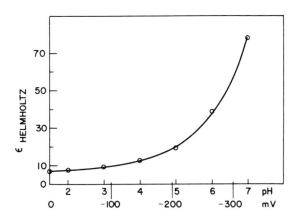

Fig. 37. Variation in the dielectric constant of the Helmholtz
region of the quartz-solution interface between pH 1.5 and 7 [242].

of the oxide surface. The ζ potential is zero at a pH where $\varphi^0 = 0$, but not at $pH_{\psi^0=0}$, because of the presence of χ potentials. In order to compensate the χ potentials, a potential V_χ, as given by Eq. (82), has to be applied. The dipoles of the surface groups are then rearranged until the dielectric constant of the Helmholtz region reaches a saturation limit. Alternatively, as the pH of the solution is increased from 1.5 to 7, the corresponding potential drop ($-V_\chi$) is used in the reorientation of the surface groups until the acidic dissociation of the surface groups begins to occur at pH 7 (see Figs. 1 and 4 in Ref. [203]).

The same conclusions concerning the interface structure may be drawn from Eq. (3), which for small Galvani potentials may be written as

$$\varphi_{sol} = \varphi_{sd}\left(\frac{L_{sol}}{L_{sd}}\right)\left(\frac{\epsilon_{sd}}{\epsilon_{sol}}\right).$$

When $\psi^0 = 0$, the $\varphi_{sol} = V_\chi$; therefore, from Eq. (82),

$$-V_\chi = \frac{2.3\,RT}{F}\,(pH_{\psi^0=0} - pH_{zpc}) = \varphi_{sd}\left(\frac{L_{sol}}{L_{sd}}\right)\frac{\epsilon_{sd}}{\epsilon_{sol}}. \qquad (83)$$

For a given composition of oxide and solution, all the terms in Eq. (83) are constant except pH_{zpc} and ϵ_{sol}. Hence in all polarization studies of the oxide-solution interface the Galvani potential φ_{sol} will continue to change even after reaching a stage at which $\psi^0 = 0$ because of changes in the dielectric constant of the Helmholtz region until a saturation limit is reached.

In the case of amphoteric oxides both χ_+ and χ_- potentials would have to be considered.

G. Variation in the zpc with the Bulk and
Surface Properties of Oxides

The zpc of oxides is found to depend on several macroscopic properties of the oxide phase, such as the following:

1. The stoichiometry (of the surface in particular) and perhaps the semiconducting properties of oxides.

2. The crystal structure.

3. The degree of surface hydration, which also depends on the heat treatment if any.

4. The difference in the acidic and basic dissociation constants of the surface hydroxyl groups.

The acid- and base-dissociation constants and the zpc of oxyacids are related to the electrostatic free energy required for proton transfer in the dissociation process. Therefore, by using the Born equation [241], the zpc can also be related to the dielectric constant of the aqueous phase and to such microscopic properties of oxides (or oxyacids) as ionic size, charge, and the coordination number. This approach for calculating the zpc is valid strictly for oxyacids in solutions, whereas in considering the zpc of oxide surfaces the contribution of the solid-state properties should also be considered. Thus, in a modified approach, Parks [231] has taken into account the crystal-field-stabilization energy (CFSE) for each transition-metal ion and also the effect of varying the coordination number of cations in the oxide phase. The general equation derived by Parks is of the form

$$zpc = A_{eff} - B[Z/R + 0.0029\,(CFSE) + a] , \tag{84}$$

where Z is the ionic charge of the various species (H^+, O^{2-}, and cations) involved, $R = 2r_O + r_+$, r_O and r_+ being the radii of O^{2-}

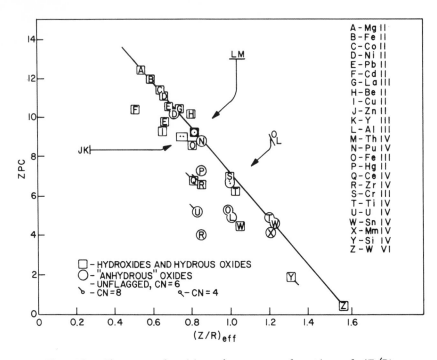

Fig. 38. The zpc of oxides shown as a function of $(Z/R)_{eff} =$ [Z/R + 0.0029 (CFSE) + a]. Reprinted from Ref. [231], p. 191, by courtesy of the American Chemical Society.

(1.4 Å) and of the cation, respectively. According to Eq. (84), the zpc is a linear function of Z/R. The values of constants A (18.6 for coordination number 6) and B (11.5) were obtained from a linear plot of Z/R against the known zpc of hydrous Al_2O_3 and hydrous MgO. The constant A is related to the ratio of the nonhydroxyl oxygens to the hydroxyl oxygens of oxides; that is, A represents the degree of hydration of the oxide. However, the degree of hydration of oxides can vary drastically with their heat treatment, and hence the hydration effect is difficult to estimate. Thus the comparison of the predicted values of zpc from Eq. (84) with the experimental values was restricted to oxides in two forms, those fully hydrous and those fully anhydrous (e. g. , ZrO_2). Suitable values of constants A and a have been evaluated tentatively for the hydrous and anhydrous oxides for coordination numbers 4, 6, and 8 in order to obtain agreement between the calculated and the experimental values. The results of these comparative studies of the zpc for a series of oxides are shown in Fig. 38. This work is discussed in greater detail by Parks [231], who has also listed values of the zpc (and isoelectric point), obtained by different experimental methods, for many oxides.

In another approach, Healy and Fuerstenau [232] have correlated the zpc of some oxides with their heats of immersion, which are related to the electrostatic field strength of oxides. In a subsequent paper [233] Fuerstenau et al. have also shown that the zpc of MnO_2 in different structural forms varies linearly with the calculated values of their electrostatic field strengths. In calculating the field strength corrections had to be made for differences in the unit-cell volumes of the different forms of MnO_2 .

In all these attempts to calculate the zpc from fundamental physical parameters a major uncertainty arises from the surface not being representative of the bulk in stoichiometry, structure, and hydration. The condition of the oxide surface mainly depends on the history of the individual oxide. On this basis Eq. (84) can at best be regarded as empirical in nature.

There are other variables that can affect the zpc. The ratio of the cationic to oxygen site densities usually differs from one crystal face to another, and hence the zpc is also expected to vary accordingly. The pzc of germanium (with a hydroxide surface) is found to be different for each of the (111), (110), and (100) planes [15, 128], and this variation in the pzc may probably be attributed to the difference in the zpc of the different crystallographic planes, which have different oxygen-to-metal ratios. Although a space charge may originate in a semiconductor oxide by electron or hole transfer between the surface states and the oxide bulk, the effect of the electronic space charge and of the Fermi level on the zpc of oxides, at room temperature, has already been shown to be negligible (Section III. B. 6) in comparison with the effects so far considered. At elevated temperatures the effective concentrations of the surface cationic and anionic charges in semiconductor oxides can be altered by increasing the electron or hole transfer between the bulk and the surface region. In such a case the zpc of the oxide may also be slightly affected [243]. The presence of a space charge at the oxide surface may, however, influence the degree of ion adsorption on oxides from the aqueous phase (e. g., chemisorption of protons) and may also affect the χ potentials by modifying the polarizability of the surface groups. Thus, as already mentioned, although the zpc of both n- and p-type germanium with an oxide surface is identical ($\sim$ pH 2), the ζ potential of p-type germanium was found to be much less than that of n-type germanium [147]. However, it is also possible that the higher potential of n-type germanium was due to contributions from the higher solid-state surface conductivity of n-type germanium.

A shift in the zpc to a lower pH has been observed after heat treatment of Al_2O_3 [236, 237] and TiO_2 [212, 235] and has been attributed to the partial dehydration of the oxide surface. However, it is not certain how far the semiconducting properties of oxides in general (particularly of TiO_2) are responsible for such changes in the zpc. The effect of temperature on the zpc of the rutile-solution interface was also studied by Bérubé and De Bruyn [213] from potentiometric titrations carried out at various temperatures. The zpc of rutile was found to shift from pH 6.0 to 5.35 on increasing the temperature from 25 to 95°C. The temperature coefficient of the change in the zpc of rutile (0.65 pH unit/70°C) was somewhat less than the temperature coefficient of the $1/2pK_w$ (0.88/70°C) of water. This difference in the temperature coefficients has been attributed to a decrease in the relative affinity of the H^+ and OH^- ions for rutile surfaces with increasing temperature. A standard entropy change of 2 cal/degree-mole has been calculated from a thermodynamic analysis of the experimental data, compared with a corresponding value of 42 cal/degree-mole for the AgI system [244]. The effect of temperature on the solid-state properties of TiO_2 and on the dissociation constants of the soluble aquo-complexes of titanium have not been considered in the foregoing work. Certain discrepancies in the above treatment have been pointed out by Ball [214].

In a similar investigation [243] the decrease in the zpc of Al_2O_3 (pH 9.06) and in the $1/2pK_w$ on increasing the temperature from 30 to 90°C was found to have the same value (0.7 unit). The value of (zpc - $1/2pK_w$) was independent of temperature and equal to 2.1. Hence the change in the zpc of Al_2O_3 with temperature was entirely due to a corresponding change in the neutral point of water. However, in the case of magnetite the variation in the zpc with temperature was considerably different from that of the $1/2pK_w$ with temperature, the value of ($1/2pK_w$ - zpc) increasing from 0.45 at 25°C to 0.8 at 90°C. It is not certain to what extent the solid-state properties of magnetite were responsible for this shift in the zpc with temperature. The behavior of n-type magnetite in the pH range below the zpc (cf. Fig. 52) has also been reported [206] to be inconsistent with many other oxides.

Specific adsorption of counterions on oxides can also shift the zpc of oxides and will be discussed together with the experimental results.

H. Experimental Methods of Studying the Primary Oxide-Solution Equilibrium

It is seen from Eqs. (61) that addition of an oxide to a solution in the region of acidic dissociation (pH > zpc) will decrease the pH

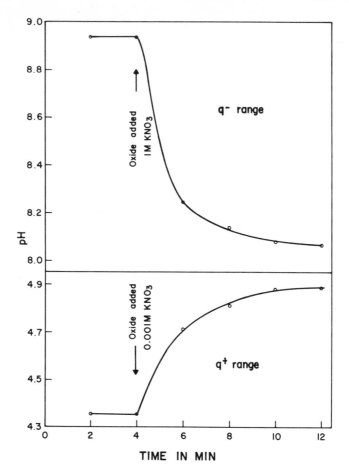

Fig. 39. Variation in pH with time before and after adding about 3g of magnetite (surface area 790 cm^2/g) in 10 ml of 0.001 and 1.0 M KNO$_3$ solutions. Reprinted from Ref. [206], p. 3842, by courtesy of the National Research Council of Canada.

of the bulk solution (Fig. 39), and the oxide will acquire a negative surface charge. In the region of basic dissociation, however, the pH of the bulk solution will increase (Fig. 39) on adding an amphoteric oxide, and the surface will acquire a positive surface charge. It is the magnitude of these changes in the pH of the bulk solution, at a constant ionic strength, that are measured in potentiometric studies of the primary oxide-solution equilibrium in order to obtain information on the properties of the double layer on oxides. The experimental procedure described here is based on the work of Ahmed et al. [146, 202, 203, 206].

The surface-charge density of any ionic oxide is equivalent to a surface concentration of about 1×10^{-9} mole of monovalent ions per square centimeter. For an oxide sample with a total surface area of about 2500 $cm^2/25$ ml of solution there is an equivalent of about 2.5×10^{-6} mole of univalent ions available to react with 25 ml of solution. If the total amount of material dissolved in solution is kept below 10^{-6} mole/l (2.5×10^{-8} mole/25 ml), the secondary effects arising from the solubility of the material will be less than 1% of the total pH variation observed during the primary oxide-solution equilibrium. In order to satisfy this requirement, coarse and crystalline oxide samples of low surface areas were used. The total surface area used in the experiments was varied depending on the solubility of the material and the pH change obtained in the reaction. Also each individual value of the surface-charge density was obtained by using a fresh sample of oxide that was in contact with the solution for not more than 10 min. In an improved, batch-wise method [242] the variation in pH with time, before and after adding the oxide to the solution, was recorded to a high precision by using a Beckman Research pH meter and a strip-chart recorder (Mosely 7100 BM). Because of the null-balancing facilities available in the Beckmann Research pH meter, the off-balance potential can be "bucked" by using the potentiometer, and the entire range of pH (3 to 11) can be accurately recorded in units of 0.25 to 0.5 pH, full scale. In spite of the various uncertainties in the absolute values of pH [245] the relative accuracy in these pH measurements is significant. A water-jacketed Metrohm titration flask of 25-ml capacity, as used in these studies of the oxide-solution equilibrium, is shown in Fig. 40. Further details concerning

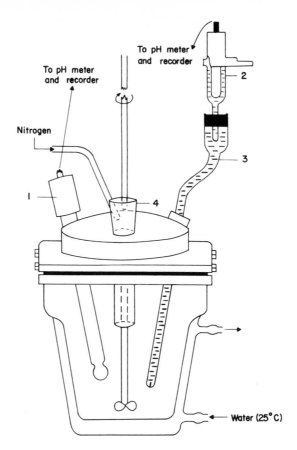

Fig. 40. Apparatus for potentiometric studies of the primary oxide-solution equilibrium: 1, glass electrode; 2, calomel electrode; 3, the solution bridge; 4, glass tube for adding oxide samples to the electrolyte solution [242].

the preparation of material, calibration of electrodes with liquid junctions, and determination of the experimental activity coefficients $(f_{\pm})_{exp}$ of H^+ and OH^- in different electrolyte solutions of different concentrations have been described elsewhere [146, 203].

Figure 41 shows the experimentally recorded pH variations before and after the addition of quartz to a 0.1 M KNO_3 solution. Similar data that were obtained manually in previous studies of the magnetite-

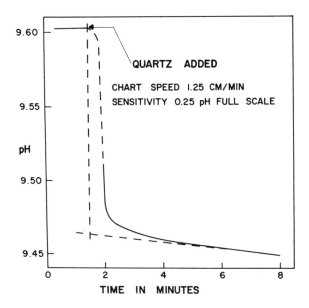

Fig. 41. Variation in pH with time before and after adding about 4g of quartz (surface area 429 cm^2/g) in 20 ml of 0.1 M KNO$_3$ solution (recorded data) [242].

solution interface are shown in Fig. 39. It is seen from these typical measurements that almost all of the pH change, corresponding to the primary oxide–solution equilibrium, occurred during the first few minutes after the addition of the oxide to the solution. Under these experimental conditions no dissolved metal ions of the oxide could be detected by standard colorimetric methods (sensitivity 1 micromole/l). However, in the case of iron oxides, considerable concentrations of dissolved iron could be detected [206] after stirring the mixture for 1 h at pH 5 to 6. The slow changes in pH with time that are apparent after about 6 min following the addition of the oxide to the solution (Figs. 39 and 41) were attributed to the secondary reactions of type 2 mentioned in Section VI. B. Corrections for these side effects were made by extrapolating the almost linear portion of the pH–versus–time plot as shown in Fig. 41. The initial and final pH values (corrected) in the experiment were converted into the corresponding a_{H+} (for pH < 7) or a_{OH-} (for pH > 7) terms. The activities were then converted into the initial and final concentration terms (C_{H+} or C_{OH-}) by multiplying with the corresponding $(1/f_{\pm})_{exp}$ values. From the resulting data one can calculate ΔC_{H+} or ΔC_{OH-} in the 25 ml of solution used and hence the charge densities $q^{\pm}$ per square centimeter of oxide. The surface areas were measured [246] by the krypton-gas-adsorption method.

I. Experimental Results and Discussion

1. The Zero Point of Charge

The zpc of a large number of oxides, as measured by various methods, has been listed by Parks [231]. In Table 1 the zpc of additional oxides has been listed, including those oxides in particular for which a double-layer analysis has been attempted.

The dependence of zpc on such oxide properties as structure, ionic size and the nature of metal coordination, heat treatment and the degree of hydration, stoichiometry, and semiconducting properties has already been discussed. In most cases variations in the zpc of oxides, as also seen in Table 1, can be explained on the basis of the differences in one or more of these properties. In addition, the zpc of oxide precipitates is also reported [209, 210, 249] to depend on the method of preparation and the impurity inclusions in the precipitate. Thus an observed variation in the experimental zpc of ZnO from a pH of less than 8 to pH 10 was attributed to the incorporation of foreign ions, particularly Cl^- and NO_3^-, in the precipitate [209, 210].

It is further seen in Table 1 (columns 6 and 7) that the zpc of specular hematite, as determined by the streaming-potential method, is in close agreement with the zpc obtained from studies (method A) of the primary oxide-solution equilibrium. However, the zpc of the same sample of hematite, but in a finely ground form (surface area 26.4 m^2), was measured by Smith and Salman [247] by the prolonged acid-base titration method (method B_f) and found to be the same as the isoelectric point ($\sim$ pH 8.3 to 8.5) of the dissolved iron complexes. The zpc of precipitated iron oxide [191, 205, 224] (samples 10 through 12, Table 1), as measured by the B_f or B_s methods (in solutions saturated with metal complexes), is also seen to be identical with the isoelectric point of the dissolved iron complexes. Hence from these studies it appears that the zpc of hydrated iron oxide is the same as the isoelectric point of the iron hydroxy complexes. However, it is also possible that the zpc of the solid phase in these measurements, made by method B, is masked by the isoelectric point of soluble $Fe(OH)_3$, whose solubility is appreciable ($\sim 10^{-6}$ to 10^{-7} mole/l). Depending on the experimental conditions, the zpc of oxides may also vary due to adsorption of dissolved metal complexes. Joy and Watson (samples 8 and 9, Table 1) also found marked differences in the zpc of different forms of hematite, the zpc also depending on the particle size and on the chemical method of pretreatment [249]. Wide variations have also been found by O'Connor and co-workers [236] and Fuerstenau et al. [237] in the zpc of different forms of Al_2O_3, depending on the crystalline form, stoichiometry, heat treatment, and surface hydration. The zpc of uranium oxides was also shown to depend on the oxide stoichiometry [234].

On increasing the KNO_3 concentration from 0.001 to 1 M in a blank experiment, the half-cell potential was found to shift to a lower pH by 0.3 ± 0.05 pH unit near the zpc, although this effect was negligible around pH 4 [202, 206]. In order to compensate for this shift in the half-cell potential, a corresponding shift in the zpc of oxides to a higher pH is expected. Such a shift in the zpc of Fe_2O_3 and Al_2O_3 may be seen in Table 1 (samples 5a and 18). However, similar shifts in the zpc of ZrO_2 and ThO_2 (samples 2 and 3) to a

Table 1

The Zero Point of Charge (zpc) of Oxides

Sample No.	Oxide	Conditions and surface area g^{-1}	Electrolyte concentration (mole/l)	pH iep$^{\underline{a}}$	Mechanism suggested$^{\underline{b}}$	pH zpc	Method$^{\underline{c}}$	Refs.
1	SiO$_2$	Quartz, 429 cm²	1.0, KNO$_3$		D	< 3.6	A, ζ_{str}	[203, 228, 236]
2	ZrO$_2$	Baddeleyite, 1438 cm²	0.001, KNO$_3$ 1.00, KNO$_3$	5.2 [247]	B	5.5±0.05 6.2±0.05	A A	[203]
3	ThO$_2$	Thorianite, 192 cm²	0.001, KNO$_3$ 1.0, KNO$_3$		B	5.9±0.05 6.8±0.05	A	[203]
4	ThO$_2$	Fired ~1600°C Fired ≤1400°C				6.8±0.05 9.5	ζ_{str} ζ_{str}	[248]
5a	$\underline{d}$Fe$_2$O$_3$	Specular hematite I$^{\underline{e}}$, 310–450 cm²	0.001, KNO$_3$ 0.10, KNO$_3$ 1.0, KNO$_3$	8.5 [191]	A or C	5.3±0.05 5.4±0.05 5.7±0.1	A	[206]
5b	$\underline{d}$Fe$_2$O$_3$	Specular hematite I$^{\underline{e}}$, 310–450 cm²	KNO$_3$			~ 6.0	ζ_{str}	[238]
6	$\underline{d}$Fe$_2$O$_3$	Specular hematite I$^{\underline{e}}$, 26.4 m²	0.001–1.0, KCl			8.7±0.1	B$_s$	[247]
7	Fe$_2$O$_3$	Specular hematite II				5.8 5.3	B$^{\underline{f}}$ ζ_{str}	[249, 250] [249, 250]
8	Fe$_2$O$_3$	Red fines (all hematites, ground)				~ 8.0	ζ_e	[249]

No.	Solid	Other forms	Electrolyte	IEP[a]	Method	pH	Method	Ref
9	Fe_2O_3					3.4–6.7	ζ_{str}	[249]
10	Fe_2O_3	Precipitate, 60 m²	NO_3^-	8.5 [191]		8.5	B_s	[191]
11	Fe_2O_3	Precipitate, 21 m²	ClO_4^-	8.3 [205]		8.3	B_f	[205]
12	Fe_2O_3	Precipitate, 44.6 m²	0.002–1.0, KCl			8.5–9.3	B_f	[224]
13	Fe_3O_4	Magnetite,$\underline{}^e$ 800 cm²	0.001, KNO_3		C	6.4±0.1; ~6.0	A; ζ_{str}	[206]
14	SnO_2	Cassiterite, 420 cm²	0.001, KNO_3; 0.1–1.0, KNO_3		A; C	5.5±0.1; 5.4±0.1	A; A	[202]
15	TiO_2	Rutile, 2300 cm²	0.001, KNO_3; 0.1, KNO_3; 1.0, KNO_3		A	5.3±0.05; 5.0±0.05; 4.8±0.1	A; A; A	[202]
16	TiO_2	Rutile, anatase, 43 m²	ClO_4^-, Cl^-, I^-, NO_3^-			5.9±0.1	B_f	[212]
		Rutile, anatase, heated >425°C	ClO_4^-, Cl^-, I^-, NO_3^-			4.0	B_f	[212]
17	TiO_2	Rutile (q too high)	0.001–0.1, KCl	6.9 [251]		7.1	B_s	[251]
18	α-Al_2O_3	Calcined, 4414 cm²	0.001, KNO_3; 0.1, KNO_3; 1.0, KNO_3	7.7 [252]	A	4.7±0.1; 4.8±0.1; 5.0±0.1	A	[146, 218]
19	$\underline{}^g\alpha$-Al_2O_3	15 m²	KCl	7.7 [252]		9.1±0.1	B_s, ζ_e	[253]
20	ZnO	0.5–13.1 m²	NO_3^-, ClO_4^-, Cl^-, Br^-, I^-	9.7		<8–10.0	B_f	[210]

[a] Isoelectric point. [b] See Sections VI.B, VI.C, and VI.D.2. [c] The experimental methods are designated as follows: A, studies of the primary oxide-solution equilibrium by pH

[continuation of footnotes to Table 1]

measurements (see Section VI. H); B_g and B_f, slow and fast continuous acid-base titrations (see Section VI. C); ζ_{str} and ζ_e, ζ-potential measurements by streaming-potential and electrophoretic methods, respectively.

$\underline{d}$Specular hematite from the same source, but different techniques used.

$\underline{e}$An n-type semiconductor.

$\underline{f}$Not specified whether slow or fast continuous acid-base titration.

$\underline{g}$The zpc of Al_2O_3 depends mainly on previous heat treatment and the surface stoichiometry of the oxide and is found to vary [236, 237, 253] from pH 2.2 to 9.1.

higher pH (exceeding 0.3 pH unit) and in the zpc of SnO_2 and TiO_2 (samples 14 and 15) in KNO_3 solutions to a lower pH indicate specific adsorption of NO_3^- on ZrO_2 and ThO_2, and of K^+ on SnO_2 and TiO_2. Such shifts in the experimental values of the zpc of oxides to lower (or higher) pH depending on the specific adsorption of cations (or anions) may be explained on the basis that it requires higher H^+ (or OH^-) concentrations to displace the adsorbed cations (or anions) in the potentiometric titrations. The zpc of solid surfaces is also known to shift toward a more positive or a more negative potential when cations or anions, respectively, are specifically adsorbed on the surfaces. This direction of the shift in the zpc of oxides is the reverse of that postulated by Parks [254] and by Blok and De Bruyn [210].

2. Positive Surface Charge at the Oxide-Solution Interface

The behavior of all oxide surfaces in the cathodic region (pH > zpc) is qualitatively similar. However, depending on the nature of the anion used, the variation in q^+ with the electrolyte concentration at a given pH shows wide differences between one oxide and another in the anodic region (pH < zpc). Different mechanisms that were proposed to account for the variation in q^+ with pH

and with electrolyte concentration have already been presented in Section
VI. D. The behavior of Al_2O_3 (Fig. 42) and TiO_2 (Fig. 44) in NO_3^-, Cl^-,
and ClO_4^- solutions, and of SnO_2 (Fig. 45) in NO_3^- and ClO_4^- solutions,
as studied by method A, indicates mechanism A (nonspecific
adsorption of anions), whereas the behavior of quartz surfaces
(Fig. 55) follows mechanism D. Experimental data that follow mechanisms other than A and indicate specific adsorption of anions are
summarized in Table 2. Column 2 in Table 2 also lists the figures
in which the experimental data have been presented. These figures
are self-explanatory when studied with the help of the data in Table 2
together with the general mechanisms (A to D) already discussed.
Further details about the oxides and their zpc may be found in Table 1.
The experimental data obtained from the slow and prolonged titrations
of oxide surfaces (method B_s) and also from the appreciably soluble
oxides, such as ZnO, are difficult to interpret. Actually more than
one mechanism probably occur simultaneously in such cases, involving solubility effects and ion-exchange [209, 210, 255] processes, and
these cases have been indicated by "E" in column 3 of Table 2. Some
apparently anomalous cases are marked by an asterisk and are discussed here.

The variation in q^+ with electrolyte concentration, as obtained
from method A, for naturally occurring rutile (Fig. 44) and specular
hematite (Figs. 50 and 51A) (n-type semiconductor) indicates mechanism A, that is, nonspecific adsorption of anions, with
$(\partial q^+/\partial \mu_{KNO_3})_{pH} \approx 1$ to $2\,\mu C/cm^2$ for a double-layer potential of about
100 mV. However, by using method B, substantial increments in q^+
values with increasing KNO_3 concentrations, at a given pH, have
been reported for precipitated forms of TiO_2 (Fig. 48) and Fe_2O_3 (Fig. 54)
in ClO_4^- solutions, thus indicating mechanism B. These variations in
the behavior of oxides in different physical forms may be due to differences in their bulk and surface properties. However, the large

Table 2

Mechanism of Anion Adsorption on Oxides

Sample No.[a]	Figure No.	Mechanism[b]	Oxide[c]	Anions	zpc pH	Method[d]	Refs.
19	43	A (or E)	Al_2O_3	Cl^-	9.1	B_s	[253]
14	45	C	SnO_2	Cl^-	–	A	[202]
2	46	B	ZrO_2	NO_3^-, Cl^-, ClO_4^-	5.5–6.2	A	[146, 203]
3	47	B	ThO_2	NO_3^-, Cl^-, ClO_4^-	5.9–6.8	A	[146, 203]
16	48	B(E)	e*TiO_2	NO_3^- (ClO_4^-, Cl^-, I^-)	5.9 0.1	B_f	[213]
20	49	B(E)	ZnO	Cl^- (NO_3^-, Br^-, I^-)	8.7	B_f	[210]
5a	50, 51A	A + C ?	*Fe_2O_3, specular hematite[f]	NO_3^-, ClO_4^-	5.3–5.7	A	[206, 218]
5a	51B	B	*Fe_2O_3, specular hematite[f]	Cl^-	5.3	A	[206, 218]
13	52		Fe_3O_4, magnetic[f]	NO_3^-	6.4	A	[206]
12	53	B	e-Fe_2O_3	Cl^-	9.3	B_f	[224]
11	54	B(E)	e*Fe_2O_3	ClO_4^-	8.3	B_f	[205]

[a] See Table 1.

[b] See Sections VI.B, VI.C, and VI.D.2. The letter "E" denotes cases in which more than one mechanism

[continuation of footnotes to Table 2]

may occur, involving solubility effects and ion-exchange processes.

$\underline{c}$Apparently anomalous cases are marked by an asterisk and are discussed in text.

$\underline{d}$Experimental methods: A, studies of the primary oxide-solution equilibrium by pH measurements; B_S and B_f, slow and fast continuous acid-base titrations.

$\underline{e}$Precipitate. $\underline{f}$An n-type semiconductor.

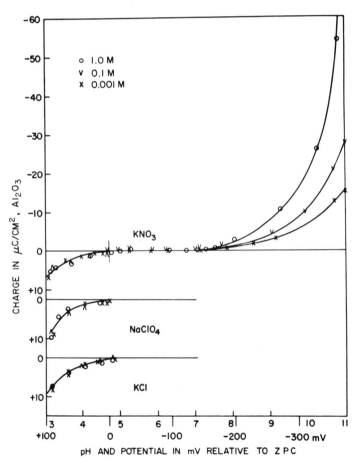

Fig. 42. Variation in the surface-charge density $q^{\pm}$ of Al_2O_3 with final pH or the potential difference relative to the zpc in KNO_3, $NaClO_4$, and KCl solutions at 25°C (sample 18 in Table 1). Reprinted from Ref. [146], p. 3549, by courtesy of the American Chemical Society.

444   S. M. AHMED

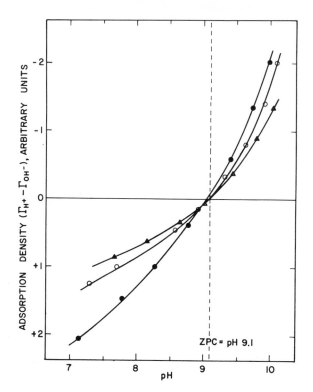

Fig. 43. Variation in the adsorption density of α-alumina obtained by slow acid-base titrations (method B_S) with pH and ionic
strength in potassium chloride solutions at 25°C (sample 19, Tables
1 and 2). Γ_{H+} and Γ_{OH-} are the surface excess of H+ and OH− in
moles per square centimeter. Reprinted from Ref. [253], p. 66, by
courtesy of Academic Press, Inc.

increments in the q^+ values of precipitated oxides (Figs. 48 and 54)
with KNO3 concentrations could perhaps also arise from several secondary effects, including an ion-exchange process (mechanism E).
These secondary reactions become predominant in the continuous-
titration method, which lasts more than 1 h even for a fast titration
(method B_f).

De Bruyn et al. [255] have proposed an ion-exchange process to
account for the long-term slow changes in pH in the oxide-solution

system. The pH changes in the slow ZnO–solution equilibria (sample 20, Tables 1 and 2) have also been interpreted [209] in terms of an ion-exchange process involving OH⁻ and the anions originally incorporated into the oxide during its preparation. Thus a wide variation

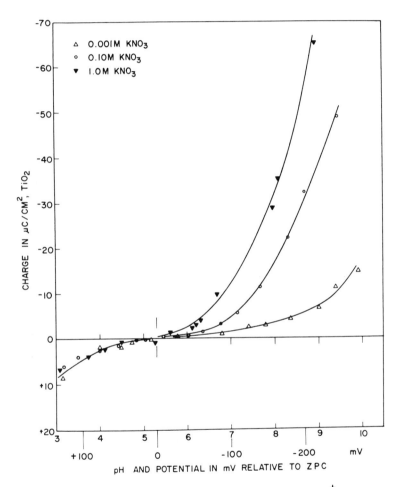

Fig. 44. Variation in the surface-charge density $q^{\pm}$ of TiO_2 with final pH or the potential difference relative to the zpc in KNO_3 solutions at 25°C. The $q^{\pm}$ values of TiO_2 in KCl solutions were almost the same as for KNO_3 solutions (sample 15, Table 1). Reprinted from Ref. [202], p. 101, by courtesy of Academic Press, Inc.

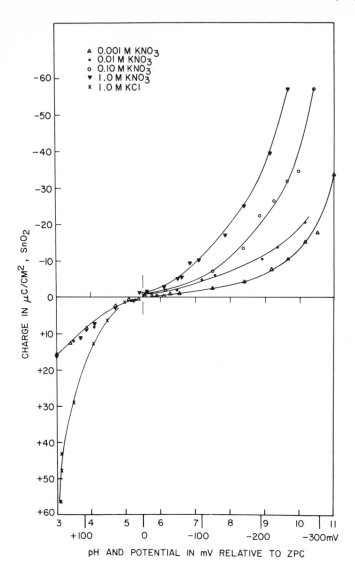

Fig. 45. Variation in the surface–charge density $q^{\pm}$ of SnO_2 with final pH or the potential difference relative to the zpc in KNO_3 or KCl solutions at 25°C (sample 14, Tables 1 and 2). Reprinted from Ref. [202], p. 100, by courtesy of Academic Press, Inc.

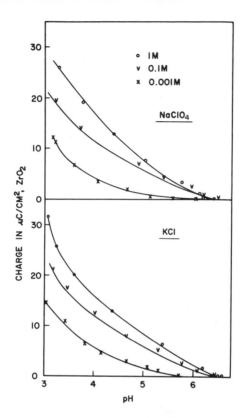

Fig. 46. Variation in the surface-charge density q^+ of ZrO_2 with final pH in $NaClO_4$ and KCl solutions at 25°C (sample 2 in Tables 1 and 2). Reprinted from Ref. [146], p. 3550, by courtesy of the American Chemical Society.

in the zpc of ZnO (pH less than 8 to 10 in NO_3^-) depending on the method of preparation has been observed and attributed to the incorporation of foreign ions (NO_3^- and Cl^-) in the precipitate [209, 210].

   In the case of ZnO, even with the use of a fast-titration method (B_f), substantial corrections had to be made [210] for the oxide solubility in order to obtain the surface-charge densities. Because of the experimental limitations due to solubility effects, charge-density data for ZnO could be obtained only in the pH range 8.5 to 10.

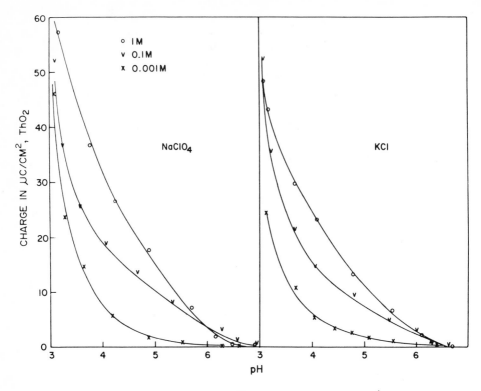

Fig. 47. Variation in the surface-charge density q^+ of ThO_2 with final pH in $NaClO_4$ and KCl solutions at 25°C (sample 3, Tables 1 and 2). Reprinted from Ref. [146], p. 3551, by courtesy of the American Chemical Society.

From an analysis of the differential-capacitance values of TiO_2 [213] and ZnO [211], obtained from the differentiation of the q versus pH plots, the order of specific interaction of anions with the oxides is given as $Cl^- \simeq ClO_4^- \simeq NO_3^- > I^-$ for TiO_2 and $Cl^- > Br^- > I^- > NO_3^- \gtrsim ClO_4^-$ for ZnO. These interactions of anions with TiO_2 and ZnO surfaces will be discussed further together with their C values in Section VI. I. 5.

Healy and Jellett [215] found only a minimum in the electrophoretic mobility of ZnO suspensions (with a negative surface charge) in KCl solutions of pH 8.0 to 8.5, but no reversal of charge on ZnO has

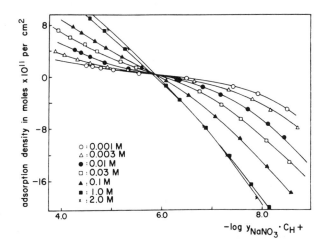

Fig. 48. Variation in the adsorption density of TiO$_2$ obtained by fast acid–base titrations (method B$_f$) in NaNO$_3$ solution; γ_{NaNO_3} is the mean activity coefficient (sample 16, Tables 1 and 2). Reprinted from Ref. [213], p. 93, by courtesy of Academic Press, Inc.

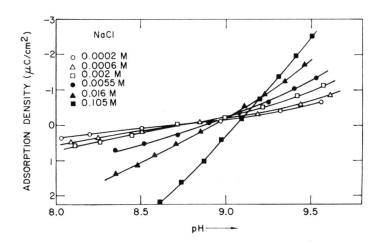

Fig. 49. Variation in the adsorption density of ZnO with pH, obtained by a fast acid–base titration (method B$_f$) in NaCl solution (sample 20, Tables 1 and 2). Reprinted from Ref. [210], p. 524, by courtesy of Academic Press, Inc.

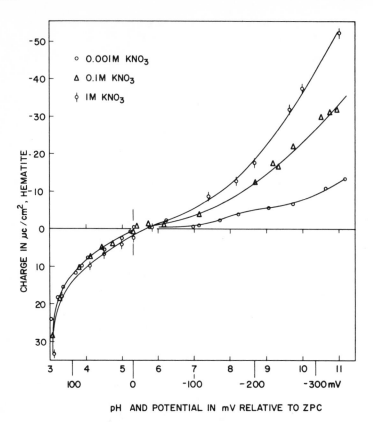

Fig. 50. Variation in the surface-charge $q^{\pm}$ of specular hematite with final pH or the potential relative to the zpc, in KNO_3 solution at 25°C (sample 5a, Tables 1 and 2). Reprinted from Ref. [206], p. 3843, by courtesy of the National Research Council of Canada.

been reported in the pH range investigated by these authors. Healy and Jellett have further pointed out that Zn^{2+} ions derived from the dissolution of ZnO are hydrolyzed, and the hydrolyzed species are readsorbed on the oxide, giving rise to a secondary ZnO–solution interface. The hydrolyzed species of Zn^{2+} resulting from the solubility of ZnO were said to interfere with the oxide surface attaining a zero charge.

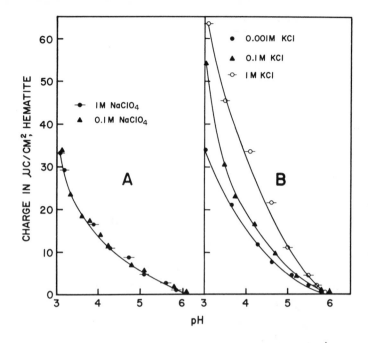

Fig. 51. Variation in the surface-charge density q^+ of specular hematite with final pH in (A) $NaClO_4$ solutions, (B) KCl solutions at 25°C (sample 5a, Tables 1 and 2). Reprinted from Ref. [206], p. 3843, by courtesy of the National Research Council of Canada.

In another recent study [256] by an electrophoretic method a charge reversal of ZnO suspensions in KNO_3 solution was found to occur at pH ~ 8, and the zpc, as found by method A, occurred at pH ~ 8.5. This result is in close agreement with the zpc of ZnO obtained by Blok and De Bruyn [210] from fast acid-base titrations.

The behavior of semiconductor and magnetic oxides in the q^+ region was found to be exceptional in comparison with that of many other oxides [206]. Thus, although the q^+ values of specular hematite (n-type semiconductor, Figs. 50 and 51A) in NO_3^- and ClO_4^- solutions were almost independent of the anion concentrations, the q^+ values are seen to be much higher (~ 30 $\mu C/cm^2$ at 130 mV relative to

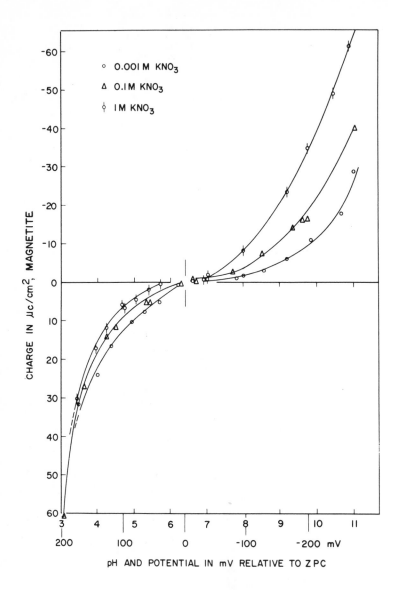

Fig. 52. Variation in the surface-charge density $q^{\pm}$ of magnetite with final pH in KNO_3 solutions at 25°C (sample 13, Tables 1 and 2). Reprinted from Ref. [206], p. 3843, by courtesy of the National Research Council of Canada.

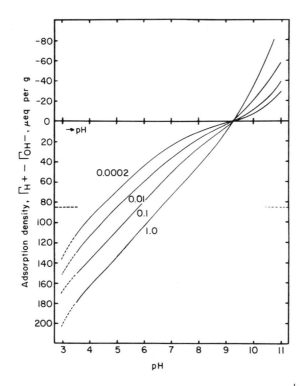

Fig. 53. Variation in the adsorption density of H^+ and OH^- on α-Fe_2O_3 (hematite) precipitate obtained by fast acid-base titrations (method B_f) in KCl solutions at $19.5\pm0.6°C$. The BET surface area $= 44.6$ m^2/g (sample 12, Tables 1 and 2). Reprinted from Ref. [224], p. 553, by courtesy of the American Chemical Society.

zpc) than the q^+ values of other oxides. These high q^+ values indicate an excessive transfer of protons from the acid solutions to the solid phase (mechanism A) and also a much closer packing of anions (NO_3^- and ClO_3^-) on Fe_2O_3 than for other oxides. Adsorption on, and penetration of protons into, the oxide films of many valve metals have also been established [74, 154-157]. The behavior of magnetite [206] (Fig. 52) in the q^+ region was also found to be inconsistent with the general behavior of the other oxides.

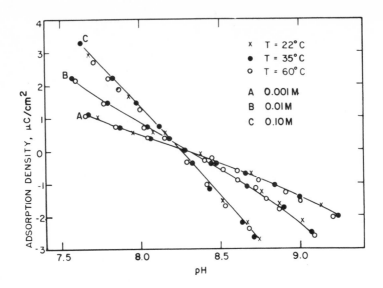

Fig. 54.   Variation in the surface-charge density of α-Fe$_2$O$_3$ (hematite) precipitate in NaClO$_4$ solutions obtained by fast acid-base titrations (method B$_f$) at various temperatures (sample 11, Tables 1 and 2).   Reprinted from Ref. [205], p. 56, by courtesy of the North-Holland Publishing Company.

3. Negative-Surface-Charge Densities of Oxides

The negative-surface-charge density q^- of all oxides, as shown in Figs. 42 through 55, originates from the acidic dissociation of the surface hydroxyl groups and is found to increase with increasing pH as well as with the electrolyte concentration.   The maximum q^- values of oxides found at pH ~ 11 are 60 to 65 $\mu C/cm^2$ under conditions where no significant dissolution of the lattice occurs [203].   This value is in agreement with the maximum surface-charge densities generally encountered in other electrochemical systems [18, 60].   These high q^- values have been attributed [146, 203] to the specific adsorption of cations on oxides (Section VI. D. 1).   Much higher experimental surface-charge densities were obtained for Fe$_2$O$_3$ by Parks and De Bruyn [191] by following the slow oxide-solution equilibria in

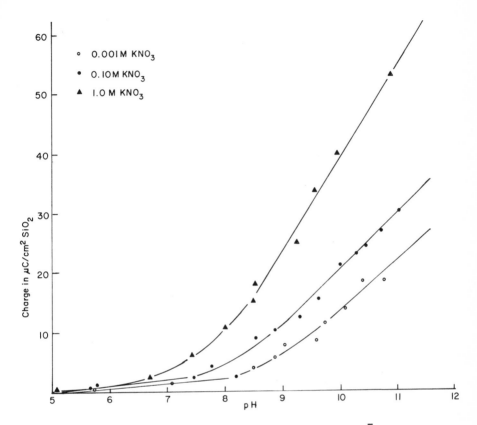

Fig. 55.   Variation in the surface-charge density q^- of quartz in KNO₃ solutions at 25°C (sample 1, Table 1) [203, 242].

prolonged-titration studies.   However, on the basis of subsequent work [209, 210, 255], De Bruyn and co-workers have attributed the slow changes of pH in aqueous Fe_2O_3 suspensions (and hence the high q^+ values) to both an ion-exchange mechanism and also partly to a diffusion of protons in the solid phase of the hydrated Fe_2O_3 precipitate.

High surface-charge densities q^-, which even exceed the surface density of the silanol groups, have also been reported by Tadros and Lyklema [257] in recent studies of the double layer on porous silica of large surface area (~ 56 m^2/g, particle diameter 500 to 1000Å, pore radius < 20Å). To account for these high surface-charge densities, penetration of OH^- and cations (counterions) into the subsurface structure of the porous oxide was postulated. On the basis of these experiments a model of the double layer on porous surfaces has been developed [258] by calculating the potential distribution inside the surface layer of the solid according to a Poisson-Langmuir equation. High q^+ values (~ 50 μC/cm^2) have also been reported in the acid medium for the sample of BDH silica that was used without purification. Hence, it is possible that the sample investigated contained silicates or cationic impurities that were leached out by acids.

4. Adsorption Isotherms and Distribution of Potential at the Oxide-Solution Interface

Attempts to analyze the double-layer structure at the oxide-solution interface from electrokinetic measurements alone have been only partly successful. However, based on studies of the reversible oxide-solution interface, recent attempts to analyze the double layer on oxides are promising.

Herczyńska and Prószyńska in their early work [179] on a series of oxides and oxide-covered metals observed a strong dependence of the zpc on the electrolyte concentration, which was attributed to the specific adsorption of cations or anions in the inner Helmholtz layer at the oxide-solution interface. These authors derived isotherms containing several constants to fit their experimental data, and they also incorrectly assumed that the inner-Helmholtz-layer potential was approximately equal to the ζ potential. As the surface-charge densities are not known in this work, it is hard to compare the results and the isotherms obtained with the more recent work in this field.

Atkinson, Posner, and Quirk [224] assumed the absence of specific adsorption of monovalent anions or cations on Fe_2O_3 and derived a form of the Langmuir isotherm in which the electrical potential terms were replaced by the electrochemical potentials of H^+ and of counterions (Cl^-). Their final equation was

$$\frac{\theta_H}{1 - \theta_H} = k_{H^+} \sqrt{[H^+][Cl^-]} \; \exp\left[(-\frac{F}{2RT} (\psi_H - \psi_{Cl})\right] , \qquad (85)$$

where k_{H^+} is a constant,

$$k_{H^+} = \exp\left[-\frac{(\Delta\mu_H + \Delta\mu_{Cl})}{2RT}\right],$$

$[H^+]$ and $[Cl^-]$ are molar concentrations of H^+ and Cl^- (KCl, supporting electrolyte), ψ_H and ψ_{Cl} are the potentials of the surface, relative to the bulk solution, due to adsorbed H^+ and Cl^-, and θ is the surface coverage. By further assuming that $1 - \theta_{H^+} \sim V_{H^+}$ (the maximum number of sites available on the oxide for H^+ adsorption) and also that Γ_{H^+} is a linear function of the potential difference $\psi_H - \psi_{Cl}$, Eq. (85) was given in an approximate form as

$$\Gamma_{H^+} = k_{H^+} V_{H^+} \sqrt{[H^+][Cl^-]} \; \exp(-K_1\Gamma_{H^+}) , \qquad (86)$$

or

$$\log \Gamma_{H^+} + \tfrac{1}{2} pH = \log k_{H^+} V_{H^+} + \tfrac{1}{2} \log [Cl^-] - \frac{K_1\Gamma_{H^+}}{2.303} . \qquad (87)$$

The exponential term in Eq. (86) contains free-energy parameters. By plotting $\log \Gamma_{H^+} + \tfrac{1}{2} pH$ against Γ_{H^+} per gram of oxide, straight lines were obtained for each KCl concentration. From the slope and intercept of these plots the values of constants K_1 and $k_{H^+}V_{H^+}$ were obtained. A similar equation that was derived for the pH range greater than the zpc is

$$\Gamma_{OH^-} = k_{OH^-} V_{OH^-} \sqrt{[K^+][OH^-]} \exp(-K_2\Gamma_{OH^-}) , \qquad (88)$$

and values of the constants k_{OH^-} V_{OH^-} and K_2 were also evaluated. Interaction constants for the adsorbed layer of counterions on oxides, that are equivalent to the experimental constants K_1 and K_2, were also calculated independently. In order that these interaction constants be comparable to the experimental constants that were obtained by using the BET surface areas for the oxide precipitates, an average distance of less than $0.5\mathring{A}$ between the plane of surface charge and the plane of adsorbed counterions had to be assumed. This small distance of separation between the two planes, together with the high values of surface-charge densities, is not consistent with the conclusion derived in the above work that the adsorbed counterions remain solvated as electrostatically bound ion pairs outside the primary coordination shell of the surface iron atoms. In a medium of low dielectric constant ($\epsilon < 10$ at the interface) Cl^- ions are known to enter readily the first coordination shell of many metal ions [222, 223, 225-227].

Some of the basic assumptions made by Posner et al. [224] in the derivations of their isotherms (Eqs. (86) through (88)), such as (a) the nonspecific adsorption of counterions, particularly cations, on the cathodic surfaces of oxides and (b) $1 - \theta_{H^+} \simeq V_{H^+}$ even at high charge densities, are not in agreement with other approaches to the same problem, which are discussed next.

a. Specific Adsorption Densities of Cations on Oxides. The following analysis of the double layer on oxides, proposed by the present author [146], is based on the concept of the specific adsorption of cations on oxides described in Section VI. D. 1.

The relative surface excess $\eta^i_{K^+}$ due to the specific adsorption of K^+ on oxide surfaces when pH > zpc was obtained for various oxides from Eq. (76) by subtracting values of $\eta^d_{K^+}$ (the cation excess in the diffuse double layer) from the total q^- values. Values of η_{K+}^d were obtained from the theory of the diffuse double layer [221] and from a knowledge of the oxide-electrode potentials relative to zpc and of the ionic strength of the solution. In calculating the η_{K+}^d for quartz (from Fig. 55) the electrostatic potential ψ was considered to become effective from pH 7, for reasons already discussed concerning the χ potential on quartz surfaces. These η_{K+}^i values of various oxides are shown plotted against the activity a_{KNO_3} in Figs. 56, 57, and 58.

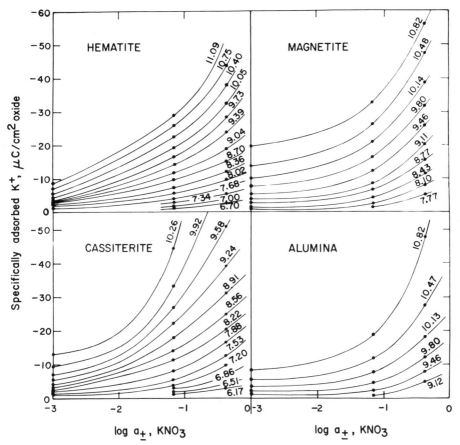

Fig. 56. Variation in the density of specifically adsorbed K^+ on Fe_2O_3, Fe_3O_4, SnO_2, and Al_2O_3 with the log of the mean activity of KNO_3 at various pH values. Reprinted from Ref. [146], p. 3552, by courtesy of the American Chemical Society.

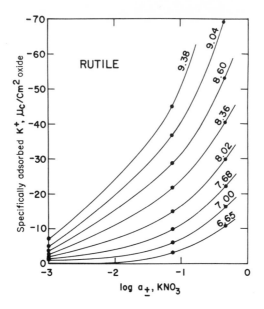

Fig. 57.   Variation in the density of specifically adsorbed K^+ on
TiO_2 with the log of the mean activity of KNO_3 at various pH values.
Reprinted from Ref. [146], p. 3552, by courtesy of the American
Chemical Society.

These adsorption isotherms for K^+ on oxides are seen to be comparable in shape and magnitude to the isotherms for the specific adsorption of anions [259-262] on anodic surfaces of mercury and of Tl^+
[263] on cathodic surfaces of mercury.

b. The Standard Free Energy for the Specific Adsorption of Cations on
Oxide Surfaces. The calculation of standard free-energy values for
the specific adsorption of K^+ from the foregoing experimental data
has been described [146] by using two approaches. In one approach,
as the pH of the solution is increased relative to the zpc of oxides
(cathodic region), the OH^- ions from the solution may be considered,
as a first step, to neutralize the surface H^+ ions to form water molecules, which are initially adsorbed on the oxide surface. The K^+ ions
from solution subsequently replace the water molecules and are

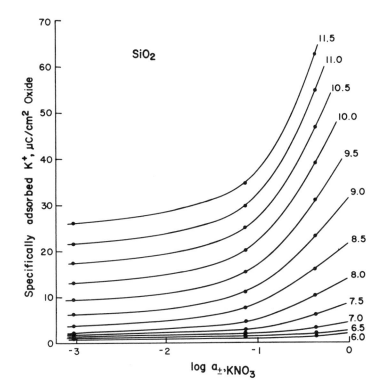

Fig. 58. Variation in the density of specifically adsorbed K^+ on quartz with the log of the mean activity of KNO_3 at various pH values [242].

specifically adsorbed on the oxide. The treatment is somewhat similar to that given by Blomgren and Bockris [264] for the adsorption of
organic ions on mercury surfaces.

If $\bar{\mu}_{K^+,sur}$ refers to the electrochemical potential of the specifically adsorbed K^+ ions on the surface, then, with reference to the
electrochemical potential $\bar{\mu}^0_{K^+,sur}$ of unity surface coverage as a
standard state and also assuming zero interaction between the
adsorbed species, we have

$$\bar{\mu}_{K^+,sur} = \bar{\mu}^0_{K^+,sur} + RT \ln \theta , \qquad (89)$$

where θ is the fractional surface coverage due to adsorbed K^+. Similarly the electrochemical potential $\bar{\mu}_{K^+,sol}$ of K^+ in solution, with reference to a standard electrochemical potential $\bar{\mu}^0_{K^+}$ of K^+ of unit mole fraction, is given by

$$\bar{\mu}_{K^+,sol} = \bar{\mu}^0_{K^+,sol} + RT \ln a_{KNO_3} \, , \tag{90}$$

where the mean activity a_{KNO_3} is used in place of a_{K^+}. Similarly for water molecules adsorbed on the surface

$$\mu_{H_2O,sur} = \mu^0_{H_2O,sur} + RT \ln (1 - \theta) \tag{91}$$

and

$$\mu_{H_2O,sol} = \mu^0_{H_2O,sol} + RT \ln a_{H_2O} \, , \tag{92}$$

where $\mu_{H_2O,sur}$ and $\mu_{H_2O,sol}$ are the chemical potentials of water in the adsorbed state and in solution. As before, $\mu^0_{H_2O,}$ in Eqs. (91) and (92) is the standard electrochemical potential of water, either in the adsorbed state (sur) or in the solution (sol) phase. Alternatively, one can also consider the direct replacement of H_3O^+ by K^+ (see later). At equilibrium,

$$\bar{\mu}_{K^+, sur} - \bar{\mu}_{H_2O, sur} = \bar{\mu}_{K^+, sol} - \mu_{H_2O, sol} \tag{93}$$

when the difference in the apparent electrochemical free energies of adsorption of K^+ and water on the oxide surface are given by

$$\Delta\bar{G} = (\bar{\mu}_{K^+, sol} - \bar{\mu}_{K^+, sur}) - (\mu_{H_2O, sol} - \mu_{H_2O, sur}) \tag{94}$$

or from Eqs. (89) through (92)

$$\Delta\bar{G} = \Delta\bar{G}^0 + 2.303\, RT \log \left(\frac{1 - \theta}{\theta} \frac{a_{KNO_3}}{a_{H_2O}} \right) . \tag{95}$$

Hence the apparent change in the standard free energy of the dissociation-adsorption process at the oxide-solution interface is given by

$$-\Delta G^0_{ap} = 2.303\, RT \, \log \left(\frac{\theta}{1-\theta} \frac{55.5}{a_{KNO_3}} \right) \tag{96}$$

and also

$$\frac{\theta}{1-\theta} = \frac{a_{KNO_3}}{55.5} \exp \left(-\frac{\Delta G^0_{ap}}{RT} \right) . \tag{97}$$

For calculating θ a surface-charge density of 65 $\mu C/cm^2$ was assumed to correspond to an effective saturation surface coverage. The values of ΔG^0_{ap} for the adsorption of K^+ on a series of oxides from 0.001 M KNO_3 solution, as calculated from Eq. (96), are shown plotted as a function of θ in Fig. 59A, in which a set of recently obtained $-\Delta G^0_{ap}$ values for quartz is also included. The $-\Delta G^0_{ap}$ values for all oxides

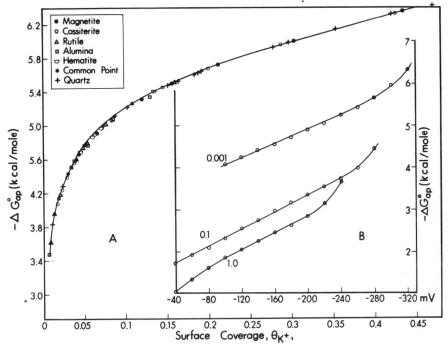

Fig. 59. Variation in ΔG^0_{ap} for the specific adsorption of K^+ on negatively charged oxide surfaces: (A) with surface coverage θ_{K^+}, of several different oxides in 0.001 M KNO_3; (B) with potential (relative to zpc), and with KNO_3 concentration, of SnO_2; numbers on curves in (B) indicate molarity of KNO_3 [146, 242].

studied fall on the same line in Fig. 59A, regardless of their specific surface area or of the total surface areas used in the experiments. The specific surface areas varied from 420 cm^2/g for SnO_2 to 4414 cm^2/g for Al_2O_3.

There is a systematic variation in $-\Delta G^0_{ap}$ with pH (or potential relative to zpc) as well as with a_{KNO_3}, as shown for a typical case of SnO_2 in Fig. 59B. Thus, whereas the $-\Delta G^0_{ap}$ values become increasingly negative with increasing pH for a given electrolyte concentration (Fig. 59A and B), the $-\Delta G^0_{ap}$ values for a given pH decrease with increasing electrolyte concentration. The increase in $-\Delta G^0_{ap}$ with increasing pH (Fig. 59A and B) is common to all oxides and is obviously due to the acidic dissociation of the surface hydroxyl groups. This dissociation increases the negative potential difference between the oxide surface and the bulk solution, whereby the adsorption of K^+ on oxide surfaces becomes increasingly easier. It is also seen in Fig. 59A that the variation of ΔG^0_{ap} with θ is not linear because of structural and induced surface-heterogeneity effects that were ignored in deriving the Langmuir isotherm. However, the variation of ΔG^0_{ap} with surface potential is almost linear (Fig. 59B) but decreases with increasing a_{KNO_3}. Hence the adsorption of K^+ on oxides becomes increasingly difficult with increasing K^+ concentrations.

From the foregoing considerations it is clear that the free-energy $(-\Delta G^0_{ap})$ term in the exponential part of Eq. (97) varies with pH and with a_{KNO_3} because of contributions from the electrostatic potentials due to the potential-determining H^+ and OH^- ions ($\psi^{sur}_{H^+/OH^-}$) and the counter electrolyte ions ($\psi^i_{K^+}$). Hence the $-\Delta G^0_{ap}$ term may be formally written in terms of its components as

$$-\Delta G^0_{ap} = zF\left[\psi_{H^+} + (\psi^i_{K^+} + \phi_{sp})\right] , \tag{98}$$

where ϕ_{sp} is the specific adsorption potential in the inner region, or

$$-\Delta G^0_{ap} = -\Delta G^0_{H^+} - \Delta G^0_{K^+} \; , \qquad (99)$$

where

$$\Delta G^0_{H^+} = RT \ln \frac{(a_{H^+})^f}{(a_{H^+})^i} \qquad (100)$$

$$\Delta G^0_{H^+} = 2.303 \, RT \, (pH_i - pH_f) \; . \qquad (101)$$

In Eqs. (100) and (101) i and f refer to the initial and final values of a_{H^+} (or pH) of the solution before and after reacting with the oxide surface, respectively. Hence, by subtracting $-\Delta G^0_{H^+}$ from $-\Delta G^0_{ap}$, one can obtain $-\Delta G^0_{K^+}$, which is independent of pH. Thus $-\Delta G^0_{K^+}$ was given in a previous publication [146] as

$$-\Delta G^0_{K^+} = 2.303 \, RT \left[\log \left(\frac{\theta}{1-\theta} \cdot \frac{55.5}{a_{KNO_3}} \right) + (pH_i - pH_f) \right] . \qquad (102)$$

In an alternative approach $\Delta \bar{G}^0$ in terms of a_{KNO_3} and $a_{H_3O^+}$ can be obtained by considering the direct replacement of H_3O^+ on the surface by K^+, and by equating the electrochemical potential of H_3O^+ and of K^+ in the adsorbed layer and in the bulk solution. Then, by substituting $\log (C_{H_2O} a_{H^+})$ for $\log a_{H_3O^+}$ and including further changes in a_{H^+} during adsorption, the isotherm obtained is identical with Eq. (102). However, the previous derivation has been more informative in understanding the potential distribution at the interface.

In Fig. 60A the $\Delta G^0_{K^+}$ for specular hematite and cassiterite are shown plotted against pH, and a similar plot for quartz is shown in Fig. 61A. As predicted, the $\Delta G^0_{K^+}$ in Figs. 60A and 61A are seen to be independent of pH within ±2%. Deviations from the predicted behavior were somewhat larger (±5%) at extreme values of pH because of the higher experimental error in measuring small changes in pH, in addition to some errors in solubility corrections. In Figs. 60B and 61B

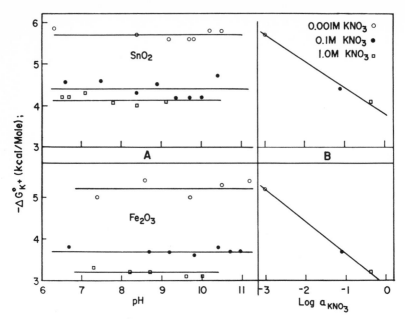

Fig. 60. (A) $\Delta G^0_{K^+}$ for Fe_2O_3 (specular hematite) and SnO_2 as a function of pH and ionic strength. (B) $\Delta G^0_{K^+}$ (pH independent), obtained from Fig. 60A, plotted as a function of a_{KNO_3}. Reprinted from Ref. [146], p. 3554, by courtesy of the American Chemical Society.

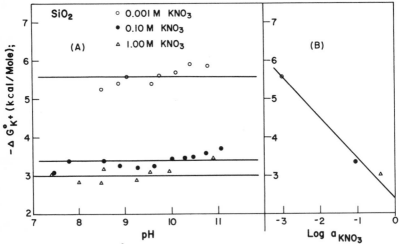

Fig. 61. (A) $\Delta G^0_{K^+}$ for SiO_2 (quartz) as a function of pH and ionic strength. (B) $\Delta G^0_{K^+}$ (pH independent), obtained from Fig. 61A, plotted as a function of a_{KNO_3} [242].

it is further seen that these $-\Delta G_{K^+}^0$ values decrease linearly, within experimental error, with increasing values of $\log a_{KNO_3}$. This decrease in $-\Delta G_{K^+}^0$ with increasing a_{KNO_3} indicates that the specific adsorption of K^+ on a negatively charged oxide surface (for a given pH) becomes increasingly difficult as the adsorption of K^+ proceeds. This behavior may be attributed to the following effects:

1. A decrease in the effective negative potential of the inner Helmholtz layer (i. e., the potential a K^+ ion encounters as it approaches the surface) due to the previous surface coverage by K^+.

2. A repulsive potential between the adsorbed K^+ and the K^+ ions about to be adsorbed.

If φ^0 (i. e., E in Eq. (68)) is assumed to represent the overall surface potential at a given pH and if ψ_{ap}^i, obtained from ΔG_{ap}^0, is the potential in the plane of specifically adsorbed K^+ (IHP) relative to the zpc, then the potential drop $\Delta(\psi^i + \chi)$, between the surface and the IHP, as obtained by subtracting ψ_{ap}^i from φ^0, is shown schematically in Fig. 62. An analysis of the potential distribution at the quartz-solution interface based on the above considerations has been carried out as follows: In Fig. 63, $-\psi_{ap}^i$ for the quartz-solution interface in 0.001, 0.1, and 1 M KNO_3 solutions is shown plotted as a function of pH. By subtracting these $-\psi_{ap}^i$ values from the potential $-\varphi^0$, relative to the zpc (pH 1.5), the resultant values of $\Delta(\psi^i + \chi)$ may also be expressed as

$$-\varphi^0 - (-\psi_{ap}^i) = -\psi^0 + \psi_{ap}^i - \chi = \Delta\psi^i - \chi \qquad (103)$$

or, in general,

$$\varphi^0 - \psi_{ap}^i = \Delta(\psi^i + \chi) \qquad (104)$$

and are shown plotted against pH in Fig. 64 for the three KNO_3 concentrations. The ψ_{ap}^i values represent the effective potential (of the

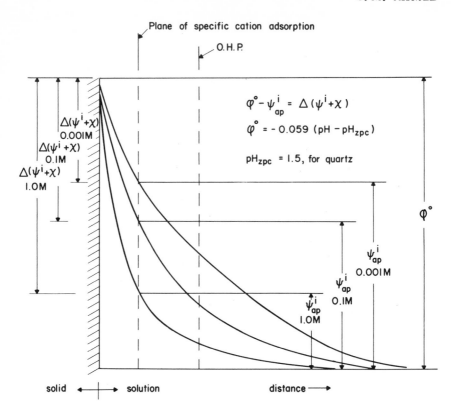

Fig. 62. Schematic representation of the potential drop in various regions of the oxide-solution interface in the cathodic region (oxide surfaces negatively charged) as a result of the specific adsorption of cations on oxides [242].

plane of K^+ adsorption, relative to the bulk solution) that a K^+ ion would encounter as it approaches the surface for adsorption from the bulk solution. The variations in ΔG^0_{ap} (Fig. 59 A, B), in ψ^i_{ap}, and in the ζ potentials of oxides with pH_s and the mean activity of the salt are qualitatively similar. The ψ^i_{ap} values are seen to be larger than the ζ potentials of quartz for a given pH and ionic strength. This difference is in qualitative agreement with the general assumption that the slipping plane in electrokinetic measurements occurs at the OHP or possibly beyond it toward the bulk solution. It is difficult to

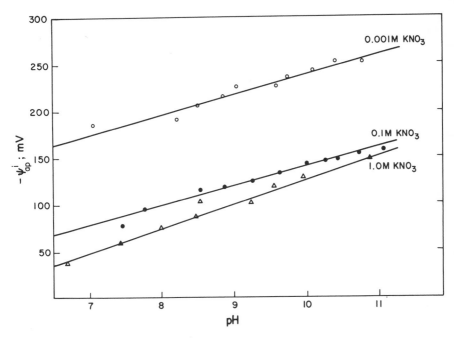

Fig. 63.   Variation in ψ_{ap}^{i} with pH at various KNO_3 concentrations [242].

bring about any quantitative comparison at this stage between the above ψ_{ap}^{i} values and the ζ potential because of uncertainties in ζ potential calculations arising from the location of the slipping plane, the interfacial dielectric constant, viscoelectric effects, and surface conductivity. Also, because of the specific adsorption of cations on oxides, the effective surface potential responsible for electrokinetic phenomena is much smaller (Fig. 62) than ψ^0 (Eq. 68).

It is further observed that the slopes of the plots of $-\Delta G_{K+}^0$ versus log a_{KNO_3} (Figs. 60B and 61B) and the extrapolated values of $-\Delta G_{K+}^0$ to unit salt activity vary for different oxides and appear to be characteristic for each oxide. Thus ΔG_{K+}^0 at unit salt activity for rutile, cassiterite, hematite, and quartz are found to be -5.7, -3.8, -2.9, and -2.4 kcal/mole, respectively. These differences in the $-\Delta G_{K+}^0$

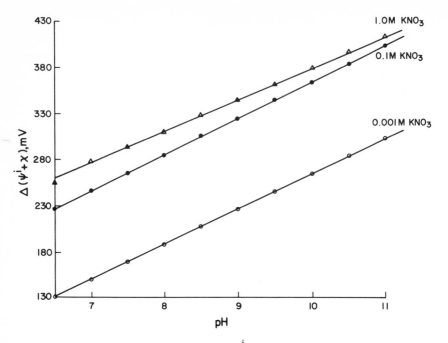

Fig. 64. The potential drop $\Delta(\psi^i + \chi)$ between the quartz surface and the IHP due to the specific adsorption of K^+ on quartz, shown as a function of pH at various KNO_3 concentrations [242].

values of different oxides may be attributed to the differences in the specific adsorption potentials ϕ_{sp} (cf. Eq. (98)) of K^+ of different oxides depending on the differences in their acidic or basic character.

A similar analysis of the double layer on positively charged surfaces of oxides may be carried out, but not without giving due considerations to the mechanisms (A through C) involved.

Blok and De Bruyn in a recent analysis [211] of the double layer on ZnO also recognized the specific adsorption of counterions on oxides as a principal factor in determining the structure of the double layer on oxides. Following Grahame's model of the double layer on metals, three planes of charge were recognized: (a) the plane of the surface charge or of the adsorbed potential-determining ions with a charge density q^s and potential ψ^s relative to the bulk solution,

(b) the plane of the centers of specifically adsorbed ions (IHP) with a charge density q^i and potential ψ^i relative to the bulk solution, and (c) the OHP with a charge density q^d and potential ψ^d. Hence, as before,

$$q^s + q^i + q^d = 0 . \tag{105}$$

The three planes, as described here, separate two regions of thickness β (surface to IHP) and γ (IHP to OHP) with dielectric constants ϵ_β and ϵ_γ, respectively. By considering the two regions as parallelplate capacitors,

$$\psi^i = \psi^d - \frac{4\pi\gamma q^d}{\epsilon_\gamma} \tag{106}$$

and similarly it can be shown [211] that

$$\psi^s = \psi^d - \left(\frac{4\pi\gamma}{\epsilon_\gamma} + \frac{4\pi\beta}{\epsilon_\beta} \right) q^d - \frac{4\pi\beta}{\epsilon_\beta} q^i . \tag{107}$$

Isotherms were constructed by calculating q^s and ψ^s using Eqs. (105) through (107) together with the other standard equations of Stern and of the diffuse double layer after assuming suitable empirical values for the physical parameters. The assumed values for the specific adsorption potentials in the Stern equation were ϕ^+ (of cations) $= 2kT$, ϕ^- (of anions) $= 4kT$, the capacitances of the inner layers $\epsilon_\gamma/4\pi\gamma =$ $\epsilon_\beta/4\pi\beta = 100 \ \mu F/cm^2$, and the total number of available sites on oxide surfaces, $N_s^+ = N_s^- = 10^{15} \ cm^{-2}$. The calculated adsorption isotherms for 0.001, 0.02, and 0.1 M solutions of a 1:1 electrolyte and the corresponding differential-capacitance curves obtained from these isotherms bear a general resemblance to the experimental isotherms (Fig. 49) for ZnO. Of course the calculated isotherms would depend on the values chosen for the physical parameters, which may not remain constant at all levels of potential and charge densities. The

specific adsorption potential ϕ^+ of ions on polarized metal surfaces
[17, 18] is known to vary with charge densities or the bias potential.
However, such variation in $\phi^{\pm}$ may not occur at the reversible oxide-
solution interface in the absence of applied potential. In the presence of specific adsorption of ions on oxides the capacitances of the
inner layers change, and the number of the available adsorption sites
is not equal to $N^{\pm}$, but equal to $(1 - \theta)N^{\pm}$ at high charge densities.
However, the foregoing approach to the problem has demonstrated
the possibility that the basic assumptions made in the double-layer
model for oxides could be valid.

The double-layer model proposed by Blok and De Bruyn [211] for
ZnO disagrees totally with the model proposed by Posner et al. [224]
for Fe_2O_3 and also disagrees in some details with a previously proposed model by Bérubé and De Bruyn [213] for the TiO_2-solution interface, but it agrees in most details with the double-layer structure
proposed previously by the present author in connection with a series
of other oxides [146, 202, 203, 206, 218]. However, compared with a
rather uniform structure of the double layer on negatively charged
oxide surfaces, the double layer on positively charged surfaces is
complex and depends largely on the nature of the coordination between
the surface metal atoms and the water molecules, the anions, and the
H^+, OH^- ions. These interactions can vary significantly from one
oxide to another. Furthermore, for the same oxide the interfacial
behavior depends on the nature of the anions, the stoichiometry, and
the degree of surface hydration as affected by heat treatment, if any.

5. The Differential Capacitance of the Reversible Oxide-Solution
    Interface

Information on the differential capacitance C of the double layer
on oxides has been obtained by graphical differentiation of the q

versus pH or potential plots of oxides. The C values first obtained
for the double layer on Fe_2O_3 (precipitate) by Parks and De Bruyn[191]
from the slow acid-base titration (B_s) data were too high (about four
times the theoretical value) to be attributed to the double layer on
oxides. These high values of C were later proved [205, 255] to be
due to the high q^+ values, at a given pH, resulting from the secondary reactions of Fe_2O_3 with electrolyte solutions. However, the minimum value of C for a number of oxides at the zpc, obtained from
studies of the primary oxide-solution equilibrium (method A) [146,
203, 206] and also by the fast-titration methods [211, 213] (B_f) were
in close agreement with the theoretically expected minimum value
of C. The C versus q or C versus pH or potential curves (henceforth
referred to as C curves) of the double layer on oxides are significantly
different from their counterparts for mercury [18, 60] and for AgI [193,
197]. The following is a summary of attempts made to explore the
significance of the C curves of oxides and also to derive information
from these C curves as to the nature of the counterion adsorption on
oxide surfaces.

The C values of a few oxides, as obtained [146, 203, 206] from
studies of the primary oxide-solution equilibrium, are shown plotted
against pH in Figs. 65 and 66. The following three main features of
these C curves have been noted [146, 206]:

1. The steep rise in the C(+) values (pH < zpc) of several oxides
was attributed qualitatively to the chemisorption of H^+ on the hydrated
oxide surfaces. This effect of H^+ adsorption and a sharp rise in C(+)
values was particularly pronounced for SnO_2 in Cl^- and for the n-type
semiconducting iron oxides, specular hematite and magnetite. The
influence of the H^+ adsorption on the double-layer properties of
oxides has also been demonstrated both from direct capacitance measurements on polarizable PbO_2 electrodes (Section V) and also by a

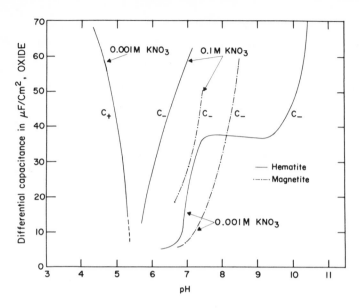

Fig. 65. Variation in the differential capacitance of the double layer on hematite and magnetite (C- only) with pH, in KNO₃ solutions. Reprinted from Ref. [206], p. 3844, by courtesy of the National Research Council of Canada.

sharp increase observed in the film capacitance of anodic oxides under a cathodic bias. In the latter case the increase in the capacitance of some oxide films (mostly of transition metals) has been attributed [4, 154–157] (Section III. C) to proton injection and to an induced space charge in the oxide film.

2. A well-defined plateau observed at 30 to 40 μF/cm² for the C(-) values (pH > zpc) of Fe₂O₃ (Fig. 65) and of quartz [203]. A similar but less pronounced effect can also be seen in the C(-) curves of TiO₂ and SnO₂ (Fig. 66), but this effect is not apparent in the case of ThO₂ [203] and Al₂O₃ (Fig. 66). This plateau probably indicates the transition from a weak adsorption to a strong specific adsorption of cations on oxides. The absence of such a plateau in the case of ThO₂ and Al₂O₃ will then indicate intense specific adsorption of cations on these two oxides. The plateau in the C(-) curves of oxides

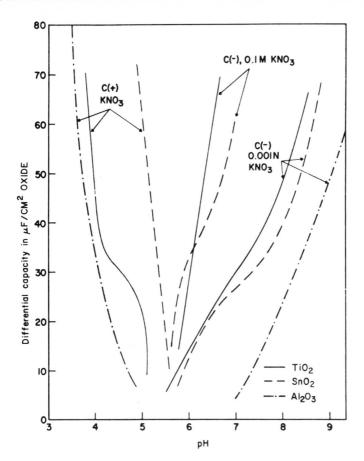

Fig. 66. Variation in the differential capacitance of the double layer on Al_2O_3, SnO_2, and TiO_2 with pH, in KNO_3 solutions. Reprinted from Ref. [146], p. 3551, by courtesy of the American Chemical Society.

has also been compared [146, 203] to a smoothed-out "hump" on mercury surfaces.

    3. The occurrence of a broad, flat minimum in the C(-) values of some oxides (quartz [203], Fe_2O_3 [206] (Fig. 65), and Al_2O_3 [146] (Fig. 66)) at the zpc. The occurrence of such flat minima in the double-layer capacitance of PbO_2 [13, 181–183] and of certain metals, such as Zn [94], has also been observed by direct capacitance

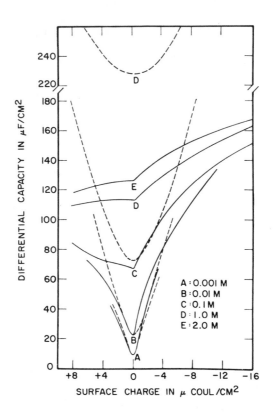

Fig. 67. Variation in the differential capacitance of the double layer on TiO$_2$ in NaNO$_3$ solutions with the surface-charge density. The capacitance values, calculated from the Gouy-Chapman theory, are shown by the broken curves. Reprinted from Ref. [213], p. 94, by courtesy of Academic Press, Inc.

measurements. For example, in the case of quartz [203] (Fig. 55) and of Al$_2$O$_3$ (Fig. 66) the broad minima in the C(-) values have resulted from the flat portions in the original q$^-$ against pH$_s$ plots (e.g., Fig. 42). In the case of the quartz-solution interface in 0.001 M KNO$_3$ solution the C(-) curve showed [203] an increase with pH above its theoretical minimum value ($\sim$ 6 μF/cm^2) only for pH > 7. This indicates that a considerable potential drop occurs (cf. variation in ϵ at quartz-solution interface, Section VI. F) within the hydration layer of the oxide ($\sim$130 mV for Al_2O_3 and > 200 mV for quartz) before proton dissociation takes place from the surface hydroxyl groups. This potential drop in the hydrated layer of the oxide may be identified as the well-known χ potential. This effect is most pronounced for oxide surfaces that are strongly hydrated — for example, quartz [203], Fe_2O_3 [206], and Al_2O_3 [146] (Figs. 65 and 66) — and does not arise on anodic oxide surfaces, probably because of the common mechanism of H^+ adsorption on oxides in this region.

The interpretation of the C(+) values is much more complex for oxides obeying mechanism B (Fe_2O_3 in Cl^- (Fig. 51B), ZrO_2 and ThO_2 in NO_3^-, Cl^-, and ClO_4^- (Figs. 46 and 47)) and mechanism C (Fe_2O_3 and Fe_3O_4 in NO_3^- and ClO_4^-, and SnO_2 in Cl^-), where specific adsorption of counterions also has to be taken into account.

An evaluation of the double-layer capacitance of TiO_2 on a semiquantitative level has been attempted by Bérubé and De Bruyn [213]. These two authors have compared the C curves of TiO_2 with the calculated capacitance curves based on the Gouy-Chapman theory of the double layer. As shown in Fig. 67, the theoretical and experimental C curves are shown to agree in shape and magnitude near the zpc, especially at low ionic strengths. At higher ionic strengths and at high positive or negative electrode potentials, the experimental C values depart significantly from the theoretical ones. The asymmetry of the C curves is usually more pronounced in the cathodic branch, due to specific adsorption of cations, than on the anodic side. This behavior is the reverse of that observed for mercury and silver iodide. The nature of the weak interactions of anions with the rutile surface ($Cl^- \simeq NO_3^- \simeq ClO_4^- > I^-$) was demonstrated by calculating the inner-layer capacity C_i from

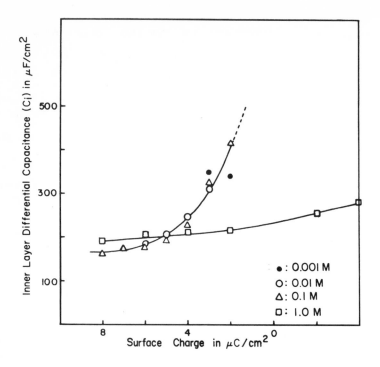

Fig. 68.   The capacitance of the inner region, C_i, of the double layer on TiO_2 in $NaNO_3$ solutions, shown as a function of the surface-charge density.   Reprinted from Ref. [213], p. 96, by courtesy of Academic Press, Inc.

$$\frac{1}{C_i} = \frac{1}{C_{exp}} - \frac{1}{C_d} , \tag{108}$$

where C_d was obtained from the diffuse-double-layer theory.   In the absence of specific adsorption of anions, $C_i(+)$ should be independent of ionic strength, as is shown in Fig. 68 when $q^+ > 4$ $\mu C/cm^2$.   For ionic strengths less than 1 M, $C_i(+)$ is seen (Fig. 68) to approach infinity at the zpc, as expected from the Gouy-Chapman model of the double layer.   A finite or a saturation value of about 200 $\mu F/cm^2$ in 1 M solutions and of 175 $\mu F/cm^2$ in more dilute solutions, was obtained for $q^+ > 4$ $\mu C/cm^2$.   This analysis does not consider the specific adsorption of anions on rutile.   This is in agreement with the other work on rutile where mechanism A was proposed [146, 202].

However, it is shown in Fig. 67 that the experimental values of C(-) (pH > zpc) for rutile lie much below the diffuse-double-layer capacitance, thus indicating that an assumption of nonspecific adsorption of cations on oxide surfaces is not valid. This is also in agreement with the conclusion derived previously concerning the specific adsorption of cations on oxides from the C curves of a series of oxides [146, 203, 206]. From an analysis of the C versus pH curves of TiO_2, obtained in the presence of different cations, the degree for specific adsorption on TiO_2 was shown by Bérubé and De Bruyn [213] to be in the order $Li^+ > Na^+ > Cs^+$. This is the reverse of the order known for the specific adsorption of cations (if any) on mercury [18, 60] and on silver iodide [193, 197]. The tendency for specific adsorption to decrease from Cs^+ to Li^+ on mercury has been attributed [18, 63] to the increasingly greater hydration and smaller size of ions from Cs^+ to Li^+. For oxide factors other than ion hydration (e.g., surface hydration, covalent forces, bond strength, and solubility) may also have to be considered for the specific adsorption of cations. Bérubé and De Bruyn [213] have proposed a complex model of the double layer on TiO_2, by taking into account the structure-promoting role of the potential-determining (H^+ and OH^-) ions, which can stay in the outer hydration layer without necessarily being in direct contact with the surface proper. Specific adsorption of inorganic ions in the double layer is related to their "structure-promoting or disrupting influence on the structural order in the surface region" [213]. This model of the double layer on TiO_2 is different in many respects from that proposed by Blok and De Bruyn for ZnO [211].

Compared to the above order of variation in the adsorption affinity of cations for TiO_2, the absorption of cations by porous silica has been found to vary by Tadros and Lyklema [257] as $(C_2H_5)_4N^+ < Li^+ <$

$Na^+ < K^+ < Cs^+$. However, much of this cation absorption by porous silica has been attributed to the penetration of cations into the pore structure of silica.

6. The $\gamma - \gamma_0$ Values

The significance of the $\gamma - \gamma_0$ values of the oxide-solution inter-face was discussed in Section VI. E, and these values are shown plotted against pH (or potential) in Fig. 69 for the typical case of Al_2O_3 in ClO_4^-, NO_3^-, and Cl^- solutions. Similar plots for other oxides have been presented elsewhere [146, 203]. The $\gamma - \gamma_0$ versus pH curves are similar in shape to the electrocapillary curves for metals and represent the integral values of the electrochemical work done in transferring the ions from the bulk solution to the interface in the process of adsorption. Except for a small correction for the contribution of the diffuse double layer, the $\gamma - \gamma_0$ values of oxides examined are of the same order of magnitude and have the same mean-ing as the "equivalent effective pressure" calculated by Parry and Parsons [265] and by Payne [266, 267] for the specific adsorption of anions on mercury. The single anodic branch of the $(\gamma - \gamma_0)$ versus pH plot for Al_2O_3 (Fig. 69) and similar plots for TiO_2 in NO_3^-, Cl^-, and ClO_4^- [218] and for SnO_2 in NO_3^- and ClO_4^- [218] represent nonspecific adsorption of anions on the metal atoms of the oxide surfaces. In this respect the single anodic branch of the $(\gamma - \gamma_0)$ versus pH plot resembles the cathodic branch of the electrocapillary curves of mercury, where the cations remain nonspecifically adsorbed.

The variation of interfacial energy at the oxide-solution interface sometimes results in a noticeable variation in contact angles of liquid drops on oxide surfaces. This problem is discussed in the next section.

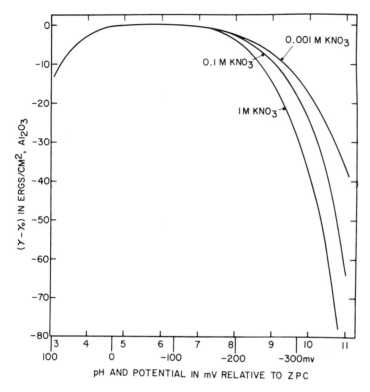

Fig. 69. Variation in $(\gamma - \gamma_0)$ at the Al_2O_3-solution interface with the final pH, and the potential difference relative to the zpc. Reprinted from Ref. [146], p. 3551, by courtesy of the American Chemical Society.

J. Contact Angles and the Free Energy of the Double Layer at the Oxide–Solution Interface

Contact angles of liquid drops on oxide surfaces can be some-times changed either by the application of an electric field across the interface or by changes in the composition of the solution phase. Attempts have been made [242, 268] to correlate these changes in contact angles with variations in the interfacial energy at the oxide-solution interface.

In the absence of specific adsorption of ions on oxides at the zpc the interfacial energy at the oxide-solution (sl) interface is given from Eq. (73) as

$$\gamma_{sl} = \gamma_{sl}^0 - \int_{\psi_{zpc}}^{\psi} q\, d\psi \; , \tag{109}$$

where ψ^0 is the electrostatic potential of the oxide surface relative to the bulk solution. By combining Eq. (109) with Young's equation [269, 270]

$$\gamma_{sg} = \gamma_{sl} + \gamma_{lg} \cos\theta \tag{110}$$

and assuming γ_{sg} and γ_{lg} to be independent of the integral term in Eq. (109), we have

$$\cos\theta - \cos\theta^0 = \frac{1}{\gamma_{lg}} \int_{\psi_{zpc}}^{\psi} q\, d\psi \; . \tag{111}$$

In Eqs. (109) through (111) sg, sl, and lg refer to the solid-gas, solid-liquid, and liquid-gas interfaces and θ is the angle of contact at the solid-liquid interphase. Equation (111) relates the change in the contact angle θ (relative to θ^0, the contact angle at the zpc) to the change in the free energy of the double layer in the solution phase relative to that at the zpc.

The possibility of the contact angle at an oxide-solution inter-face being changed by the application of an external emf to the sys-tem Ge/Ge oxide/solution has been examined by Sparnaay [268]. Since most of the potential drop on applying a field to this system occurs within the oxide layer, any change in the contact angle with the applied voltage is related to changes in the electrostatic free energy in the oxide phase as

$$\cos\theta_{II} - \cos\theta_I = \frac{1}{\gamma_{lg}} \int_I^{II} q_{ss}\, d\psi_s \; , \tag{112}$$

where q_{ss} is the density of slow surface states and ψ_s is the potential of the oxide surface relative to the bulk solid. The limits of the
integral in Eq. (112) are ψ_s^{II} and ψ_s^I, which are the values of ψ_s at
the applied potential V_{II} and V_I, respectively. It is debatable [268]
whether ψ_s or ψ^0 (solution phase, Eqs. (109) and (111)) should be
used in Eq. (112), since any change in the contact angle should
result from a change in the density of surface groups that are in equilibrium with the electrolyte solution. In any case, it is seen from
Eqs. (111) and (112) that for any change in the contact angle θ to be
detectable, the change in the integral terms of Eqs. (111) and (112)
should be significant in relation to the value of γ_{lg}, which is about
72 ergs/cm^2 for water at room temperature. The integral term in
Eq. (112) (for the solid phase) is equal to $e(\psi_s^{II} - \psi_s^I)N_A$ for $\psi_s^{II} > \psi_s^I$
and $-e(\psi_s^{II} - \psi_s^I)N_D$ for $\psi_s^{II} < \psi_s^I$, N_A and N_D being the densities of
the acceptor and donor states at the surface. The value of this inte-
gral (the contribution of the solid phase) was estimated to be only on
the order of 1 erg/cm^2 if N_A and N_D are on the order of 10^{13} cm^{-2}
[268]. However, if the value of $|e\Delta\psi_s|$ is 10 kT ($\sim 4.2 \times 10^{-13}$ erg
at room temperature) and the density of surface states (slow) is
10^{14} cm^{-2}, then the value of the integral is about 40 ergs/cm^2 [30],
which is comparable to γ_{lg}. Attempts were made by Sparnaay [268]
to measure electrocapillary curves for the Ge/Ge oxide/solution
system by using the experimental arrangement shown in Fig. 70. At
a certain potential (pzc) the contact angle is expected to have a
maximum value and to decrease at other values of applied potential.
This effect was observed only qualitatively and was time dependent.
This time dependency was attributed to the time constants of the slow
surface states.

On the other hand, the change in the value of the integral in
Eq. (111), due to changes in the free energy of the aqueous double
layer, has been shown by Ahmed [146, 203, 218] to be as high as

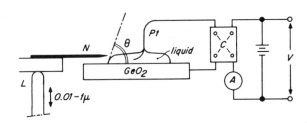

Fig. 70. Experimental arrangement for measuring contact angle θ at the germanium oxide-solution interface as a function of the applied voltage V. Part L is a sensitive lever system for a controlled and known displacement of a needle L, A is an ammeter, and C is a commutator. Image was projected with a thirtyfold magnification for measuring θ. Reprinted from Ref. [268], p. 219, by courtesy of the North-Holland Publishing Company.

10 to 100 ergs/cm^2 (Fig. 69).* Hence, if the original contact angle θ^0 at the zpc of an oxide surface is sufficiently high (i.e., if the surface is sufficiently hydrophobic), a definite change in the contact angle should be observed by changing the pH relative to the zpc and also using organic liquids, such as alcohol, that have a lower dielectric constant than water. A suitable oxide of this type was found to be specular hematite [206, 218], which has a hexagonal crystal structure, the coordination number of Fe^{3+} being 6. It is weakly magnetic, and the sample investigated was an n-type semiconductor. Crystals of specular hematite possess several smooth, lustrous planes, which, when cleaned chemically without any polishing, show native hydrophobicity in the vicinity of the zpc (pH 5.3 to 5.7 in 0.001 and 1 M KNO_3). Hence the surface forces of specular hematite at the zpc appear to be essentially molecular in nature. A maximum contact angle of 71°±3.5° (Fig. 71) of specular hematite with water was measured at a pH (5.3) close to the zpc in the absence of any supporting electrolyte. In

---

*The $\gamma - \gamma_0$ values in Figs. 5-7 of Ref. [203] are 10 times smaller due to a graphical mistake. See Refs. [146] and [218].

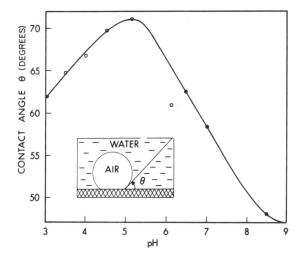

Fig. 71. Variation in contact angle θ at the specular hematite-solution interface as a function of pH [242].

order to avoid surface-heterogeneity effects, these contact angles were measured on a naturally cleaved surface of a single oxide sample. The contact angles decreased significantly with change in pH relative to the zpc due to interfacial reactions that decrease the free energy of the double layer (Fig. 69) and make the oxide surface hydrophilic. The addition of KCl to the solution also decreased the contact angle (~ 53° at the zpc in 0.1 M KCl).

The effect of the contact angle on specular hematite being a maximum at the zpc was also demonstrated by passing air bubbles through a bed of specular hematite in a 0.001 M KNO_3 solution in the absence of any surface-active agent. In Fig. 72 the oxide is seen to show maximum native floatability in the pH range 5 to 7, where the contact angle θ is also maximum.

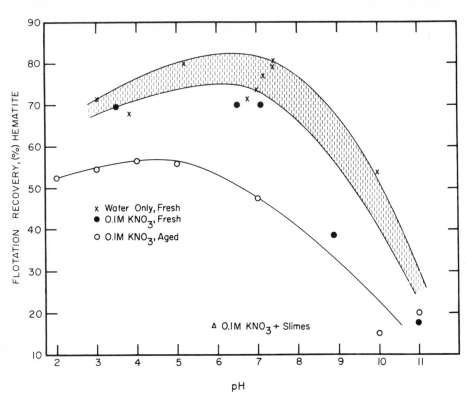

Fig. 72. Variation in native floatability of specular hematite as a function of pH and varying conditions of the solution and of the oxide surface [242].

K. The ζ Potential at the Oxide–Solution Interface

1. General Considerations

Under interfacial conditions other than the zpc, and provided that the surface charge of the solid is not neutralized by chemisorption of counterions, a potential is developed whenever a relative movement occurs between a solid phase and a solution phase. The potential difference between the plane of shear (or slipping plane) and the bulk solution is generally known as the ζ potential, which is

responsible for the following interrelated electrokinetic effects:
(a) streaming potential or streaming current, (b) sedimentation potential, (c) electrophoresis, and (d) electroosmosis. Although the ζ
potential at the oxide-solution interface can be determined from
measurements of any of these effects, most experimental determinations of the ζ potential have been made by streaming-potential and
by electrophoretic methods. Streaming-potential measurements are
carried out with particles of at least 0.1 to 0.2 mm in diameter,
whereas electrophoretic measurements are made with particle suspensions in liquids. As the ζ potential is closely related to the double
layer on oxides, some essential features of these two experimental
methods and the state of the present knowledge of this subject are
presented here, with particular reference to oxides. Further details
may be found in a number of recent publications [271-274].

2. Streaming Potentials

a. Theoretical. The streaming potential E_s is the potential difference
developed between the ends of a capillary tube or a porous plug
through which a liquid is forced to flow under a pressure P. The
liquid flow should be laminar. Knowing the specific conductance of
the liquid in the plug, the dielectric constant ϵ, and the viscosity η
of the liquid, the ζ potential may be calculated from the measured
values of E_s and P by using the classical Smoluchowski equation.
For this equation to be valid the material should be nonconducting,
and the capillary or pore radius should be much greater than the
double-layer thickness τ (particle size ≈ 0.2 mm [275]). The Smoluchowski equation has to be further corrected for the surface conductivity of the solution due to accumulation of ions in the double layer
and for the variation in ϵ and η of the liquid in the double layer [273,
276-278]. Corrections for the surface conductivity have been

calculated for single capillaries [273]. However, for such corrections to be made in the case of porous plugs, the plug conductance
has to be determined experimentally for each streaming-potential
measurement. Furthermore, if the solid-state conductivities of
semiconducting oxides are comparable to the plug conductance in
dilute solutions, additional corrections will have to be made in the
measured E_s. Such complications were noted by O'Connor et al.
[279] in streaming-potential measurements on several oxides. In
such cases of noninsulating oxides the measurement of streaming
current [280-282] has definite advantages over streaming-potential
measurements. In another investigation Holmes et al. [248] found
that the specific surface conductivities of nonconducting ThO_2 in
aqueous solutions were much larger than those predicted from the
theory of surface conductivity based on excess ions in the double
layer. This excess surface conductivity was ascribed to the conductance of the ionizable surface hydroxyl groups on the oxide.

The effect on ζ potential of the variation in ϵ and η of the liquid
medium in the double layer has been examined in detail by Lyklema
and Overbeek [276] and more recently by Stigter [277] and by Hunter
[278]. Although the dielectric constant of the Helmholtz layer is
known to be considerably lower than that of the bulk solution, corrections in ζ potential resulting from the variation in ϵ with field in
the aqueous bulk region of the slipping plane are considered to be
insignificant. However, as the position of the slipping plane or the
slipping layer [276] at the interface varies with values of the Debye-
Hückel constant $\kappa = 1/\tau$ and of the potential of the OHP, the viscosity
of the medium in the region of the shear can also change. Under these
conditions the viscoelectric corrections were calculated and found to
be significant by Lyklema and Overbeek [276]. But in later investigations of this problem Stigter [277] and Hunter [278] have shown
that the viscoelectric corrections under most conditions are much

smaller than those previously estimated by Lyklema and Overbeek. Corrections in ζ for the variation in both ϵ and η in the double layer have been provided in graphical form [276, 278]. The same conclusions concerning ϵ and η should be valid in electrophoretic measurements.

b. Experimental. A noteworthy improvement made by Schulman and Parreira [283, 284] in the measurement of E_s is the use of a pressure transducer, which enables direct plotting of $\Delta E_s/\Delta P$ on an x-y recorder. This method takes only a few seconds for a reading compared with the manual measurements, which are laborious and time consuming. However, a minor difficulty in the new experimental arrangement was the necessity to correct for the backpressure developed in the solution reservoir. This difficulty has been eliminated by Dibbs [285] by connecting the two ports of the pressure transducer to the two ends of the porous diaphragm and measuring the pressure difference across the cell directly. A Teflon cell, enclosed in a metal casing, with a pressure transducer and a demodulator were used by Dibbs in measuring $\Delta E_s/\Delta P$. The ζ potential of a quartz sample obtained as above is shown in Fig. 73. The same quartz sample was later used by the present author in studies of the reversible double layer (Section I. 4).

A major experimental source of error in streaming-potential measurements can arise from the variable and random differences in the rest potentials of platinum electrodes due to polarization effects [274, 286]. Freshly prepared Ag-AgCl electrodes have been found to give more reproducible results than platinum electrodes.

3. Microelectrophoresis

a. Theoretical. In electrophoresis, particles suspended in a liquid medium move under the influence of an applied electric field, of

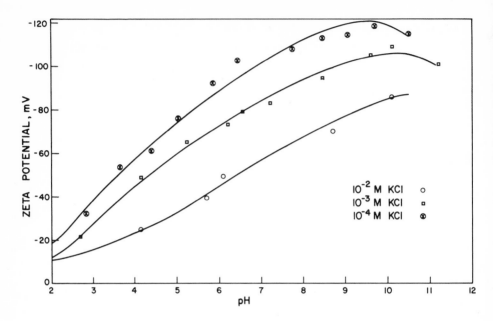

Fig. 73. The ζ potential of quartz (used in the double-layer studies) as a function of pH in KCl solutions of various concentrations [238].

strength E_e, as a result of the force exerted on the surface charge of the particle. The general expression for the electrophoretic velocity v_E of a nonconducting particle, for any ratio of particle size to the double-layer thickness (not including surface conductivity), as given by Henry [287] is

$$v_E = f_H (\epsilon \, E_e \, \zeta / \pi \eta) \, , \tag{113}$$

where f_H is a numerical factor varying [273] from 1/4 to 1/8 depending on the particle shape, the direction of the field, and on the product κa (a being the radius of curvature of the particle). It has been shown that for $\kappa a \gg 1$ (~ 1000 times) $f_H = 1/4$ regardless of the shape of the particle, as was also substantiated by Morrison [288] in a theoretical analysis of this subject. For spherical particles with small values of κa, $f_H = 1/6$. Equation (113) with $f_H = 1/4$ or 1/6

has been known as Smoluchowski's or Hückel's equation, respectively. Other values of f_H have been discussed by Overbeek [273].

Accumulation of excess charge surrounding the particle can give
rise to significant surface-conductivity effects, which can also
deform the double layer in an applied electric field and hence retard
the particle velocity considerably. This electrophoretic-retardation
is similar to relaxation phenomena in strong electrolytes. Corrections for both surface conductivity and relaxation effects have been
derived mathematically [273, 289] and discussed by a number of
workers [272, 273, 290-292]. For small as well as for large values
of κa, the relaxation effects are reported to be unimportant, but κa
even in colloids is not sufficiently small (e. g., $\kappa a = 10^5 \times 10^{-6} = 0.1$
even in a 10^{-5} M solution) for the relaxation effects to be negligible.
For intermediate values of κa (~ 0.1 to 100) relaxation effects are
considerable.

b. Experimental. Apparatus suitable for microelectrophoretic studies
of suspended particles has been described by a number of workers
[292-295]. A laterally oriented flat observation chamber for micro-
electrophoresis has been described by Neihof [293] and by Parreira
[294] for the particular use with organic liquids of low dielectric constants. The Cytopherometer, supplied by Carl Zeiss Inc. , has all
the special features described by Neihof and Parreira, and several
additional advantages, including a water jacket around the observation chamber. However, care should be taken with this apparatus
because, due to an uncorrected optical-inversion effect, the observed
direction of particle movement is the reverse of the actual movement.

Another apparatus available commercially for electrophoretic
studies of particles is the Zeta Meter [295], which employs a cylin-
drical capillary cell with nonreversible metal electrodes and has no
provision for temperature control of the cell. The Cytopherometer

employs $Cu/CuSO_4$ reversible electrodes, which are difficult to pre-
pare in duplicate with identical behavior. These electrodes can be
replaced by high-purity palladium electrodes with electrochemically
occluded hydrogen, as described recently by Neihof [293]. However,
due to electrolysis a substantial change in the solution pH was found
to occur [242] in the electrode compartments while using palladium
electrodes. The change in the solution pH can be reduced substantially by reversing the current for equal lengths of time in each measurement, and also by proper design of the cell compartments.

4. Analysis of the Double Layer on Oxides from ζ-Potential Data

On the basis of ζ-potential measurements several attempts have
been made to evaluate the potential distribution and the double-layer
charge densities for oxides [228, 279, 293-295] and for several non-
oxide systems [296, 297], including AgI and AgBr [298-300]. By
assuming the interface to consist of a compact Stern layer followed
by a diffuse double layer, attempts have been made to correlate the
electrostatic surface potential ψ^0 to the potential of the Stern plane
ψ_δ ($\sim \zeta$) of AgI and AgBr. By assuming that the χ potential at the
interface does not change,

$$\frac{d\varphi}{d(pAg)} = \frac{d\psi^0}{d(pAg)} = -\frac{2.3\,RT}{F} = -59 \text{ mV at } 25°C \qquad (114)$$

However, at the zpc in the absence of specific adsorption, it is found
experimentally that the slope $(d\zeta/dpAg)_{\zeta \to 0}$ near the zpc is only
-40 mV. By assuming further that the slipping plane coincides with
the Stern plane ($\psi_\delta \approx \zeta$), the above discrepancy between the rate of
change of ψ^0 and of the ζ potential with pAg, is attributed to the
contribution from the double-layer capacitances, so that

to be adsorbed on oxides the oxide surfaces have to be oppositely charged relative to the adsorbate ion, and that the ζ potential of oxides approaches zero when monovalent inorganic ions are adsorbed on oxides. However, when multivalent ions are specifically adsorbed at the oxide-solution interface, the ζ potential can not only approach zero but also change sign on further increase of the ion concentration in solution [230, 279, 301, 303-307].

Adsorption of cationic [250, 308-313] and anionic [312-316] surface-active compounds on oxides has also been investigated by ζ-potential measurements in connection with research into flotation and agglomeration of oxide minerals and the wetting of surfaces. In the case of long-chain organic electrolytes as adsorbates, the ζ potential of oxides has been shown to change its sign even in the presence of monovalent surfactants, depending on their concentration and chain length. Thus Gaudin and Fuerstenau [308] and Fuerstenau [309] found that, unlike in the case of inorganic monovalent cations, adsorption of dodecylammonium ions (RNH_3^+) on quartz from a solution with a critical concentration of RNH_3^+ can reduce the negative ζ potential of quartz to zero, whereas with further increase in the RNH_3^+ concentration the ζ potential was reversed to large positive values. This sign reversal for the ζ potential indicates molecular association of RNH_3^+ on the surface with a considerable concentration of the cationic groups facing the solution. It was therefore inferred [308, 309] that the adsorption of individual RNH_3^+ ions occurs on the negative sites of quartz until a critical concentration is reached at the surface. With further increase in the concentration of RNH_3^+, the ammonium ions, possibly together with some RNH_2 molecules, begin to associate into patches called hemi-micelles (giant polyvalent ions). The adsorption of RNH_3^+ and RNH_2 on quartz has also been demonstrated [317] by the measurement of contact angles, which were found to be a maximum at pH 10.

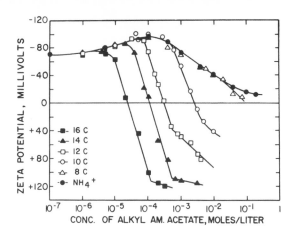

Fig. 74. Variation in the ζ potential of quartz with varying concentrations of ammonium acetate and alkyl ammonium acetates of various chain lengths. Reprinted from Ref. [312], p. 186 (based on Ref. [309]), by courtesy of the Society of Mining Engineers of AIME.

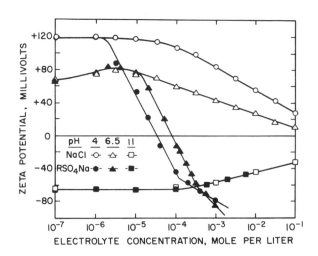

Fig. 75. Variation in the ζ potential of corundum in aqueous solutions of sodium dodecyl sulfate and sodium chloride of various concentrations at pH 4, 6.5, and 11. Reprinted from Ref. [312], p. 187, (based on Ref. [313]), by courtesy of the Society of Mining Engineers of AIME.

Similar conclusions have been reached by Furerstenau and Modi [313], who studied the adsorption on corundum of 12-carbon cationic (RNH_3^+ as chloride) and anionic (RSO_3^-, RSO_4^- and $RCOO^-$) surface-active agents by ζ-potential measurements. Joy and Watson [250] have reported considerable adsorption of RNH_2, together with RNH_3^+ (12-carbon) on hematite surfaces. These surface-active agents were found to be adsorbed appreciably only when the oxide surface is oppositely charged relative to the surfactant, as shown in Figs. 74 to 76. The adsorption potentials for the adsorption of organic ions in the Stern layer have been calculated and discussed [312, 313].

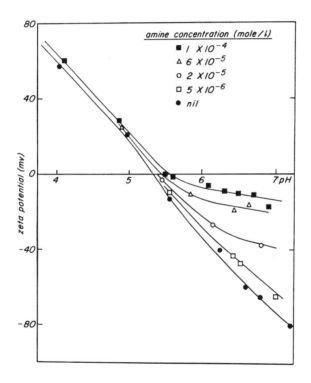

Fig. 76.   Effect of dodecylamine on the ζ potential of hematite at various pH values (ionic strength 6×10^{-4}).   Reprinted from Ref. [250], p. 330, by courtesy of the Institution of Mining and Metallurgy, London, England.

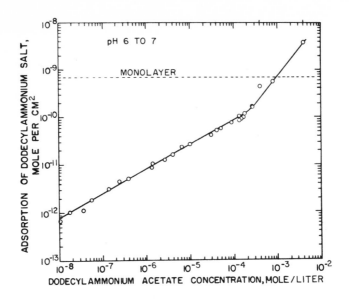

Fig. 77. Variation in the adsorption density of dodecylammonium acetate on quartz with the concentration of the adsorbate in solution at pH 6 to 7. Reprinted from Ref. [312], p. 185 (based on Ref. [310]), by courtesy of the Society of Mining Engineers of AIME.

Direct measurements of the adsorption of dodecylammonium acetate on quartz have also indicated [310] two types of adsorption, as shown from a change in the slope of the adsorption isotherm in Fig. 77. The tendency for molecular association of the adsorbed RNH_3^+ and RNH_2 on oxide surfaces has been shown [309, 312] to become significant and increase with increasing chain length longer than eight carbon atoms (Fig. 74).

The effect of molecular association in the adsorption of surface-active compounds on oxides appears to be particularly strong in the case of oleic, linoleic, and linolenic acids (18-carbon unsaturated acids with one, two, and three double bonds, respectively). Purcell and Sun [316] found from ζ-potential measurements that these acids, when adsorbed on rutile at a concentration of 10^{-3} mole/l, can reverse the surface charge from positive to negative and thus shift the zpc of rutile from pH 6.8 to about pH 3. The surface activity of these acids was also shown to decrease somewhat with increasing unsaturation in the organic acids.

In the presence of polyvalent cations like Ba^{2+} and Ca^{2+} anionic surface-active agents, such as carboxylates ($RCOO^-$) can also be adsorbed on the initially negatively charged surfaces of oxides. When Ba^{2+} [230] and Ca^{2+} [229] are specifically adsorbed on negatively charged oxide surfaces, positive sites are available for the adsorption of $RCOO^-$ ions from solution [314]. Such activation of mineral surfaces for the adsorption of surfactants is commonly used in the selective flotation of minerals.

ACKNOWLEDGMENTS

The author is grateful to the Director, Mines Branch, Department of Energy, Mines, and Resources, Ottawa, Canada, for granting permission and providing facilities for making this contribution to this series. The author is particularly grateful to Dr. H. P. Dibbs, Head, Surface Science Group, Mineral Sciences Division, Mines Branch, Ottawa, Canada, for his continuous support and assistance throughout the preparation of this review.

REFERENCES

[1] R. Parsons, in Modern Aspects of Electrochemistry (J. O'M. Bockris, ed.), Vol. I, Butterworths, London, 1954, p. 103.

[2] B. E. Conway, Theory and Principles of Electrode Processes Ronald Press Co., New York, 1965, p. 13.

[3] H. Gerischer, in Advances in Electrochemistry and Electrochemical Engineering (P. Delahay and C. W. Tobias, eds.), Vol. 1, Interscience, New York, 1961, p. 142.

[4]   P. F. Schmidt, J. Electrochem. Soc. , 115, 167 (1968).

[5]   N. J. Harrick, Ann. N. Y. Acad. Sci. , 101, 928 (1963).

[6]   G. W. Poling, J. Colloid Interface Sci. , 34, 365 (1970).

[7]   J. W. Strojek and T. Kuwana, J. Electroanal. Chem. , 16, 471 (1968).

[8]   G. C. Grant and T. Kuwana, J. Electroanal. Chem. , 24, 11 (1970).

[9]   F. P. Kober, J. Electrochem. Soc. , 114, 215 (1967).

[10] W. Paik, M. A. Genshaw, and J. O'M. Bockris, J. Phys. Chem. , 74, 4266 (1970).

[11] J. O'M. Bockris, A. Damjanovic, and W. E. O'Grady, J. Colloid Interface Sci. , 31, 387 (1970).

[12] N. A. Balashova and V. E. Kazarinov, in Electroanalytical Chemistry (A. J. Bard, ed.), Vol. 3, Marcel Dekker, New York, 1969, p. 135.

[13] J. P. Carr, N. A. Hampson, and R. Taylor, J. Electroanal. Chem. , 27, 109 (1970).

[14] W. N. Hansen, T. Kuwana, and R. A. Osteryoung, Anal. Chem. , 38, 1810 (1966).

[15] P. J. Boddy, J. Electroanal. Chem. , 10, 199 (1965).

[16] V. A. Myamlin and Yu. V. Pleskov, Electrochemistry of Semiconductors (English transl. by C. G. B. Garrett from the revised Russian edition), Plenum, New York, 1967, p. 414.

[17] P. Delahay, Double Layer and Electrode Kinetics, Interscience, New York, 1965.

[18] M. A. V. Devanathan and B. V. K. S. R. A. Tilak, Chem. Rev. , 65, 635 (1965).

[19] C. A. Barlow, Jr. , in Physical Chemistry, an Advanced Treatise (H. Eyring, D. Henderson, and W. Jost, eds.), Vol. IX A, Academic Press, New York, 1970, p. 167.

[20] J. O'M. Bockris, K. Müller, H. Wroblowa, and Z. Kovac,
      J. Electroanal. Chem., 10, 416 (1965).

[21] A. N. Frumkin, O. Petry, and B. Damaskin, J. Electroanal.
      Chem., 27, 81 (1970).

[22] A. N. Frumkin, N. A. Balashova, and V. E. Kazarinov, J. Electro-
      chem. Soc., 113, 1011 (1966).

[23] H. Wroblowa and K. Müller, J. Phys. Chem., 73, 3528 (1969).

[24] J. O'M. Bockris and A. K. N. Reddy, Modern Electrochemistry,
      Vol. 2, Plenum, New York, 1970, p. 623.

[25] H. Gerischer, in Physical Chemistry, an Advanced Treatise
      (H. Eyring, D. Henderson, and W. Jost, eds.), Vol. IX A,
      Academic Press, New York, 1970, p. 463.

[26] M. Green, in Modern Aspects of Electrochemistry (J. O'M.
      Bockris, ed.), Vol. 2, Butterworths, London, 1959, pp. 343–407.

[27] J. F. Dewald, in Semiconductors (N. B. Hannay, ed.), ACS Mono-
      graph No. 140, Reinhold, New York, 1959, p. 727.

[28] H. F. Göhr, in The Electrochemistry of Semiconductors (P. J.
      Holmes, ed.), Academic Press, New York, 1962, p. 1.

[29] R. Memming and G. Schwandt, Angew. Chem. Intern. Ed., 6,
      851 (1967).

[30] M. J. Sparnaay, Adv. Colloid Interface Sci., 1, 277 (1967).

[31] J. M. Hale, J. Electrochem. Soc., 115, 208 (1968).

[32] W. Mehl and J. M. Hale, in Advances in Electrochemistry and
      Electrochemical Engineering (P. Delahay, ed.), Vol. 6, Inter-
      science, New York, 1967, pp. 399–458.

[33] W. Mehl, J. M. Hale, and F. Lohmann, J. Electrochem. Soc.,
      113, 1166 (1966).

[34] F. Lohmann, Surface Sci., 14, 431 (1969).

[35] S. Levine, G. M. Bell, and D. Calvert, Can. J. Chem., 40,
      518 (1962).

[36] S. Levine, J. Mingins, and G. M. Bell, J. Electroanal. Chem.,
      13, 280 (1967).

[37] C. A. Barlow, Jr., and J. R. Macdonald, in Advances in Electro-
      chemistry and Electrochemical Engineering (P. Delahay, ed.),
      Vol. 6, Interscience, New York, pp. 2-199.

[38] F. Beck and H. Gerischer, Z. Elektrochem., 63, 500 (1959).

[39] Reference [3], p. 164.

[40] P. J. Boddy, J. Electroanal. Chem., 10, 231 (1965).

[41] P. J. Boddy and W. H. Brattain, J. Electrochem. Soc., 110, 570
      (1963).

[42] P. J. Boddy, J. Electrochem. Soc., 115, 199 (1968).

[43] P. J. Boddy, D. Kahng, and Y. S. Chen, Electrochim. Acta, 13,
      1311 (1968).

[44] A. H. Wilson, Theory of Metals, 2nd ed., Cambridge, London,
      1954.

[45] L. D. Loch, J. Electrochem. Soc., 110, 1081 (1963).

[46] D. Elliot, D. L. Zellmer, and H. A. Laitinen, J. Electrochem.
      Soc., 117, 1343 (1970).

[47] A. T. Howe and P. J. Fensham, Quart. Rev., 21, 507 (1967).

[48] A. K. Vijh, J. Electrochem. Soc., 117, 173C (1970).

[49] D. Adler, in Solid State Physics (F. Seitz, D. Turnbull, and
      H. Ehrenreich, eds.), Vol. 21, Academic Press, New York,
      1968, pp. 1-111.

[50] C. N. R. Rao and G. V. Subba Rao, Phys. Stat. Sol. (a), 1, 597
      (1970).

[51] F. J. Morin, Bell System Tech. J., 37, 1047 (1958).

[52] G. J. Hyland, J. Solid State Chem., 2, 318 (1970).

[53] F. J. Morin, in Semiconductors (N. B. Hannay, ed.), ACS Mono-
      graph No. 140, Reinhold, New York, 1959, p. 600.

[54] N. B. Hannay, in Semiconductors (Proceedings of the Inter-

national School of Physics, Vienna), Academic Press, New York, 1963, p. 409.

[55]   J. B. Goodenough, J. Appl. Phys., 37, 1415 (1966).

[56]   G. W. Rathenau and J. B. Goodenough, J. Appl. Phys., 39, 403 (1968).

[57]   J. H. De Boer and E. J. W. Verwey, Proc. Phys. Soc. (London), 49 (1937).

[58]   W. D. Johnston, J. Chem. Educ., 36, 605 (1959).

[59]   S. Van Houten, J. Phys. Chem. Solids, 17, 7 (1960).

[60]   D. C. Grahame, Chem. Rev., 41, 441 (1947) (see p. 483).

[61]   Reference [16], p. 36.

[62]   E.J.W. Verwey and J.Th.G. Overbeek, Theory of the Stability of Lyophobic Colloids, Elsevier, New York, 1948, p. 29.

[63]   J. O'M. Bockris, M. A. V. Devanathan, and K. Müller, Proc. Roy. Soc. (London), A274, 55 (1963).

[64]   B. E. Conway and L. G. M. Gordon, J. Electroanal. Chem., 15, 7 (1967).

[65]   E. Gileadi and B. E. Conway, in Modern Aspects of Electro-chemistry (J. O'M. Bockris and B. E. Conway, eds.), Vol. 3, Butterworths, London, 1964, p. 347.

[66]   D. C. Grahame, Comptes Rendus de la IIIe Réunion, CITCE, Berne, Aug. 1951 (C. Mandredi, ed.), Milan, 1952, p. 330.

[67]   B. B. Damaskin, Russ. Chem. Rev., 30, 78 (1961).

[68]   G. D. Robbins and C. G. Enke, J. Electroanal. Chem., 12, 102 (1966).

[69]   G. M. Schmid, J. Electrochem. Soc., 115, 1033 (1968).

[70]   L. Ramaley and C. G. Enke, J. Electrochem. Soc., 112, 943, 947 (1965).

[71]   B. D. Cahan, J. B. Ockerman, R. F. Amlie, and P. Rüetschi, J. Electrochem. Soc., 107, 725 (1960).

[72] J. J. McMullen and N. Hackerman, J. Electrochem. Soc., 106, 341 (1959).

[73] J. S. Riney, G. M. Schmid, and N. Hackerman, Rev. Sci. Instr., 32, 588 (1961).

[74] G. M. Schmid and N. Hackerman, J. Electrochem. Soc., 109, 243 (1962).

[75] H. S. Isaacs and J. S. Llewellyn Leach, J. Electrochem. Soc., 110, 680 (1963).

[76] R. G. Barradas and E. M. L. Valeriote, J. Electrochem. Soc., 112, 1043 (1965); ibid., 114, 593 (1967); ibid., 117, 650 (1970).

[77] E. M. L. Valeriote and R. G. Barradas, J. Electroanal. Chem., 12, 67 (1966).

[78] E. M. L. Valeriote and R. G. Barradas, Chem. Instrument., 1(2), 153 (1968).

[79] A. N. Frumkin, Proc. Symp. Electrode Processes, Philadelphia, Wiley, New York, 1959, p. 2.

[80] P. Delahay and G. G. Susbielles, J. Phys. Chem., 70, 647 (1966).

[81] P. Delahay, J. Electrochem. Soc., 113, 967 (1966).

[82] J. O'M. Bockris, S. D. Argade, and E. Gileadi, Electrochim. Acta, 14, 1259 (1969); J. O'M. Bockris and S. D. Argade, J. Chem. Phys., 50, 1622 (1969); see also, Surface Sci., 30, 237 (1972).

[83] R. S. Perkins and T. N. Anderson, in Modern Aspects of Electro-chemistry (J. O'M. Bockris and B. E. Conway, eds.), Vol. 5, Plenum, New York, 1969, p. 203.

[84] S. D. Argade and E. Gileadi, in Electrosorption (E. Gileadi, ed.), Plenum, New York, 1967, p. 87.

[85] L. Campanella, J. Electroanal. Chem., 28, 228 (1970).

[86] R. Parsons, in Proc. 2nd Intern. Congr. Surface Activity, Vol. III, Interfacial Phenomena, Solid-Liquid Interface, Butter-worths, London, 1957, p. 38.

[87] A. N. Frumkin, O. Petry, A. Kossaya, V. Entina, and V. Topolev, J. Electroanal. Chem., 16, 175 (1968).

[88] T. R. Beck, J. Phys. Chem., 73, 466 (1969).

[89] R. A. Fredlein, A. Damjanovic, and J. O'M. Bockris, Surface Sci., 25, 261 (1971).

[90] T. N. Anderson, J. L. Anderson, and H. Eyring, J. Phys. Chem., 73, 3562 (1969).

[91] D. Armstrong, N. A. Hampson, and R. J. Latham, J. Electroanal. Chem., 23, 361 (1969).

[92] N. A. Hampson and D. Larkin, J. Electrochem. Soc., 114, 933 (1967); ibid., 115, 612 (1968).

[93] N. A. Hampson and D. Larkin, J. Electrochem. Soc., 114, 817 (1967).

[94] D. S. Brown et al., J. Electroanal. Chem., 17, 421 (1968).

[95] P. Caswell, N. A. Hampson, and D. Larkin, J. Electroanal. Chem., 20, 335 (1969).

[96] B. J. Piersma, in Electrosorption (E. Gileadi, ed.), Plenum, New York, 1967, p. 19.

[97] M. A. Genshaw, in Electrosorption (E. Gileadi, ed.), Plenum, New York, 1967, p. 73.

[98] Reference [22]; discussion section, pp. 1019-1023.

[99] A. N. Frumkin, O. Petry, and R. Marvet, J. Electroanal. Chem., 12, 504 (1966).

[100] E. Gileadi, S. D. Argade, and J. O'M. Bockris, J. Phys. Chem., 70, 2044 (1966).

[101] Reference [65], p. 368.

[102] E. Gileadi, in Electrosorption (E. Gileadi, ed.), Plenum, New York, 1967, p. 1.

[103] B. E. Conway and L. G. M. Gordon, J. Phys. Chem., 73, 3523 (1969).

[104] J. F. Dewald, Bell System Tech. J., 39, 615 (1960).

[105] J. Bardeen, Phys. Rev., 71, 717 (1947).

[106] W. Shockley and W. T. Read, Phys. Rev., 87, 835 (1952).

[107] W. H. Brattain and J. Bardeen, Bell System Tech. J., 32, 1 (1953).

[108] C. G. B. Garrett and W. H. Brattain, Phys. Rev., 99, 376 (1955).

[109] N. B. Hannay, in Semiconductors (N. B. Hannay, ed.), ACS Monograph No. 140, Reinhold, New York, 1959, p. 1.

[110] A. Many, Y. Goldstein, and N. B. Grover, Semiconductor Surfaces, Wiley, New York, 1965.

[111] E. Spenke, Electronic Semiconductors, McGraw-Hill, New York, 1958.

[112] C. Kittel, Introduction to Solid State Physics, 2nd ed., Wiley, New York, 1956.

[113] J. E. Tamm, Physik Z. Sowjetunion, 1, 733 (1932).

[114] W. Shockley, Phys. Rev., 56, 317 (1939).

[115] J. Koutecky, J. Phys. Chem. Solids, 14, 233 (1960).

[116] T. B. Grimley, J. Phys. Chem. Solids, 14, 227 (1960); P. Mark, Surface Sci., 25, 192 (1971).

[117] G. Heiland, Fortschr. Physik, 9, 393 (1961).

[118] A. Many, J. Phys. Chem. Solids, 8, 87 (1959).

[119] G. Rupprecht, Ann. N.Y. Acad. Sci., 101, 960 (1963).

[120] P. Handler, J. Phys. Chem. Solids, 14, 1 (1960).

[121] R. Memming and G. Schwandt, Surface Sci., 5, 97 (1966).

[122] W. H. Brattain and P. J. Boddy, in Proc. Intern. Conf. Phys. Semiconductors, Institute of Physics and Physical Society, London, 1962, p. 797.

[123] P. J. Boddy and W. H. Brattain, J. Electrochem. Soc., 109, 812 (1962).

[124] R. Memming and G. Neumann, Surface Sci., 10, 1 (1968); Phys. Letters, 24A, 19 (1967).

[125] H. Gerischer, in The Surface Chemistry of Metals and Semi-conductors (H. C. Gatos, ed.), Wiley, New York, 1959, p. 177.

[126] R. H. Kingston and S. F. Neustadter, J. Appl. Phys., 26, 718 (1955).

[127] H. Gerischer, J. Electrochem. Soc., 113, 1174 (1966).

[128] P. J. Boddy, Surface Sci., 13, 52 (1969).

[129] G. Brouwer, J. Electrochem. Soc., 114, 743 (1967).

[130] P. J. Boddy and W. H. Brattain, Surface Sci., 4, 18 (1966).

[131] P. J. Boddy and W. H. Brattain, Ann. N. Y. Acad. Sci., 101, 683 (1963); Surface Sci., 3, 348 (1965).

[132] R. M. Hurd and P. T. Wrotenbery, Ann. N. Y. Acad. Sci., 101, 876 (1963).

[133] P. J. Boddy and W. J. Sundburg, J. Electrochem. Soc., 110, 1170 (1963).

[134] W. H. Brattain and P. J. Boddy, Proc. Natl. Acad. Sci. U. S., 48, 2005 (1962).

[135] H. J. Engell and K. Bohnenkamp, in The Surface Chemistry of Metals and Semiconductors (H. C. Gatos, ed.), Wiley, New York, 1959, p. 225; Z. Elektrochem., 61, 1184 (1957).

[136] Yu. V. Pleskov and M. D. Krotova, in Proc. Intern. Conf. Phys. Semiconductors, Institute of Physics and Physical Society, London, 1962, p. 807.

[137] W. H. Brattain and P. J. Boddy, J. Electrochem. Soc., 109, 574 (1962).

[138] H. Gerischer et al., Surface Sci., 4, 431, 440 (1966).

[139] H. A. Laitinen, C. A. Vincent, and T. M. Bednarski, J. Electrochem. Soc., 115, 1024 (1968).

[140] T. O. Rouse and J. L. Weininger, J. Electrochem. Soc., 113, 184 (1966).

[141] M. Hoffman-Perez and H. Gerischer, Z. Elektrochem., 65, 771 (1961).

[142] Reference [16], p. 73.

[143] R. Memming, Surface Sci., 2, 436 (1964).

[144] H. Yoneyama and H. Tamura, Bull. Chem. Soc. Japan, 43, 350 (1970).

[145] M.J. Sparnaay, Surface Sci., 1, 102 (1964).

[146] S. M. Ahmed, J. Phys. Chem., 73, 3546 (1969).

[147] M.J. Sparnaay, Rec. Trav. Chim., 79, 950 (1960).

[148] G.T. Wright, Solid State Electronics, 2, 165 (1961); R.H. Tredgold, Space Charge Conduction in Solids, Elsevier, Amsterdam, 1966; R.M. Hill, Thin Solid Films, 1, 39 (1967).

[149] T. Freund, J. Phys. Chem., 73, 468 (1969).

[150] H.K. Kiess, J. Phys. Chem. Solids, 31, 2379, 2391 (1970).

[151] H. Gerischer, Surface Sci., 18, 97 (1969); ibid., 13, 265 (1969).

[152] W.P. Gomes, T. Freund, and S.R. Morrison, J. Electrochem. Soc., 115, 818 (1968).

[153] T. Freund and S.R. Morrison, Surface Sci., 9, 119 (1968); S.R. Morrison, ibid., 15, 363 (1969).

[154] D.A. Vermilyea, Surface Sci., 2, 444 (1964).

[155] D.A. Vermilyea, J. Electrochem. Soc., 112, 1232 (1965).

[156] D.A. Vermilyea, J. Phys. Chem. Solids, 26, 133 (1965).

[157] D.A. Vermilyea, J. Electrochem. Soc., 115, 177 (1968).

[158] F. Huber, J. Electrochem. Soc., 110, 846 (1963).

[159] Y. Sasaki, J. Phys. Chem. Solids, 13, 177 (1960).

[160] H.E. Haring, J. Electrochem. Soc., 99, 30 (1952).

[161] A. Middelhoek, J. Electrochem. Soc., 111, 379 (1964).

[162] W. Ch. Van Geel, Physica, 27, 761 (1951).

[163] P.F. Schmidt, F. Huber, and R.F. Schwarz, J. Phys. Chem. Solids, 15, 270 (1960).

[164] J. H. Anderson and G. A. Parks, J. Phys. Chem., 72, 3662 (1968).

[165] Reference [4], p. 169, and references cited therein.

[166] M. M. Egorov and V. F. Kiselev, Russ. J. Phys. Chem., 40, 1069 (1966).

[167] A. V. Kiselev, Russ. J. Phys. Chem., 40, 1073 (1966).

[168] W. Hertl and M. L. Hair, J. Phys. Chem., 72, 4676 (1968).

[169] M. M. Bhasin, C. Curran, and G. S. John, J. Phys. Chem., 74, 3973 (1970).

[170] J. W. Diggle, T. C. Downie, and C. W. Goulding, Chem. Rev., 69, 365 (1969).

[171] D. A. Vermilyea, in Advances in Electrochemistry and Electrochemical Engineering (P. Delahay, ed.), Vol. 3, Interscience, New York, 1963, p. 211.

[172] F. Huber, J. Electrochem. Soc., 115, 203 (1968).

[173] R. A. Marcus, Ann. Rev. Phys. Chem., 15, 155 (1964).

[174] V. G. Levich, in Physical Chemistry, an Advanced Treatise (H. Eyring, D. Henderson, and W. Jost, eds.), Vol. IX B, Academic Press, New York, 1970, p. 985.

[175] J. N. Butler, J. Chem. Phys., 35, 636 (1961).

[176] F. F. Vol'kenshtein, V. S. Kuznetsov, and V. B. Sandomirskii, Kinetika i Kataliz, 3, 712 (1962); English translation, Kinetics and Catalysis, 3, 619 (1962).

[177] A. M. Kuznetsov and R. D. Dogonadze, Izv. Akad. Nauk SSSR, Ser. Khim., No. 12, 2140 (1964); English translation, Consultants Bureau, New York, 1964, p. 2042.

[178] E. Herczyńska and K. Prószyńska, Naturwissenschaften, 9, 351 (1963).

[179] E. Herczyńska and K. Prószyńska, J. Inorg. Nucl. Chem., 26, 1429, 2127 (1964).

[180] B. N. Kabanov, I. G. Kiseleva, and D. I. Leikis, Dokl. Akad. Nauk SSSR, 99, 805 (1954).

[181] J. P. Carr, N. A. Hampson, and R. Taylor, J. Electroanal. Chem.,
      27, 201, 466 (1970).

[182] R. T. Angstadt, C. J. Venuto, and P. Rüetschi, J. Electrochem.
      Soc., 109, 177 (1962); P. Rüetschi and R. T. Angstadt, ibid.,
      111, 1323 (1964).

[183] J. P. Carr and N. A. Hampson, J. Electroanal. Chem., 28, 65
      (1970).

[184] D. J. G. Ives and G. J. Janz, Reference Electrodes, Theory and
      Practice, Academic Press, New York, 1961.

[185] Reference [17], p. 56.

[186] B. E. Conway, in Modern Aspects of Electrochemistry (J. O'M.
      Bockris and B. E. Conway, eds.), Vol. 3, Butterworths, London,
      1964, p. 43.

[187] B. E. Conway, in Physical Chemistry, an Advanced Treatise
      (H. Eyring, D. Henderson, and W. Jost, eds.), Vol. IXA,
      Academic Press, New York, 1970, p. 1.

[188] C. S. Fuller, in Semiconductors (N. B. Hannay, ed.), ACS Mono-
      graph No. 140, Reinhold, New York, 1959, p. 92.

[189] Reference [2], p. 92.

[190] J. O'M. Bockris and D. B. Mathews, J. Electroanal. Chem., 9,
      325 (1965); Proc. Roy. Soc. (London), A292, 479 (1966).

[191] G. A. Parks and P. L. de Bruyn, J. Phys. Chem., 66, 967 (1962).

[192] E. Herczyńska and K. Prószyńska, Report No. 372/V, Institute
      of Nuclear Research, Warsaw, Poland, 1962.

[193] E. L. Mackor, Rec. Trav. Chim., 70, 663, 747, 763 (1951).

[194] J. Th. G. Overbeek, Semicentennial Symposium, Electrochemical
      Constants, Natl. Bur. Stand. U. S., Circ. No. 524, 1953,
      p. 213.

[195] T. B. Grimley and N. F. Mott, Discussions Faraday Soc., 1, 3
      (1947).

[196] T. B. Grimley, Proc. Roy. Soc. (London), A201, 40 (1950).

[197] J. Lyklema and J. Th. G. Overbeek, J. Colloid Sci., 16, 595 (1961).

[198] B. H. Bijsterbosch and J. Lyklema, J. Colloid Sci., 20, 665 (1965).

[199] E. P. Honig, Trans. Faraday Soc., 65, 2248 (1969).

[200] E. P. Honig, Nature, 225, 537 (1970).

[201] E. P. Honig and J. H. Th. Hengst, J. Colloid Interface Sci., 31, 545 (1969).

[202] S. M. Ahmed and D. Maksimov, J. Colloid Interface Sci., 29, 97 (1969).

[203] S. M. Ahmed, Can. J. Chem., 44, 1663, 2769 (1966).

[204] E. J. W. Verwey, Rec. Trav. Chim., 60, 625 (1941).

[205] G. Y. Onoda, Jr., and P. L. de Bruyn, Surface Sci., 4, 48 (1966).

[206] S. M. Ahmed and D. Maksimov, Can. J. Chem., 46, 3841 (1968).

[207] G. Biedermann and J. T. Chow, Acta Chem. Scand., 20, 1376 (1966).

[208] A. A. Van der Giessen, Philips Res. Repts. Suppl., No. 12, 1-93 (1968).

[209] L. Blok and P. L. de Bruyn, J. Colloid Interface Sci., 32, 527 (1970).

[210] L. Blok and P. L. de Bruyn, J. Colloid Interface Sci., 32, 518 (1970).

[211] L. Blok and P. L. de Bruyn, J. Colloid Interface Sci., 32, 533 (1970).

[212] Y. G. Bérubé and P. L. de Bruyn, J. Colloid Interface Sci., 27, 305 (1968).

[213] Y. G. Bérubé and P. L. de Bruyn, J. Colloid Interface Sci., 28, 92 (1968).

[214] B. Ball, J. Colloid Interface Sci., 30, 424 (1969).

[215] T. W. Healy and V. R. Jellett, J. Colloid Interface Sci., 24, 41 (1967).

[216] J. W. Fulton and D. F. Swinehart, J. Amer. Chem. Soc., 76, 863 (1954).

[217] L. Blok, The Ionic Double Layer on Zinc Oxide in Aqueous Electrolyte Solutions, Ph. D. thesis, State University of Utrecht, 1968, p. 24.

[218] S. M. Ahmed and D. Maksimov, Mines Branch Research Report, R 196, Dept. of Energy, Mines, and Resources, Ottawa, Canada, 1968.

[219] G. Kortüm and J. O'M. Bockris, Text Book of Electrochemistry, Vol. 1, Elsevier, London, 1961, p. 293.

[220] Reference [2], p. 30.

[221] C. D. Russell, J. Electroanal. Chem., 6, 486 (1963).

[222] D. W. Fong and E. Grunwald, J. Amer. Chem. Soc., 91, 2413 (1969).

[223] A. Takahashi, J. Phys. Soc. Japan, 24, 657 (1968).

[224] R. J. Atkinson, A. M. Posner, and J. P. Quirk, J. Phys. Chem., 71, 550 (1967).

[225] A. Fratiello, V. Kubo, R. E. Lee, and R. E. Schuster, J. Phys. Chem., 74, 3726 (1970).

[226] A. Fratiello, S. Peak, R. E. Schuster, and D. D. Davis, J. Phys. Chem., 74, 3730 (1970).

[227] A. Fratiello, R. E. Lee, V. M. Nishida, and R. E. Schuster, J. Chem. Phys., 50, 3624 (1969).

[228] H. C. Li and P. L. de Bruyn, Surface Sci., 5, 203 (1966).

[229] S. M. Ahmed and A. B. Van Cleave, Can. J. Chem. Eng., 43, 23 (1965).

[230] A. M. Gaudin and C. S. Chang, Trans. AIME, 193, 193 (1952).

[231] G. A. Parks, Chem. Rev., 65, 177 (1965).

[232] T. W. Healy and D. W. Fuerstenau, J. Colloid Sci., 20, 376 (1965).

[233] T. W. Healy, A. P. Herring, and D. W. Fuerstenau, J. Colloid Interface Sci., 21, 435 (1966).

[234] V. Pravdić and S. Sotman, Croat. Chem. Acta, 35, 247 (1963).

[235] T. Morimoto and M. Sakamoto, Bull. Chem. Soc. Japan, 37, 719 (1964).

[236] D. J. O'Connor, P. G. Johansen, and A. S. Buchanan, Trans. Faraday Soc., 52, 229 (1956).

[237] M. Robinson, J. A. Pask, and D. W. Fuerstenau, J. Amer. Cer. Soc., 47, 516 (1964).

[238] A. Marchetti, Internal Report MS 70-79, Mines Branch, Dept. of Energy, Mines, and Resources, Ottawa, Canada, 1970.

[239] Reference [219], p. 176.

[240] T. Moeller, Inorganic Chemistry, an Advanced Text Book, Wiley, New York, 1952, pp. 312-316, 318-321.

[241] Reference [219], p. 325; Ref. [24], Vol. 1, p. 510.

[242] S. M. Ahmed, to be published.

[243] P. H. Tewari and A. W. McLean, private communication.

[244] J. Lyklema, Discussions Faraday Soc., 42, 81 (1966).

[245] R. G. Bates, Determination of pH, Theory and Practice, Wiley, New York, 1964; J. Res. Natl. Bur. Stand., 66A, 179 (1962).

[246] S. M. Ahmed, Tech. Bull. TB 84, Mines Branch, Dept. of Energy, Mines, and Resources, Ottawa, 1966.

[247] G. W. Smith and T. Salman, Can. Met. Quart., 5, 93 (1966).

[248] H. F. Holmes, C. S. Shoup, Jr., and C. H. Secoy, J. Phys. Chem., 69, 3148 (1965).

[249] A. S. Joy and D. Watson, in Proc. 6th Intern. Mineral Processing Congr., Cannes, 1963, Pergamon, New York, 1965, p. 355.

[250] A. S. Joy and D. Watson, Bull. Instn. Min. Metall., 73, No. 687, 323 (1964).

[251] G. W. Smith and T. Salman, Can. Met. Quart., 6, 167 (1967).

[252] K. H. Gayer, L. C. Thompson, and O. T. Zajicek, Can. J. Chem.,
36, 1268 (1958).

[253] J. A. Yopps and D. W. Fuerstenau, J. Colloid Sci., 19, 61 (1964).

[254] Reference [231], p. 186.

[255] Y. G. Bérubé, G. Y. Onoda, Jr., and P. L. de Bruyn, Surface
Sci., 8 (or 7, there is a typographical error in the journal),
448 (1967).

[256] S. M. Ahmed, unpublished work.

[257] Th. F. Tadros and J. Lyklema, J. Electroanal. Chem., 17, 267
(1968).

[258] J. Lyklema, J. Electroanal. Chem., 18, 341 (1968).

[259] D. C. Grahame and R. Parsons, J. Amer. Chem. Soc., 83, 1291
(1961).

[260] D. C. Grahame, J. Amer. Chem. Soc., 80, 4201 (1958).

[261] J. Lawrence, R. Parsons, and R. Payne, J. Electroanal. Chem.,
16, 193 (1968).

[262] H. Wroblowa, Z. Kovac, and J. O'M. Bockris, Trans. Faraday
Soc., 61, 1523 (1965).

[263] G. G. Susbielles, P. Delahay, and E. Solon, J. Phys. Chem.,
70, 2601 (1966).

[264] E. Blomgren and J. O'M. Bockris, J. Phys. Chem., 63, 1475
(1959).

[265] J. M. Parry and R. Parsons, Trans. Faraday Soc., 59, 241 (1963).

[266] R. Payne, J. Chem. Phys., 42, 3371 (1965).

[267] R. Payne, J. Electrochem. Soc., 113, 999 (1966).

[268] M. J. Sparnaay, Surface Sci., 1, 213 (1964).

[269] J. J. Bikerman, in Proc. 2nd Intern. Congr. Surface Activity,
Vol. III, Interfacial Phenomena, Solid–Liquid Interface, Butterworths, London, 1957, pp. 125 and 187.

[270] B. A. Pethica and T. J. P. Pethica, in Proc. 2nd Intern. Congr. Surface Activity, Vol. III, Interfacial Phenomena, Solid-Liquid Interface, Butterworths, London, 1957, p. 131.

[271] P. Sennett and J. P. Olivier, in Chemistry and Physics of Interfaces, American Chemical Society, Washington, D. C., 1965, p. 75.

[272] D. J. Shaw, Electrophoresis, Academic Press, London, 1969.

[273] J. Th. G. Overbeek, in Colloid Science (H. R. Kruyt, ed.), Vol. I, Elsevier, Amsterdam, 1952, p. 194.

[274] G. L. Zucker, Sc. D. thesis, Columbia University, 1959.

[275] L. A. Wood, J. Amer. Chem. Soc., 68, 432 (1946).

[276] J. Lyklema and J. Th. G. Overbeek, J. Colloid Sci., 16, 501 (1961).

[277] D. Stigter, J. Phys. Chem., 68, 3600 (1964).

[278] R. J. Hunter, J. Colloid Interface Sci., 22, 231 (1966).

[279] D. J. O'Connor, N. Street, and A. S. Buchanan, Aust. J. Chem., 7, 245 (1954).

[280] R. M. Hurd and N. Hackerman, J. Electrochem. Soc., 102, 594 (1955); ibid., 103, 316 (1956).

[281] V. Pravdić, Croat. Chem. Acta, 35, 233 (1963).

[282] P. H. Cardwell, J. Colloid Interface Sci., 22, 430 (1966).

[283] J. H. Schulman and H. C. Parreira, Report No. 1, Office Naval Research Contract NONR-266 (64), 1964.

[284] H. C. Parreira, J. Colloid Sci., 20, 1 (1965).

[285] H. P. Dibbs, private communication, Mines Branch, Dept. of Energy, Mines, and Resources, Ottawa, 1970.

[286] H. C. Parreira and J. H. Schulman, Adv. Chem. Ser., 33, 160 (1961).

[287] D. C. Henry, Proc. Roy. Soc. (London), 133, 106 (1931); Trans. Faraday Soc., 44, 1021 (1948).

[288] F. A. Morrison, Jr., J. Colloid Interface Sci., 34, 210 (1970).

[289] F. Booth, Proc. Roy. Soc. (London), A203, 514 (1950).

[290] M. Sengupta and A. K. Bose, J. Electroanal. Chem., 18, 21 (1968).

[291] J. Th. G. Overbeek and P. H. Wiersema, in Electrophoresis: Theory, Methods, and Applications (M. Bier, ed.), Vol. 2, Academic Press, New York, 1967, p. 1.

[292] S. Ross and R. F. Long, Ind. Eng. Chem., 61, 59 (1969).

[293] R. Neihof, J. Colloid Interface Sci., 30, 128 (1969).

[294] H. C. Parreira, J. Colloid Interface Sci., 29, 432 (1969).

[295] R. Schmut, in Chemistry and Physics of Interfaces, American Chemical Society, Washington, D. C., 1965, p. 93.

[296] D. Stigter, J. Phys. Chem., 68, 3603 (1964).

[297] D. A. Haydon, Proc. Roy. Soc. (London), A258, 319 (1960).

[298] Reference [273], p. 231.

[299] R. H. Ottewill and R. F. Woodbridge, J. Colloid Sci., 19, 606 (1964).

[300] Reference [272], pp. 10-15.

[301] P. J. Anderson, in Proc. 2nd Intern. Congr. Surface Activity, Vol. III, Interfacial Phenomena, Solid-Liquid Interface, Butterworths, London, 1957, p. 67; Trans. Faraday Soc., 54, 130, 562 (1958).

[302] R. J. Hunter and H. J. L. Wright, J. Colloid and Interface Sci., 37, 564 (1971).

[303] D. P. Benton, G. A. Horsfall, and S. K. Nicol, J. Chem. Soc., 5067 (1963).

[304] J. M. W. Mackenzie, Trans. AIME, 235, 82 (1966).

[305] J. M. W. Mackenzie and R. T. O'Brien, Trans. AIME, 244, 168 (1969).

[306] D. J. O'Connor and A. S. Buchanan, Aust. J. Chem., 6, 278 (1953).

[307] H. J. Modi and D. W. Fuerstenau, J. Phys. Chem., 61, 640 (1957).

[308] A. M. Gaudin and D. W. Fuerstenau, Trans. AIME, 202, 958 (1955).

[309] D. W. Fuerstenau, J. Phys. Chem., 60, 981 (1956).

[310] P. L. de Bruyn, Trans. AIME, 202, 291 (1955).

[311] J. M. Cases, Les phénomènes physico-chémiques à l'interface, application au procédé de la flotation, Sciences de la Terre, Mémoire No. 13, Nancy, 1968.

[312] F. F. Aplan and D. W. Fuerstenau, in Froth Flotation, 50th Anniversary Volume (D. W. Fuerstenau, ed.), AIME, 1962, p. 170.

[313] D. W. Fuerstenau and H. J. Modi, J. Electrochem. Soc., 106, 336 (1959).

[314] A. M. Gaudin and D. W. Fuerstenau, Trans. AIME, 202, 66 (1955).

[315] M. J. Jaycock, R. H. Ottewill, and I. Tar, Trans. Instn. Mining and Met., 73, No. 686, 255 (1964).

[316] G. Purcell and S. C. Sun, Trans. AIME, 223, 6 (1963).

[317] R. W. Smith, Trans. AIME, 223, 427 (1963).

AUTHOR INDEX

Plain numbers indicate pages where a name appears. Numbers in parentheses indicate the reference for all preceding page numbers given. Each reference is in turn followed by an underlined number, which shows where it is listed. For example, T. N. Anderson is mentioned on page 337, in relation to reference (83), listed on page 504; also mentioned on page 338, in relation to reference (90), listed on page 505.

A

Adler, D., 328-332 (49) 502
Ahmed, S. M., 168 (38) 283;
    371, 373, 397, 406, 412-
    416, 420, 421, 425, 433,
    434, 439, 442, 443, 447,
    448, 454, 458-460, 463,
    466, 472, 473, 474, 475,
    477, 479-481, 483-484 (146)
    508; 405, 406, 412, 414-
    416, 421, 433, 437, 439,
    442, 445, 446, 472, 479
    (202) 511; 405, 406, 413,
    416, 417, 420, 426, 433,
    434, 438, 442, 454, 455,
    473-477, 479, 480, 483,
    484 (203) 511; 405, 406,
    409, 416, 431-433, 435,
    442, 450-453, 472-475,
    477, 479, 484 (206) 511;
    412-416, 420, 439, 442,
    472, 480, 483, 484 (218)
    512; 420, 499 (229) 512;
    423, 425, 434, 435, 455,
    463, 466, 468, 469, 470,
    481, 485, 486, 492 (242)
    513; 436, 461 (246) 513;
    451 (256) 514
Aida, H., 25 (102) 85
Al'tovskii, R. M., 11 (57) 82;
    44 (142) 87

Altukhov, V. K., 27 (110) 85
Ames, W. F., 131 (27) 282
Amlie, R. F., 335, 338 (71) 503
Amsel, G., 201 (53) 284;
    271 (99) 286
Anderman, G., 299 (3) 318
Anderson, J. L., 338 (90) 505
Anderson, J. H., 337 (164) 509
Anderson, P. J., 493, 495 (301) 516
Anderson, T. N., 337 (83) 504;
    338 (90) 505
Andreeva, V. V., 17, 21, 75
    (84) 84
Angstadt, R. T., 397, 399, 402,
    475 (182) 510
Aplan, F. F., 495, 496, 498
    (312) 517
Argade, S. D., 337, 338 (82, 84)
    504; 339, 402 (100) 505
Armstrong, D., 338 (91) 505
Arnold, K., 9, 20, 21, 54, 66,
    79 (35) 81
Aronowitz, G., 74 (182) 89
Asakura, S., 34, 35 (115) 85;
    34, 35 (116) 86
Atkinson, R. J., 416, 421, 437,
    439, 442, 453, 457, 472
    (224) 512
Awad, S. A., 77 (187) 89
Appleby, A. J., 24 (99) 85

Nishida, V. M. , 417, 458 (227) 512

Nobe, K. , 34, 35 (115) 85; 34, 35 (116) 86

Nordberg, R. , 45 (145) 87

Nordling, C. , 45 (145) 87

Novakovsky, V. M. , 9, 21, 61, 73 (32, 33) 81; 57 (33) 81

Nyburg, S. C. , 209 (62) 285

O

O'Brien, R. T. , 495 (305) 516

O'Connor, D. J. , 422, 430, 437, 438, 440 (236) 513; 488, 492, 493, 495 (279) 515; 495 (306) 516

Ockerman, J. B. , 335, 338 (71) 503

O'Grady, W. E. , 43 (141) 87; 323 (11) 500

Okamoto, M. , 19, 70 (92, 93) 84; 27 (93) 84

Oliver, J. P. , 487 (271) 515

Onada, G. , Jr. , 405, 407, 421, 439, 442, 454, 473 (205) 511; 441, 444, 455, 473 (255) 514

Ord, J. L. , 18 (87–90) 84; 48 (89, 90) 84; 60 (89) 84; 50, 60 (90) 84

Oshe, A. I. , 52 (173, 174) 88

Oshe, E. K. , 45 (146-148) 87

Osteryoung, R. A. , 323 (14) 500

O'Sullivan, J. P. , 275 (103) 286

Ottewill, R. H. , 492 (299) 516; 495 (315) 517

Otto, K. , 191, 192 (46) 283

Overbeek, J. Th. G. , 333, 405, 412, 419 (62) 503; 405, 412 (194) 510; 405, 473, 479 (197) 511; 487, 488, 490, 491 (273) 515; 487, 488, 489 (276) 515; 491 (291) 516; 492 (298) 516

P

Padovani, F. A. , 315 (6) 318

Paik, W. , 323 (10) 500

Paleolog, E. N. , 46 (152) 87

Park, J. A. , 422, 430, 437, 440 (237) 513

Parker, G. H. , 315 (7) 318

Parks, G. , 377 (164) 509; 405, 407, 411, 412, 421, 437, 439, 454, 473 (191) 510; 421, 422, 427–429, 436 (231) 512; 440 (254) 514

Parreira, H. C. , 489 (283, 284, 286) 515; 491, 492 (294) 516

Parry, J. M. , 480 (265) 514

Parsons, C. , 94 (12) 282

Parsons, R. , 320, 341 (1) 499; 338 (86) 504; 460 (259, 261) 514; 480 (265) 514

Pavlov, D. , 19, 77 (94, 95) 84

Payne, R. , 460 (261) 514; 480 (266, 267) 514

Peak, S. , 417, 458 (226) 512

Perkins, R. S. , 337 (83) 504

Pethica, B. A. , 482 (270) 515

Pethica, T. J. P. , 482 (270) 515

Petry, O. , 324, 339 (21) 501; 338, 339 (87) 505; 339 (99) 505

Pickering, H. W. , 12, 24, 36, 72 (63) 83

Piersma, B. , 339 (96) 505

Piontelli, R. , 12 (60) 83; 25, 26 (107) 85

Pistorius, C. A. , 262 (95) 286

Pleskov, Yu. V. , 323, 324, 344, 354, 355, 364 (16) 500; 333, (61) 503; 355, 364 (136) 507; 359 (142) 508

Polling, G. W. , 323 (6) 500

Polling, J. J. , 217 (67) 285

Popova, T. J. , 9, 64 (37) 81

Popova, R. , 19, 77 (95) 84

Porter, G. B. , 209 (63) 285

Posner, A. M. , 416, 421, 437, 439, 442, 453, 457, 472 (224) 512

Pourbaix, M. , 38, 39 (137) 87

Powers, R. W. , 78 (188) 89